The Selected Papers of
BOULTON & WATT

The Selected Papers of

BOULTON & WATT

VOLUME I

THE ENGINE PARTNERSHIP

1775–1825

EDITED BY

Dr Jennifer Tann

THE MIT PRESS

Cambridge, Massachusetts

Published in the United States by The MIT Press and in England by Diploma Press.

Printed and bound in the United States of America

Library of Congress Cataloging in Publication Data

Boulton, Matthew, 1728–1809.
The selected papers of Boulton and Watt.
Bibliography: v. 1, p.
Includes index.
Contents: v. 1. The engine partnership, 1775–1825.
1. Steam-engines—History—Sources.
I. Watt, James, 1736–1819. II. Tann, Jennifer, 1939– III. Title.
TJ461.B68 1981 621.1 81-3695
ISBN 0-262-02167-6 (v. 1) AACR2

Contents

List of Plates

Acknowledgements

It is a great pleasure to record my indebtedness and thanks to the many people who have helped me, both directly and indirectly, in compiling this volume.

I owe a great debt to the many eminent historians who have worked on the Boulton and Watt and Matthew Boulton Papers. J.P. Muirhead, John Lord, Eric Roll, H.W. Dickinson, Rhys Jenkins, T.S. Ashton, A.E. Musson, Eric Robinson, W.H. Chaloner, J.R. Harris, R.L. Hills, W.K.V. Gale, J. Rowe, G.N. von Tunzelmann and others have done so much to throw new light on the role of steam power in the industrial revolution.

My thanks are due to the Social Science Research Council and to the University of Aston Management Centre who generously supported the research on which this volume is based and to my colleague, Michael Breckin, who worked with me for four years. This series of Selected Papers could not have been undertaken without the encouragement and support of Mr. W.A. Taylor, Librarian Emeritus of the City of Birmingham, and of the present City Librarian, Mr. B. Baumfield. I thank both them and their staff. I am particularly grateful to Miss D. McCulla, Head of the Local Studies Department and to Mr. A. Andrews, Archivist, who with their staff have had to bear the brunt of numerous requests and enquiries. My debt to Mr. A.H. Westwood, formerly director of Birmingham Assay Office, is great, both for his kindness in allowing me access to the Matthew Boulton Papers and for the privilege of working in the Assay Office Library. I am grateful to the Matthew Boulton Trust for permission to reproduce some documents from the Matthew Boulton Papers in this volume. I thank my husband and sons who have uncomplainingly shared their home with Boulton and Watt for a number of years, and Mrs. Joy Tonkin who has deciphered all manner of illegible handwriting, including mine, to produce the typescript. Finally I wish to thank Diploma Press for their helpfulness at all stages in the production of this book.

Author's Introduction

Invention of the Separate Condenser and Beginning of the Engine Partnership

James Watt's interest in steam engines dates back to at least 1761 when he is known to have tried some experiments on high pressure steam with a Papin digester.[1] But he became involved in a more practical way in 1763 when Professor John Anderson asked him, as mathematical instrument maker to Glasgow university, to repair a model atmospheric engine belonging to the Natural Philosophy Class. Numerous experiments followed, during which Watt found that the boiler could not supply sufficient steam to operate the engine for more than a few strokes. (A scale model poses additional problems of volume in this respect, besides which the effect of friction is greater than in a full-sized engine.) Watt identified the problem as being twofold: that of conserving heat by keeping the cylinder hot whilst yet condensing the steam, and in May 1765 he devised the separate condenser and air pump. Watt was loaned money for his experiments by Dr. Joseph Black who also introduced him to Dr. John Roebuck of Birmingham, then engaged in developing the coalfield at Bo'ness.

Atmospheric engines had already been installed on the coalfield and in October 1765 Watt made a trial of one of them. At about the same time he produced a design for a reciprocating engine which he intended to erect for Dr. Roebuck at Kinneil. This differed from the Newcomen engine in having an inverted cylinder which actuated the pumps directly,[2] but he shortly afterwards adopted the standard beam engine design having made the necessary structural alterations for his improvements. He also devised a rotative engine on the steam wheel principle, the design for which was included in the 1775 application to Parliament,[3] but whilst an experimental wheel was constructed at Soho, the rotative engine, when it was finally produced, was an adaptation of the reciprocating beam engine.

In 1766 Watt commenced in business as a surveyor, probably retaining his instrument-making shop, and a visit to Birmingham was made in 1767 whilst *en route* to London in connection with a Canal Bill. His object had been to see Samuel Garbett, Roebuck's Birmingham partner, but he was away from home. On his return journey Watt called on Dr. Erasmus Darwin at Lichfield and described the separate condenser to him.[4] He began corresponding with Boulton and Small, continuing to modify his engine design and by January 1768 there was talk of a partnership.[1] But Roebuck had a two-thirds share in Watt's engine, having taken over Watt's debt to Joseph Black, besides agreeing to advance whatever sums were necessary for further experiments and securing a patent [2]. In August 1768 Watt took the oath in London and his first patent was sealed in January 1769.[5] [10]

On his return from London, Watt met Boulton at Soho. Boulton and Dr. William Small discussed the possibility of a connection with Watt who, on returning to Glasgow, called on Roebuck with the suggestion that Boulton might become a partner in the engine project. Watt clearly welcomed the possibility of Boulton's involvement, but once the patent was enrolled, Roebuck was less disposed to Boulton taking a share, offering him licence to make engines for the Midland counties only. Boulton was not attracted by such a limited offer: 'I cannot think it is a proper one for me to meddle with as I do not intend turning engineer. I was excited by two motives to offer you my assistance which were love of you and love of a money-getting ingenious project ... It would not be worth my while to make for these counties only, but I find it very well worth my while to make for all the world.'[6] Watt, however, continued his correspondence with Small,[7] detailing further engine experiments and indicating that he hoped a settlement between Boulton and Roebuck would be reached. By 1773 Roebuck was in severe financial difficulties and Watt became the owner of the Kinneil engine in return for a discharge of unpaid debts due to him. Roebuck, however, still owned the two-thirds' share in Watt's patent, but Boulton, through a cash settlement with his toy manufacturing partner, Fothergill, secured Roebuck's agreement [3 & 4] for a transfer of the two-thirds' share in Watt's patent to him in settlement of debts owing to Boulton and Fothergill, and to him personally.[8] The Kinneil engine was sent to Soho and in May 1774 Watt and his family followed.

The Kinneil engine was re-erected and experiments recommenced[9] but there were still many development problems to be overcome. Moreover, with nearly half the term of Watt's patent expired, there was the question of financial reward. Boulton proposed seeking an extension of the patent by Act of Parliament. Watt drew up a memorandum describing and illustrating his improvements and, with Boulton's expertise in parliamentary lobbying, the Bill to extend the patent for twenty-five years passed all its stages and received Royal Assent in May 1775, the partnership between Boulton and Watt commencing on June 1st of that year.

Engine Technology and Patents

It will be remembered that Watt had devised a steam wheel to produce rotary motion as early as 1765-9 but when, after the commencement of the Boulton and Watt partnership, there seemed on many occasions to be more orders for reciprocating engines than could be met, Watt did not pursue the matter. Boulton raised the issue in 1779,[10] however, and in 1781[11] he urged Watt to take out a patent for certain methods of producing rotative motion [14] 'There is no other Cornwall to be found and the most likely line for the consumption of our engines is the application of them to mills, which is certainly an extensive field.' In the interim period, however, Pickard had patented the crank and Watt, probably not wishing to open a debate on the question of patents and monopolies, decided not to contest it but developed, instead, the sun and planet motion,[12] amongst other methods, [13, 14 & 16] which was patented in 1781.[13] The firm used the sun and planet gear on all their rotative engines until 1794 when Pickard's patent expired, and indeed for some years afterwards.

In 1782 Watt was granted a patent[14] for several new improvements to his engine. These included the use of the expansive principle in which a valve cut off the steam before the piston stroke was complete, thereby economising on fuel; the double-acting engine in which steam pressure was utilised both above and below the piston; a toothed rack-and-sector connecting the piston rod and beam for use in double acting engines; a double engine comprising two engines that could be linked in such a way that steam admitted to the first cylinder could be employed expansively in the second; and a steam wheel or rotative engine similar to that patented in 1769. Of these, the double-acting engine was the improvement with the widest potential application. The fuel savings of the expansive engine were found, in practice, to be too small to warrant extensive development and the rack-and-sector was superseded by parallel motion which was patented in 1784[15] together with various further improvements, notably the application of engines to wheel carriages,[16] which would seem to have been included to forestall developments by others. [28] Of all his inventions, parallel motion was the one which most pleased Watt,[17] [16] but one further invention was seen by many contemporaries [158] to make an important contribution towards the improvement of industrial health in attempting to prevent the increase in the atmospheric pollution created by steam engines and coal-burning industrial processes. This was the smokeless furnace or fire place patented in 1785.[18]

Besides obtaining patent protection in Britain, the partners were keen to seek patents and privileges abroad, attempting to make the granting of a privilege a condition of the first sale in a new overseas market. Some form of exclusive privilege was obtained in France, Spain and the Netherlands and negotiations were begun with individuals in the United States of America, Austria/Bohemia, Prussia, Belgium and, possibly, Italy.[19] [29-40]

But their British patents had been under pressure since the 1780s. Their largest subcontractor, Wilkinson, [52] had begun to make Watt engines for his own use, other engine makers had patented pretended improvements and were selling their engines at highly competitive prices. When Jonathan Hornblower, one of these pirates, attempted to obtain an Act of Parliament in 1792, extending his patent of 1781, Boulton and Watt opposed him and the Bill was thrown out at the second reading. In the following year Boulton and Watt brought an action against Edward Bull in the Court of Common Pleas. This case dragged on, being overtaken in 1796 by a case against further pirates, Jabez Hornblower and his partner Maberley. [53] The Court found for the plaintiffs and, after being taken to the King's Bench, the judges unanimously affirmed the judgement of the Court of Common Pleas in January 1799.[20] During the interval between the trial against Bull and the final opinion of the judges, the Court of Chancery granted injunctions founded on the verdict of the jury, restraining further piracies. Accordingly, M.R. Boulton and Watt Junr., with the assistance of employees and friends, began an energetic anti-piracy campaign which reached a peak of activity in 1796-7. 'But for the good sense and indefatigable activity of Mr. Boulton Junr. and my son,' Watt remarked, 'we must have succumbed, for want of animal life and spirits.'[21]

Where patent protection was secured overseas, it was ineffectual. Not only did Boulton and Watt break their own exclusive sales agreements, by-passing the Soho-appointed sole-selling agencies in two out of the three countries in which privileges were obtained, but they made no attempt to take proceedings against the infringement of their rights in any country. If the aim in obtaining overseas patents was to increase their sales abroad, they were a failure. In spite of much interest in Watt's engine in France, few orders materialised. Spain could not have been expected to produce more than a very few orders and the prospects in the Netherlands must have been strictly limited. But, on the other hand, overseas patents were secured at no financial loss to the partners and, since they served to spread information about the Watt engine, they could be regarded as an inspired form of advertising.[22]

THE FIRST ENGINES

During the first year of the partnership four steam engines were ordered, the first one to be erected being the 50″ cylinder pumping engine at Bloomfield Colliery which was set to work in March 1776.[23] A 38″ cylinder blowing engine for the ironmaster, John Wilkinson, was set to work at about the same time, the first of a large number of Watt engines to be erected by Wilkinson.

During the first ten years of production Boulton and Watt generally sold fewer than ten engines per year, only once producing as many as fifteen. The annual output occasionally rose to twenty or more engines during the 1780s, although the total annual horsepower produced was not necessarily any greater. Production did not ever reach thirty engines per year, however, until after the opening of Soho Foundry (Table 1).

For the first seven years of the partnership, production consisted solely of pumping and blowing engines. Orders from the home market predominated with those for Cornwall playing an increasingly important part, for mine drainage was particularly costly in that area on account of the high coal prices. [124-131] Some of the largest reciprocating engines made by Boulton and Watt went to the Cornish mines, one with a horsepower equivalent of 95 h.p. being ordered by Poldice as early as 1778. Boulton's prediction of the extent of the market for rotative engines proved correct, however, and they rapidly outnumbered reciprocating engines in the order books. Three rotative engines were ordered in 1783 and by the end of 1795 rotative engine orders totalled 139.

The first overseas order was received in 1778, being followed by two more in the subsequent year. All three engines were for French customers. By the end of 1795, sixteen engines had been ordered by overseas customers from France, Belgium, Germany, Italy, Spain, the Netherlands, Sweden and India and by this date the total British orders for Boulton and Watt engines had risen to approximately 282.

For a number of years it was believed that the output of Boulton and

Watt engines during the period covered by Watt's patent represented 'the entire steam engine building of the country, if not the whole world'.[24] This has been proved incorrect by several recent writers;[25] indeed Boulton and Watt's share of the late 18th century British steam engine market seems to grow smaller and smaller as references to further atmospheric and pirate Watt engines are discovered. No accurate data exist on the total number of engines erected by 1800 either in Britain or in other parts of the world, although it is clear that, by then, several countries, including France, Germany, Belgium, Russia and the U.S.A., had their own indigenous engine industries besides Britain. And no attempts have hitherto been made to estimate the number of Boulton and Watt engines made after 1800.

Table 1: Estimated annual Boulton and Watt engine sales in Britain 1775-1800 (numbers of engines)

Year	*Reciprocating*		*Rotative*		*Total*	
	per year	*cumulative total*	*per year*	*cumulative total*	*per year*	*cumulative total*
1775	4	4			4	4
1776	5	9			5	9
1777	8	17			8	17
1778	9	26			9	26
1779	10	36			10	36
1780	10	46			10	46
1781	4	50			4	50
1782	2	52			2	52
1783	7	59	3	3	10	62
1784	9	68	6	9	15	77
1785	10	78	9	18	19	96
1786	13	91	8	26	21	117
1787	2	93	9	35	11	128
1788	3	96	11	46	14	142
1789	8	104	12	58	20	162
1790	11	115	10	68	21	183
1791	7	122	18	86	25	208
1792	1	123	17	103	18	226
1793	6	129	11	114	17	243
1794	6	135	7	121	13	256
1795	8	143	18	139	26	282
1796	6	149	25	164	31	313
1797	5	154	16	180	21	334
1798	5	159	18	198	23	357
1799	13	172	34	232	47	404
1800	11	183	36	268	47	451
Totals:	183	183	268	268	451	451

Source: Boulton and Watt Mss; Letter Books, Engine Books, Order Books.

The first attempt to estimate the total number of Bolton and Watt engines erected in Britain and overseas during the term of Watt's patent, was that of John Lord who suggested that 325 engines, producing c.5163 h.p., were erected by 1800. Of these, 289 were erected in England, eight in Wales, twenty-one in Scotland and seven in Ireland, France and Spain.[26] H.W. Dickinson pushed the figure to 484 British engines and twelve overseas engines, averaging them at 15 h.p. each, which is fairly near the mark, to produce an estimated total horsepower of 7440.[27] This figure found general acceptance until recently. Dickinson gave no source for his total but a count of the engines recorded in the Catalogue of Old Engines reveals that this is the basis for the figure. J.R. Harris[28] and A.E. Musson[29] both state that 'nearly 500' Boulton and Watt engines were erected whilst J.A. Robey and J. Kanefsky[30] have suggested that the figure should be increased to 512. Von Tunzelmann, however, points out that Dickinson both double-counted several engines and under-estimated the average power of each one and he estimates the total at 478 engines producing c.12,750 h.p.[31] But these figures are also too high and, whilst Lord's figures are certainly too low, the actual number of Boulton and Watt engines ordered for the British market by the end of 1800 would seem to have been about 183 reciprocating and 268 rotative engines, a total of 451, producing 11,251 h.p., an average of 25 h.p. per engine (Table 1). In addition to these, twenty-four engines producing 773.4 h.p. were ordered by overseas customers.[32] Seven of these, producing 81 h.p. were countermanded, the remaining seventeen engines producing 692.4 h.p. being despatched (Table 2).

Table 2: Overseas Orders for Boulton and Watt engines 1775-1800 inclusive

Country	*Total no. engines ordered*	*Total h.p. ordered*	*No. of engines counter-manded*	*Counter-manded h.p.*	*Total h.p. exported*
France	5	261	1	16	245
Belgium	1	10	1	10	0
Germany	3	19.4	1	8	11.4
Italy	1	28	0	0	28
Spain	5	263	0	0	263
Netherlands	3	75	0	0	75
Sweden	2	17	2	17	0
Russia	2	70	0	0	70
Austria	1	10	1	10	0
India	1	20	1	20	0
Totals:	24	773.4	7	81	692.4

Source: Boulton and Watt Mss; Letter Books, Engine Books, Order Books.

ORGANISATION OF ENGINE PRODUCTION

It has been stated that for the first twenty years, Boulton and Watt were not engine manufacturers at all, but rather, consulting engineers in that they merely provided drawings and supervised the erection of the engine whilst the customer could have the parts made where he wished. Nor, it is suggested, did they wish to become executive engineers, being prepared to lease that part of the business to local millwrights and engineers.[33] This is a rather too simple view of the organisation of the steam engine business, however, and Boulton and Watt's letters are inconsistent on many points of business policy anyway. The partners made such claims as suited them at a particular moment in time and customers rarely compared notes on the terms on which they were provided with engines. A fact which, in the circumstances, was fortunate.

Seven years before the engine partnership commenced, Boulton had declared his intention of building an engine works adjoining his toy factory, Soho Manufactory, where Boulton and Watt would be able to 'serve all the world with engines of all sizes'.[34] When Watt eventually arrived in Birmingham no further mention of the engine factory was made. [66] Watt worked mainly from his home, Regent's Place at Harper's Hill. Some of the smaller engine parts, such as nozzles, which required special skills were made at an engine shop at Boulton's Soho Manufactory, but the larger forged and cast work was subcontracted.[35] [73]

No single subcontractor was chosen to make all the engine parts, the partners selecting specialists for particular pieces.[36] And they were not necessarily obtained from the nearest, but from the best suppliers. Rolled iron plates were purchased from Burton Forge and Marston Forge, tubes from Izon of West Bromwich, piston rods from Jukes Coulson of Rotherhithe and boilers from Horton of West Bromwich. For iron castings Boulton and Watt depended largely on the Coalbrookdale Company and John Wilkinson, particularly the latter. Wilkinson had patented a machine for boring cylinders in 1774 and whilst there was some choice of suppliers for other parts, the partners insisted on Wilkinson for cylinders and such parts as related to the engine end of the beam... 'as there is no other proper apparatus in Britain for producing the parts with the truth and exactness we require'.[37] Some other founders, the Coalbrookdale Co. amongst them, installed boring machines on Wilkinson's plan but they seem to have been unable to match his quality at the outset. Close links with as eminent a caster as Wilkinson were advantageous, for his advice was frequently sought on engine design problems besides being proferred unasked. By the early 1780s Wilkinson had some kind of exclusive right to the manufacture of cylinders [69] for Watt remarked to Boulton that he felt that they should feel free to order all the engine parts from Coalbrookdale if that Company procured the engine order. Indeed there was a potential problem with all iron founders who naturally wished to cast their own engine parts. [70] Only very rarely did Boulton and Watt waver in their resolve to insist on selecting the suppliers of precision parts and they never sold manufacturing licences to other engineers despite a number of requests.

Boulton and Watt gradually increased the range of engine parts made at Soho and by the early 1790s were capable of producing over 50% of the parts by value. There were several reasons for this increase in manufacturing capacity. First, whilst in some years a loss was incurred on Soho-made parts, the partners must have wished to secure for themselves a greater share of the profit made by their subcontractors. Whilst they claimed never to surcharge subcontractors' work and, because of the high prices charged by Wilkinson probably never did his, they nevertheless made it a practice to surcharge other subcontractors' work where the market could bear it. And costing remained a problem with subcontracting.

A more important reason was the slow rate of throughput, the difficulty of co-ordinating subcontractors and assembling all the parts of an engine in a given place at a given time. Leaving aside such issues as the problems of transporting large castings by ship, and the frequent freezing of canals in winter, entrepreneurs needed to plan the installation of their engine, synchronising this with the low or dead season in their manufacturing cycle. One of Boulton and Watt's constant problems was that of estimating the waiting period between order and delivery. With the manufacturing in their own hands, a more accurate prediction was possible, at least in principle.

The greatest problems, however, were the lack of standardisation and quality control. All the early engines were custom-built and, whilst the partners sought suppliers of renown, there were frequent complaints of poor quality parts. Subcontractors, both large and small, from Wilkinson and the Coalbrookdale Co. to the Eagle Foundry, Birmingham, were accused of bad workmanship. Even the sun and planet rotative wheels made at Soho Manufactory frequently broke and Boulton and Watt adopted the policy of sending a spare set with most distant British and all overseas orders. [74-82]

Subcontracting held considerable advantages for Boulton and Watt, however. Parallels can be observed with The Proprietors of the Invention for Raising Water by Fire (1715) who charged an annual rent in exchange for their licence to erect an engine, but the latter generally required customers to order their own parts whilst Boulton and Watt almost always arranged for the manufacture of the important parts and, from the outset, produced some themselves. One of the great advantages lay in the small capital base from which the partners could operate. Little fixed capital investment was required and by sharing some employees with Boulton's other partnerships, and keeping their own employees to a minimum, the Boulton and Watt partnership was not adversely affected by Boulton's severe liquidity problems of the 1770s.[38] Moreover, Watt could continue his experimental work and practical engine developments whilst subcontractors bore a large proportion of the risks.

Soho Foundry

There can be little doubt that it was John Wilkinson's piracy, coupled with the problems of throughput and quality control, which, combined with the fact that Watt's patent would expire in 1800, prompted the partners to build their own engine foundry. Rather than extend the engine workshops at Soho Manufactory, a new canal-side site was chosen at Smethwick and Soho Foundry was begun in 1795, being opened in January 1796. But the Soho Engine Manufactory continued in existence and was redesigned between 1800 and 1803. Boulton and Watt sought the advice of their friends and employees both on the overall plan of the new foundry and on particular machines and operations:[39]

> I told you yesterday that we should make a dash at the foundery this season — 60 foot sqr is the size we propose. Please to give us all your ideas upon the subject respecting the position of the furnaces, stoves, cyl^{r}, pit &c. and the abstract of your observations upon the founderies in the North. You must let us have them soon as we are to commence our operations immediately not to lose this astonishingly fine season.

William Wilkinson provided much guidance on the foundry plan although not all his suggestions were adopted:[40]

> We have however launched out too widely in our infant establishment at Smethwick to be able to recede, having already laid out upwards of £2,000 which sum will be nearly doubled before the year expires ... I had almost forgot to thank you for your former obliging favour some parts of which we have adopted ... but in general have disposed the works differently so as to suit the peculiarities of our business.

Ewart, amongst others, contributed to the planning and design of the machinery, whilst the all-important boring operation was observed at competitors' foundries:[41]

> We went to Banks and Onions Works and saw their boring mill and some large cylinders for a blowing engine — but they are not well done, nor do I think they have aimed at accuracy. Their

> machine is the same as the Dale Co.s except perhaps ½ inch diameter of their rod which is less (B & Os is less) — I think their castings are as well done as the Dale Co.s but their notions of accuracy from what I saw appear inferior to the Dale Co. and theirs are very short of Mr. Wilkinson's performance.

Boring was still proving a problem several years later:[42]

> There is no boring rod in the kingdom (because there is not one at Soho) that can bore a cyl^{r} long enough for a 10 feet stroke, so I have calculated only on a 9 feet stroke.

In due course houses were built for some of the foundry employees, one row surviving to the present day.

Soho Engine Manufactory had, from 1794, been run by a new partnership 'Boulton, Watt & Sons' which, besides its original partners, included M.R. Boulton, James Watt Junr. and Gregory Watt. The shares were distributed in the proportion laid down in 1775 but when in 1800 the old partnership was dissolved the shares were equally divided between the Boultons and the Watts, the style being changed to 'Boulton, Watt & Co., Soho'. James Watt retired from business, handing his share to his son, and, on Boulton's death in 1809, Soho Engine Manufactory was in the sole ownership of M.R. Boulton and James Watt Junr. Soho Foundry was owned, and run, from the outset, as a separate concern, by Boulton and his son and James Watt Junr. and Gregory Watt. James Watt, although he had purchased the land on which the foundry stood and maintained a close interest in its affairs, had no financial interest in the concern. In 1800 three of the firm's most senior and able engineers were granted a share in the profits. John Southern, originally engaged as a draughtsman, was given a 2¼% commission on all goods manufactured at Soho with a guaranteed minimum of £600 and was admitted as a partner in 1810, receiving one-sixth of the profits. James Lawson, the supervising engineer, was given a 1% commission on all goods sent to Scotland and ½% on all orders procured by him in England with an initial annual guaranteed minimum of £350; whilst William Murdock was also given 1% with a salary of £300. He was offered a partnership in the firm in 1810 but declined it, accepting £1,000 p.a. instead.[43] In 1814 and 1815 some discussions took place between Watt Junr. and Peter Ewart with a view to the latter joining the engine partnership, but nothing materialised. [6-9]

With the opening of Soho Foundry, twenty years after the commencement of the Boulton and Watt partnership, the firm became, in principle, complete engine manufacturers, although in fact they continued to subcontract to some Birmingham and Black Country firms for a few years. Although the foundry was small at the outset, the partners had planned for expansion and, from its opening,

Soho Foundry was capable of producing a larger number of engines per year than was possible with the former subcontracting system.

In 1796, the year the Foundry was opened, six reciprocating and twenty-five rotative engine orders were accepted and by 1800, when Watt's patent expired, about 169 engines had been made at the foundry, nearly 60% of the number of engines produced during the twenty years of subcontracting. The opening of Soho Foundry marks an important watershed in the history of the Boulton and Watt engine. The foundry was divided into three sections: loam, dry sand and green sand casting and within each section there were subdivisions into classes which related to the premiums paid to the moulders. Much of this planning seems to have been carried out by M.R. Boulton. [108, 114]

The quality of the goods produced at the foundry in the first few years was frequently poor, however. Green sand casting techniques were inferior to those of Matthew Murray at Leeds and Watt Junr. himself engaged in industrial espionage in an attempt to learn more of Murray's methods. Murray had earlier enticed Halligan away from Soho and Watt Junr. managed to induce him both to divulge the secrets of Murray's success and to return to Soho. [59, 60] Watt seems to have seriously considered building a branch foundry alongside Murray's but, instead, the partners reorganised both Soho Foundry and the Engine Workshop at Soho Manufactory between 1800 and 1803.[44]

> It will be highly essential for us to make a general revision of our establishment both at the foundry and here. Both urgently require a reformation in the general superintendance of the workmen who especially at the foundry have got completely the upper hand of Vivian and, in some respects, of Murdock.

A month later they could report:[45]

> We have made considerable additions to the works at Soho as well as at the foundery but no part is yet completed. However we are now making rather more than an Engine per week and expect to do one half more in the course of the ensuing year.

LABOUR

In 1802 a total of fifty-four men and boys were employed at Soho Foundry. Of these, thirty-four, organised in nine teams, were piecemen, employed in various casting operations whilst the remaining twenty were day men. The piecemen comprised eleven foremen with assistants and apprentices who were employed by Boulton and Watt but paid by the foremen. [114] As the foremen's agreements demonstrate, [115] a variant of the butty system prevailed at the foundry and, in this, Soho Foundry was no different from the vast majority of coal and iron works in the West Midlands at the time. The day men were all employed and paid by Boulton and Watt and were engaged in what could be regarded as the central services at the foundry required by all piecemen: tending air furnaces, loam mixing, smithing and scrap picking.

During the early years of the Boulton and Watt partnership the majority of employees had been paid a weekly wage. But piece rates had been discussed in the later 1770s and by the 1790s piece rates had been applied to those articles that had become reasonably standardised.[46] At Soho Engine Manufactory piece rates were applied most widely in fitting. Some were flat rates payable for a certain article fitted, others varied according to the diameter of the article, whilst others varied with the horsepower of the engine. Piece rates had traditionally been employed in the iron foundry trades, being related to the weight of output and agreed to for different classes of goods. And after the opening of Soho Foundry the proportion of employees working for piece rates increased.

In order to arrive at profitable piecework rates a detailed study of output, materials' prices and previous wage rates had to be made and, where possible, compared with those of other founders. [111] The manufacture of almost every component of a steam engine was studied before any offers were made to the foremen and once piecework was introduced, output and wage bills were closely monitored for several months. Much of this work was carried out by Abraham Storey, formerly John Wilkinson's chief caster, with the close involvement of M.R. Boulton and James Watt Junr. [120]

A premium system was introduced which consisted of an agreed bonus payable to the foremen on completion of an agreed quantity of work. This, as E. Roll has observed,[47] would act as an incentive to speed up production whilst no reward was payable to the other employees in the team.

There is little material amongst the firm's papers on the subject of labour relations. The engine smiths struck in 1791 on the settlement of certain piecework rates but the dispute only lasted a couple of days. More serious unrest occurred in 1814, however, but the matter was dealt with summarily by Watt Junr:[48]

> In consequence of the attempt made by the Moulders to dictate to me by forming a combination and proceeding in a body to this place upon two successive days under a pretence of grievances which have no existence, I have given them notice that the piecework will be taken away from all those who have acted in this improper manner and not restored until I am satisfied that they are sensible of their fault and will pursue a more correct line of conduct in future. I have entirely discharged from our service

> Thomas Middleton as the ringleader in this affair and have given him notice to give up his house at the end of the ensuing week, he will therefore let his account be made out and settled tomorrow morning. There are two other men whom I suspect of having taken an active part and whom I shall also discharge when the fact is fully established. The settlement of the piecework has for several years been made annually at the periods when it was convenient to me which as far as I can recollect has always been later than this. I am aware of no reason and none has been stated to me why it should take place earlier this year than usual. If however any reason did exist it was easy for the men to have applied to me individually when I came to the foundry as they well know that I am always ready to hear and regress [sic] their complaints when properly stated. You will read this letter to the Moulders in the Foundry to impress better upon their minds the substance of what I stated personally.

Crucial to the success of the Boulton and Watt engine both at home and abroad was the existence of an efficient group of itinerant Soho-trained engine erectors. One or more men were sent from Soho, at the customer's expense, to superintend the erection of practically every engine sold in Britain and many sold abroad. Some early overseas customers made the sending of a mechanic a condition of sale, but these men usually failed to return, being offered more lucrative work abroad.[49] [101, 102] After a few years Boulton and Watt declined to accept this condition but usually managed to recruit a millwright or mechanic especially for the customer, giving him sufficient training in engine maintenance before he went. Engine erecting was not over-popular with some older Soho mechanics:[50]

> Yesterday we despatched per Coach your two intended elévés... you know of what importance it is to us to have a good stock of Engine erectors in order to render us independent of such of our old hands as are become too great men to act in a subordinate capacity and to enable us not to keep engines waiting for the want of these gentry and always to have two or three spare men in the yard to be employed in cases of emergency.

Some emigrating engineers and mechanics, previously unconnected with Soho, offered their services to Boulton and Watt on their arrival overseas. [190] Such men provided valuable links in the diffusion chain. In the home market Boulton and Watt had links with former employees in London (the engineer John Rennie and Richard Dayus, a millwright) and the West Country (George and James Haden), whilst establishing their own agencies in Scotland and Lancashire through Hutton and the Creighton brothers. West of England orders increased rapidly after the Haden brothers settled in Trowbridge to start a millwrighting business[51] and, on being suplied with current price lists, these agents filtered off many of the preliminary enquiries from prospective customers.

MARKETING

After-sales service became increasingly important from the early 1790s and in this the agents and senior mechanics performed an important role. [207-213] Moreover, when any of the partners were away from Birmingham, they took the opportunity of visiting customers in the area, realising that efficient engines and satisfied customers were the best advertisements for their engines.

An important aspect of Boulton and Watt's British marketing was the identification of key, innovating entrepreneurs in different industries. Some, like Arkwright, were approached by a partner in an endeavour to sell the engine. [140] Others came to Boulton and Watt. [142] The result was that in a number of industries prospective customers could be encouraged to follow the precedent set by the larger innovators. Samuel Whitbread, Peter Drinkwater, the Robinsons of Papplewick, John Wilkinson, the Coalbrookdale Co., the Albion Mill Co. and Josiah Wedgwood performed an invaluable advertising function in this respect. Prospective overseas customers too sought out their British counterparts in the hope of viewing an engine. [176] But such direct sales drives as were attempted overseas had little effect. Pull was more effective than push in the foreign market.

Until the expiry of Watt's patent the main source of profit to the firm was the royalty for plans and licence to work the engines payable by each customer. For the early reciprocating engines this was computed at the value of one-third of the savings in fuel effected by Watt's engine compared with an atmospheric engine. It was, therefore, not in Boulton and Watt's interest to sell engines to areas where coal was cheap and herein lay the great attraction of the Cornish market. [85] Watt made detailed calculations of the possible annual profit accruing to Boulton and Watt in areas of cheap coal before agreeing or declining to supply an engine. [83] Customers too, made their calculations and where an engine was required in a cheap coal area, or for only part of the year, a Watt engine was not the most economic solution. [160, 161] Royalties were payable in half-yearly instalments and because of both the expense and difficulty in finding suitable engines for comparative trials, Watt compiled tables giving a constant for engines of all sizes based on elaborate comparative tests at a few sites. In order to measure more effectively the actual work done by his engine, Watt devised a counter which was affixed to the engine beam and thereafter the premium was related to the actual number of strokes made.[52]

A new system of charging was introduced for rotative engines. Since their sizes were measured in terms of nominal horsepower, £5 per horsepower per year was charged in the provinces and £6 in London. In the 1790s, customers were offered the chance to commute the premium at discount by payment of a lump sum on purchasing the engine [89] and the majority of new engine purchasers did. Few actual receipts for the payment of premiums survive but if the figure of £5 per h.p. per year is applied to all engines,[53] the power of reciprocating engines being converted to horsepower equivalents, the partners should have received something of the order of £26,405 by the end of 1780 and a further £164,415 by the end of 1789.[54]

Table 3: Retail prices of Boulton and Watt[a], Murray[b] and Trevithick[b] engines

h.p.	*Boulton and Watt* *Total price excl. boiler* £	*£ per h.p.*	*Murray* *Total price excl. boiler* £	*£ per h.p.*	*Trevithick* *Total price excl. boiler* £	*£ per h.p.*
2			175	87	221	111
3					284	95
4	350	88			347	87
6	385	64	380	63	441	74
8	470	59			494	62
10	520	52	510	51	536	54
12	560	47				
14	600	43			750	54
16	710	44				
20	820	41	600	30		
24	1,020	43				
25			715	29		
30	1,160	39	830	28		
32	1,200	38				
35			938	27		
40			1,045	26		
45	1,526	34				

[a]at 1801
[b]at 1804
Sources: Boulton and Watt Mss 'Calculations of the Costs and Prices of Small Engines Oct. 9 1801'; Goodrich Mss (Science Museum) 438, 1511.

The capital cost of a Boulton and Watt engine was, on the whole, higher than that of an atmospheric or Watt-type engine of another make both before and after 1800, the cost per h.p. decreasing with engine size (Table 3). Assuming, for the moment, that every purchasing decision was entirely rational, there were certain basic advantages and disadvantages to purchasing a Watt engine. The chief advantage lay in the lower coal consumption of Watt's engine, plus the reduced chance of engine failure. On the other hand, the requirement of an annual premium, plus the fact that the engine was more complicated, were in some cases the over-riding factors which led to the adoption of an atmospheric engine. But reliability and complexity, at least in principle, were constants. So was the premium if the engine in question was a rotative one. The variable element in running costs was the price of coal. G.N. von Tunzelmann[55] has calculated the threshold at which it would have become economically rational to purchase a Watt engine at about 1790 and has suggested that if an entrepreneur was, on other grounds, indifferent between a Watt and an atmospheric engine, he would select an atmospheric engine where coals cost less than 7s 10d. per ton outside London. If, however, the entrepreneur already had an atmospheric engine installed, it would have paid him to scrap it only when its running costs rose above the *total* costs of the new Watt engine. But towards the end of the 1790s the balance changed sharply in favour of Watt engines and this may go some long way towards explaining why there was a dramatic increase in Boulton and Watt engine orders at the turn of the century.

On the expiry of Watt's patent in 1800, open and increasing competition was experienced from other manufacturers. But demand was great and Boulton and Watt's order books were full:[56]

> We are extending our manufactory of small Engines but at this time the demand for them is far greater than our means of execution and until the addition to our establishment is completely in activity we could not engage to execute any order for Engines of this construction in less than 8-10 months.

Practically all the competitors undercut Boulton and Watt's prices, some quite considerably. When, for instance, the Dutch government ordered a pair of 70 h.p. marine engines in 1825, Watt Junr. quoted £9,000, noting that Fawcett of Liverpool had quoted £8,000 for a pair of 75 h.p. engines.[57] Increasingly from the second decade of the 19th century, Boulton and Watt's home and European customers were governments or public works enterprises, or customers replacing or augmenting an earlier model.

In 1800, 47 engines producing 1083 h.p. were ordered by British customers. Fifty-four engines producing 1017 h.p. were ordered in 1802 and forty-nine producing 1044 in 1803. In 1804 the figure fell to thirty-two engines producing 590 h.p., the figures for the previous years not being exceeded until 1824 (Table 4). In 1813 the first two pairs of marine engines were ordered by a British customer and by 1825 marine engines producing a total of 3084 h.p., valued conservatively at £201,674, had been ordered. Watt Junr. was closely involved in the promotion of the shipping interest and with their agent, James Brown, based in London, obtained a number of important and lucrative contracts for the Post Office, the Navy and some independent shipping lines. [197-206] In the early years of steam shipping, constant attention had to be paid to engine and hull design as the many letters that passed between James Brown and William Creighton demonstrate. Creighton's letters are particularly refreshing with their frank comments, often written in pencil and sometimes illustrated. Marine engines played an increasingly important role in the overseas market too, forty engines producing 970 h.p. being ordered, or nearly 40% of the total h.p. exported, between 1775 and 1825,[58] West Indies excluded.

Table 4: Estimated annual Boulton and Watt engines sales in Britain 1801-1825

Year	*Number of engines per year*	*Cumulative total of engines from 1775*	*h.p. per year*	*Cumulative total h.p. from 1775*
1801	47	498	1,352	12,603
1802	54	552	1,017	13,602
1803	49	601	1,044	14,664
1804	32	633	590	15,254
1805	28	661	697	15,951
1806	25	686	596	16,547
1807	18	704	359	16,906
1808	20	724	575	17,481
1809	23	747	549	18,030
1810	35	782	943	18,973
1811	21	803	556	19,529
1812	20	823	607	20,136
1813	13	836	173	20,309
1814	19	855	499	20,808
1815	23	878	394	21,202
1816	9	887	258	21,460
1817	10	897	172	21,632
1818	11	908	263	21,895
1819	19	927	664	22,559
1820	20	947	374	22,933
1821	19	966	478	23,411
1822	17	983	542	23,953
1823	22	1,005	704	24,657
1824	36	1,041	1,571	26,228
1825	54	1,095	2,078	28,306
Totals:	644	1,095	17,055	28,306

Source: Boulton and Watt Mss; Letter Books, Engine Books, Order Books.

The estimated cumulative fixed capital formation in Boulton and Watt engines in Britain was £937,323 by 1800. It had exceeded £1m by 1802 and £1.5m by 1821. By 1825 an estimated £1,791,432 had been invested in 28,306 h.p. of stationary and marine Boulton and Watt engines in Britain.[59]

Partington estimated that a stock of 10,000 engines of all makes producing at least 200,000 h.p. existed in Britain in 1822.[60] This figure probably includes marine engines but Musson suggests that the figure is too high even allowing for this. He suggests that by 1824, 5000 stationary engines producing 100,000 h.p. may be a more reasonable figure.[61] By this date 978 stationary Boulton and Watt engines (out of a total of 1041 British engines manufactured by them) producing 24,144 h.p. had been installed and, whilst data does not exist on the number of Boulton and Watt engines that had been scrapped during the fifty-year period to 1825, it is known that the majority of them had a long working life and second-hand prices remained high.[62] It may, therefore, not be too fanciful to suggest that Boulton and Watt engines could account for between one-sixth and one-fifth of the engines and from one-fifth to one-quarter of the horsepower in Musson's estimated total.

As with stationary engines, the data for an estimate of the total stock of marine engines are sparse. But it is known that Brunel and Fawcett, besides several Clydeside engineers, were successfully competing in this expanding market. Moreover, in the land engine market Boulton and Watt had active and successful rivals. In 1824, for instance, of 129 engines producing 2318 h.p. known to have existed in Leeds, only seven were Boulton and Watt engines compared with seventy made by Fenton Murray and Wood.[63]

It may, therefore, not be unreasonable to suppose that the cumulative total power of all steam engines erected in Britain between 1750 and 1824 lay somewhere in the region of 150,000 h.p. If this figure is accepted, Boulton and Watt manufactured 1 h.p. for every 4.7 h.p. supplied by all other makers during the period. With the growth in both the number and scale of rival establishments, Boulton and Watt's share of the total engine stock must have fallen fairly rapidly in the 1820s, however. It is not unlikely, therefore, that a cumulative total of £10,000,000, or more, fixed capital had been invested in steam power in Britain by the end of 1825, to which Boulton and Watt had contributed nearly 18%.

THE BOULTON AND WATT PAPERS

The core of this magnificent collection of business papers was presented to the City of Birmingham in 1911 by Mr. George Tangye and deposited in the Central Library in 1915. The documents had been acquired by George Tangye when Soho Manufactory was demolished in 1863 and they were subsequently displayed in his Cornwall Works. The collection was considerably augmented in 1936 and again in 1946 by Messrs. W. & T. Avery who had acquired Soho Foundry in 1896. In 1921 the Muirhead Collection of Boulton and Watt papers was presented to the Central Library by Mr. J.B.C-L. Muirhead. These papers had been used by J.P. Muirhead for his *Mechanical Inventions of James Watt* (1854) and *Life of James Watt* (1859) and probably originated from Doldowlod, the Watt family's country house, where a number of private papers of James Watt and James Watt Junr. still exist.[64]

Other deposits have subsequently been made, including two scrapbooks collected by Samuel Timmins, a volume of letters written between 1796 and 1808 by Gilbert Hamilton, an employee, to Robert Hamilton and a volume of topographical drawings made by William Creighton, an employee.

The Boulton and Watt Papers consist, primarily, of the firm's own business records, including incoming letters relating to engine sales; together with some personal papers of James Watt and his family. Vast though some of the series of business records are, no one series before 1825 is complete, and while the history of a single engine transaction may be reconstructed, it is not always possible. The Land Engine Books, apart from one for Cornish engines, covering the period 1779-86, run from 1797 to the 1840s (23 volumes) and Boat Engine Books (13 volumes) from 1815 to 1853 and these contain a record of the majority, but not all, of the engines made during these years. They are also valuable as a source of information on costing and profit levels. Cost, not retail prices, are entered for each customer and these can frequently be compared with retail estimates contained in letters.

There are 156 volumes of financial records including nineteen Account Books and thirty-six Cash Books running from 1795, fourteen Inventory volumes from 1797, twenty-one Day Books from 1795 and ten Piecework Books from 1825. There are sixty-two volumes of Order Books, Castings Books and miscellaneous Supplies Books and thirty-two Packing Books, whilst eight volumes cover despatch and transport. Rentals of houses and other property cover six volumes and there are four volumes of Watt family papers.

But without doubt, the most outstanding section of the written records generated by the firm is the collection of 183 letterbooks in five main series, most of them press copies. These contain copies of letters between the partners as well as to their agents, but the majority of the letters consists of replies to customers, both potential and actual. In many cases the original incoming letter has been lost but it is possible from these very full replies to deduce the subject of the enquiry. The Office Letterbooks run from 1775 to 1828 in forty-eight volumes, whilst the Foundry Letterbook series commences in 1791 and runs to 1847 in fifty-seven volumes. There is considerable overlap between them, an engine transaction frequently appearing in both series. Firm Letterbooks commence in 1828, running to 1850 in seventeen volumes; letter books begin in 1848 and run to 1886 in twenty-three volumes and London Letters run from 1876 to 1893 in ten volumes. In addition to these there are twenty-eight miscellaneous volumes including letters of James Watt, Cornish letters and others, amongst which there are a number of original letters. In spite of such wealth of material, something in the region of 37,000 letters covering the period 1775-1825 alone, the late 1780s to early 1790s are only imperfectly covered by the letterbooks and unfortunately the other business records are less than complete for this period also.

Complementing the letterbooks is a series of some 15,000 letters and papers originating from customers and, to a lesser extent, from the Soho partners and their employees. These range from early enquiries for steam engines from potential British and overseas customers to discussions on business policy between the partners and instructions to their agents and mechanics. Since Boulton and Watt corresponded, at some time or other, with many of the greatest manufacturers, scientists and engineers of their day, besides many of whom very little is known, these incoming papers, together with the letterbooks, provide a unique insight into the development of industrial society in the late 18th and early to mid-19th centuries.

In addition to the written records there are approximately 36,000 engine drawings in 1,339 numbered portfolios, besides some miscellaneous drawings in a few un-numbered portfolios.

These drawings consist of complete views of engines, some of them coloured for the customer, some of them 'reverse', being press copies of the originals; besides many drawings of parts, some on a large scale for the foundry pattern maker. Most of the engine drawings have the customer's name on, but where a customer had several engines within a short space of time and the drawings were not numbered identification is not always easy. Boulton and Watt generally supplied drawings for the engine and boiler houses and these often survive in the appropriate portfolio. Moreover, in order to assist the engineers in planning the dimensions, customers were often requested to send plans, sections and, where appropriate, elevations of their mills, mines and factories. Many of these survive and provide a unique record of the craft of the engineer, the millwright and the draughtsman in the industrial revolution.[65]

There are four catalogues to parts of the Boulton and Watt Papers. Harry Hazleton, a one-time apprentice of Soho Foundry, compiled a catalogue of engine drawings, in which he listed engines by type or use and county, giving a portfolio number for most of them. This Catalogue of Old Engines, often inaccurately referred to as The Engine Book, is extremely valuable but must be used with caution. It is not a complete record of all engines sold by Boulton and Watt and the subdivision by 'type' is inconsistent. Moreover, the list contains a number of second-hand Boulton and Watt engines which are entered as if they were new, besides certain 'pirates' which were not made by Boulton and Watt at all. A subject name and geographical index to this catalogue has been compiled by Miss Waveney Payne who also produced a two-volume name index to the letterbooks covering the years 1775-1800; and a name index to some of the incoming letters and papers. It is difficult to assess how much incoming material exists in addition to that listed, but it is quite considerable. One further major aid to researchers is the contemporary name index bound in at the front of most of the letterbooks which enables particular transactions to be traced after 1800.

MATTHEW BOULTON PAPERS

The Matthew Boulton Papers are, in many ways, complementary to the Boulton and Watt Papers. For many years cared for by the Birmingham Assay Office, they are now owned by the Matthew Boulton Trust, being placed on loan in Birmingham Reference Library. The collection comprises Matthew Boulton's and Matthew Robinson Boulton's personal papers together with papers relating to their various business enterprises including the mint, the merchant house, plate, buttons, and the Boulton and Watt engine partnership.

Incoming material and business papers are housed in boxes. Where there are more than thirty or so documents devoted to a particular topic, family or person, they are housed in one or more of the 113 'Accounts' boxes. Many of these contain lists of contents made by Mr. W.O. Westwood. Other incoming material is arranged in alphabetical order in forty-eight boxes and this is indexed in eleven folio catalogue volumes compiled by Mr. Westwood. There are, moreover, ninety-six notebooks and a number of other miscellaneous boxes, besides long runs of Letterbooks.

Material relating to the engine partnership is to be found in the nine boxes of James Watt and James Watt Junr. letters and in the Small, Roebuck, Wilkinson and Matthews letters as well as being scattered throughout the collection, for many prospective engine customers corresponded with Boulton rather than Boulton and Watt. Moreover it is the Matthew Boulton Collection which provides the documentation on the founding of the Boulton and Watt partnership.

SELECTION AND ARRANGEMENT OF THE DOCUMENTS

Some of the great riches of the records of the Boulton and Watt partnership have been known to scholars for many years but, after J.P. Muirhead's *Origin and Progress of the Mechanical Inventions of James Watt* published in 1854, it was not until 1969 that further selections from these archives were published. In that year Eric Robinson and A.E. Musson's *James Watt and the Steam Revolution* appeared in celebration of the bi-centenary of Watt's first patent and in the following year Eric Robinson and Douglas McKie's book *Partners in Science: Letters of James Watt and Joseph Black* was published. The originals of many of the documents printed in these two scholarly volumes are at Doldowlod, home of Major D. Gibson-Watt and they are not generally available to researchers. Moreover, the aims of these editors were, in the first volume, to select papers illustrative of Watt's patents and the law suits which followed them; and in the second, to print all the known letters between Watt and Black and Watt and Robson.

The aim of the present series is rather different. A selection of letters and papers has been made from the mass of material surviving from the partners, their employees, friends and customers, to illustrate themes in the organisation of industry in Britain and overseas during the early industrial revolution. The overseas material in the Boulton and Watt Papers is probably the least known part of the collection. The present volume covers the steam engine partnership, its beginnings, the patents at home and abroad, production problems, marketing policies and the leading innovators of steam power. Subsequent volumes will cover the application of steam power to different industries in Britain and overseas.

It has been observed that: 'It is an exceedingly difficult problem to select the letters that deal with any one theme from the mass of surviving correspondence relating to James Watt.'[66] To attempt this from the entire Boulton and Watt Collection has indeed been a daunting task. Wishing to avoid exercising the editor's prerogative too harshly by cutting passages from the papers, the selected letters and papers have, with two or three exceptions, been printed in their entirety. In the few cases where deletions have been made they are indicated thus...; spelling and contractions remain unchanged, as does punctuation save where the addition of a full-stop makes the document more intelligible. The date is printed at the top of the letters, even when it appears at the end in some of the originals and a full locational reference is given at the end of each document. Where the original is in the Boulton and Watt Collection the box or letterbook reference alone is given. If the original is in the Matthew Boulton Papers, the location has the prefix MB.

The documents printed in the present volume have been arranged in sections to illustrate themes in the organisation of the engine partnership. Each document is numbered and where reference is made to a particular document elsewhere in the volume it is denoted by the number in square brackets.

NOTES AND REFERENCES TO INTRODUCTION

[1]Much of this section is based on 'Letter proposed to be addressed to Mr. Prony by J. Watt in the year 1803 upon his Inventions relative to the Steam Engine, but the war breaking out it was never finished.' B & W Papers MIV Misc. 3. Further references to papers in the Boulton and Watt Collection, Birmingham Reference Library, will have no prefix. References to the M. Boulton Papers, formerly held at the Assay Office but now on loan to Birmingham Reference Library, will have the prefix M.B. Where no box or parcel number is given the document quoted will be found in the Letterbook of the appropriate date. Numbers in brackets in the Introduction refer to the documents which follow. H.W. Dickinson, *A Short History of the Steam Engine,* 1938, pp. 66-77; ibid, *James Watt, Craftsman and Engineer,* 1936, pp. 32-66.
[2]Illustrated in H.W. Dickinson, *James Watt* ..., p.58.
[3]'An Act for vesting in James Watt, Engineer ... The sole use and property of certain steam Engines ...' is printed in Eric Robinson and A.E. Musson, *James Watt and the Steam Revolution,* 1969, pp. 76-80.
[4]H.W. Dickinson, *James Watt* ..., pp. 67-82; D. King-Hele, *Doctor of Revolution, The Life and Genius of Erasmus Darwin,* 1977, pp. 68-69.
[5]Printed in Eric Robinson and A.E. Musson, *James Watt* ..., pp. 56-61. See also Eric Robinson, James Watt and the Law of Patents, *Technology & Culture,* 13, 1972.
[6]M.B. – M. Boulton to J. Watt, 7 Feb. 1769.
[7]See M.B. – Small Box.
[8]M.B.-J. Roebuck to M. Boulton, 17 May 1773, Roebuck/23.
[9]Notebook of engine experiments Parcel A.
[10]H.W. Dickinson, *James Watt* ..., p. 123.
[11]M.B. – M. Boulton to J. Watt, 21 June 1781.
[12]'Rotative Motions' notes by James Watt, 10 Nov. 1808, MIV Misc. 3.
[13]Printed in Eric Robinson and A.E. Musson, *James Watt* ..., pp. 89-95.
[14]Printed in ibid, pp. 96-108.
[15]Printed in ibid, pp. 111-122.
[16]M. Boulton to J. Watt, 6 Aug. 1784, Box 20.
[17]'Rotative Motions' ... MIV Misc. 3.
[18]Printed in Eric Robinson and A.E. Musson, *James Watt* ..., pp. 122-125.
[19]Jennifer Tann, Marketing Methods in the International Steam Engine Market, *Journ. Econ. Hist.,* 38, 1978.
[20]The cases are succinctly explained in J. Watt Junr. to F. Arago, 13 Oct. 1834, printed in Eric Robinson and A.E. Musson, *James Watt* ..., pp. 155-158. See also Jennifer Tann, Mr. Hornblower and his Crew: Pirates of Watt's engine in the late 18th Century, *Trans. Newcomen Soc,* 50, 1980.
[21]J. Watt to J. Black, 1 June 1796, quoted in Eric Robinson and Douglas McKie, *Partners in Science* ..., pp. 224-25.
[22]Jennifer Tann, Marketing Methods ...
[23]A computer catalogue of Boulton and Watt engines 1775-1825, based on entries in letterbooks, incoming letters, engine books, drawings etc., has been compiled at the University of Aston. This will be published in due course.

[24]John Lord, *Capital and Steam Power,* 1923, p. 166.
[25]See below.
[26]John Lord, *Capital* ..., pp. 167-175.
[27]H.W. Dickinson, *Matthew Boulton,* 1937, p. 176.
[28]J.R. Harris, The Employment of Steam Power in the Eighteenth Century, *History,* 52, 1967.
[29]A.E. Musson, Industrial Motive Power in the United Kingdom, 1800-70, *Econ. Hist. Rev.,* 29, 1976.
[30]J.A. Robey and J. Kanefsky, Appendix in L.T.C. Rolt and J.S. Allen, *The Steam Engine of Thomas Newcomen,* 1977.
[31]G.N. von Tunzelmann, *Some Economic Aspects of Steam Power in the British Isles to 1856 with Special Reference to Textile Industries,* Oxford D.Phil. Thesis, 1974, p.33; ibid *Steam Power and British Industrialisation,* 1978, p.29.
[32]Jennifer Tann and M.J. Breckin, The International Diffusion of the Watt Engine, *Econ. Hist. Rev.,* 31, 1978.
[33]Eric Roll, *An Early Experiment in Industrial Organisation, Being a History of the Firm of Boulton and Watt,* 1775-1805, 1930, p.24.
[34]M. Boulton to J. Watt, 7 Feb. 1769, quoted in ibid, p.23.
[35]Jennifer Tann, Suppliers of Parts: The Relationship between Boulton and Watt and the Suppliers of Engine Components, *Centenary Trans. Birm. & Warwicks. Arch. Soc.,* 1974.
[36]Much of the following section is based on ibid, Boulton and Watt's Organisation of Steam Engine Production Before the Opening of Soho Foundry, *Trans. Newcomen Soc.,* 48, 1978.
[37]Quoted in T.S. Ashton *Iron and Steel in the Industrial Revolution,* 1924, p.65.
[38]J.C. Cule, Finance and Industry in the 18th Century: The Firm of Boulton and Watt, *Economic History,* Feb. 1940.
[39]J. Watt Junr. to J. Southern, 15 Sept. 1795.
[40]J. Watt Junr. to M. Wilkinson, 31 Aug. 1795.
[41]J. Southern to J. Watt, 29 May 1795.
[42]J. Southern to Gregory Watt, 21 Oct. 1800.
[43]Eric Roll, *An Early Experiment* ..., pp. 163-164. Throughout this Introduction the firm is, for convenience, consistently styled Boulton and Watt, however.
[44]M.R. Boulton to J. Watt Junr., 3 Oct. 1801.
[45]Boulton Watt & Co. to Lawson, 28 Nov. 1801.
[46]Eric Roll, *An Early Experiment* ..., pp. 189-236.
[47]ibid.
[48]J. Watt Junr. to Ward, 1 April 1814.
[49]Jennifer Tann, Marketing Methods ...
[50]Watt Junr. to Lawson, 1 Nov. 1796.
[51]Jennifer Tann, The Employment of power in the West of England Wool Textile Industy, 1790-1840, N.B. Harte and K.G. Ponting (Eds), *Textile History and Economic History, Essays in Honour of Miss Julia de Lacy Mann,* 1973.
[52]But since there were different stroke lengths and rates with different sizes and types of engine, this system of measurement did not go unchallenged by customers
[53]And this will produce an under-estimate of the sum due.
[54]But see 'Memorandums relative to Boulton and Watt's Engines in Cornwall, 20 April 1799'. Printed in Eric Robinson and A.E. Musson, *James Watt* ..., pp. 208-209, where the partners claim to have received c.£100,000 from Cornwall between 1778 and 1798 and were owed c.£42,000, having made voluntary concessions of the order of £35,000.
[55]G.N. von Tunzelmann, *Steam Power and British Industrialisation,* op. cit. pp. 74-79.
[56]Box 53.
[57]J. Watt Junr. to B & W, 26 April 1825, Box 33.

[58]Jennifer Tann and M.J. Breckin, International Diffusion ..., op. cit., Appendix.
[59]Jennifer Tann and M.J. Breckin, Fixed Capital Formation in Steam Power ..., C. Feinstein and S. Pollard (Eds), *Fixed Capital Formation in the Industrial Revolution,* 1980.
[60]C.F. Partington, *An Historical and Descriptive Account of the Steam Engine,* 1822.
[61]A.E. Musson, Industrial Motive Power in the United Kingdom ..., p. 435.
[62]There was great demand for second-hand engines and it was not uncommon for Boulton and Watt engines to last a century which, as Peter Mathias has observed, whilst being eloquent testimony to British engineering, speaks less well of British management.
[63]Brotherton Library Mss 18.
[64]A. Andrews, Special Collections in West Midlands Libraries: The Boulton and Watt Collection, Birmingham Reference Library, *Open Access,* 13, No. 3, 1965.
[65]Many of them are reproduced in Jennifer Tann, *The Development of the Factory,* 1970.
[66]Eric Robinson and Douglas McKie, *Partners in Science ...*, p.6.

Section A

The Engine Partnership

The Engine Partnership

Documents 1-9

The negotiations which led to the formation of the Boulton and Watt partnership extended over a period of seven years. For much of this period William Small, Boulton's friend and personal physician, acted as intermediary. He offered friendship to Watt besides proffering advice on the intricacies of partnership, finance, and applied science. From the outset Small, like Boulton, had high expectations of Watt's engine[1]. But the debts incurred by Watt in the development of his engine had been taken over by Dr. John Roebuck (whose business interests extended from coal mining to the lead chamber process of sulphuric acid manufacture) in return for a share in Watt's engine. It was this delicate situation which accounted for the protracted nature of the Boulton and Watt negotiations. When in 1773 Roebuck agreed in principle to assign his share in Watt's engine to Boulton, four years of Watt's patent had already expired[3 & 4]. The partnership finally commenced in 1775 with the passing of an Act of Parliament extending Watt's patent to the year 1800. Certain changes in the allocation of shares took place within the Boulton and Watt families before 1800, however. Boulton's son, Matthew Robinson, and Watt's sons, James and Gregory, were admitted to the Soho Engine Manufacturing partnership in 1794, the sons entirely owning and managing Soho Foundry from its building in 1796.

James Watt formally retired in 1800 and in the year following Boulton's death in 1809, James Watt jnr., and M.R. Boulton discussed admitting two of their most senior employees to partnership. Only

John Southern accepted[5]. M.R. Boulton would appear to have been less committed to the engine partnership than Watt, jnr. When writing to his cousin Z. Walker, jnr., in 1797 he alluded to his aversion to the business[103] and between 1814 and 1815 Watt, jnr., was actively seeking a new partner in the person of Peter Ewart, a millwright. Ewart had been employed at Soho for a short while but had subsequently been in partnership with Samuel Gregg the cotton spinner, at Styal, Cheshire. The projected partnership did not materialise and from this period Watt, jnr., took the major management decisions in the firm.

DOCUMENT 1

To James Watt Birmingham 7 Jany 1768.

Dear Sir

Our friend Boulton will by this post send letters both to you and to D^{r} R[1]. I know not well how to resolve without seeing you. I have not the pleasure of being enough acquainted with Dr R. to judge whether we should all suit one another. His integrity and generosity every body agrees are great. You certainly know the proposal he has made to Boulton, who will tell you his determination about it. Before I knew your connexion wth Dr R. my idea was that you should settle here, and that Boulton and I should assist you as much as we could, which in any case we will most certainly do. I have no kind of doubt of your success, nor of your acquiring fortune, if you proceed upon a proper plan as to the manner of doing business, which if you do, you will be sole possessor of the affair even after your patent has expired. I had not thoroughly considered this part of the matter when you left me. In a partnership that I liked, I should not hesitate to employ any sum of money I can command, in your scheme, & I am certain it may be managed with only a moderate capital. Whether it would be possible to manage the wheel & reciprocating engines by separate partnerships without their interfering I am not certain. If it is, Boulton & I would engage with you in either, provided you will live here. Thank you for the boat varnish.

One Experiment was shewed to me, and two were described to me since you left me which will certainly soon lead the operators, altho they are not great cymists to your alcali secret.[2]

This is what I alluded to, and I beg you will attend to it. I am in haste

Dear Watt

Your affectionate humble Servt

W. Small

MB — Small/1

---

[1]John Roebuck, first promoter of Watt's engine.

[2]Watt, James Keir and Joseph Black were working independently on the production of alkali. They later formed an informal partnership.

DOCUMENT 2

To Matthew Boulton — Glasgow Oct 20th 1768

Dear Sir

I got safe home on Wednesday last week. When you were so kind as to express a desire to be concerned in My fire engine I was sorry I could not Imediately make you an offer. The case is thus. By several unsuccessfull projects & expensive experiments I had Involved myself in a considerable debt before I had brought the theory of the fire engine to its present State. About three years ago a Gentleman[1] who was concerned with me dyed. As I had at that time conceived a very Clear Idea of my present Improvements & had even made some tryal of them, tho' not so satisfactory as has been done since, Doctor Roebuck agreed to take my debts upon him & to lay out whatever more money was necessary either for Experiments or securing the Invention, for which cause I made over to him two thirds of the property of the Invention, the debt & expences are now about £1200. I have been since that time employed in Constructing several working fire engines on the common principles as well as in trying experiments to verify the theory. As the doctor from his engagements at Bo'ness & other business cannot pay much attention to the executive part of this, tho' Greatest part of it must devolve on one who am from my natural Inactivity, want of health & resolution Incapable of it, It gave me great Joy when you seemed to think so favourably of our scheme as to wish to engage in it. I therefore made it my business as soon as I got home to wait on the doctor & propose you as one I wished he would make an offer to which he agreed to with a great deal of pleasure & will write you in a few days that, if agreeable, you may be a third part Concerned on paying the half of the cost & whatever you may think the risque he has run deserves, which last he leaves to yourself. If you should not chuse to engage on these terms, we will make you an Offer when the whole is more perfect which I hope it will soon be. The Objections against the engine before the tryals made Last winter were that it might be difficult to make a piston sufficiently steam tight without water or some other fluid above it. Water cannot be used because if any of it got in it would boil & occasion a loss of heat — Oil of some kind was the next thought of but possibly it might be destroyed in time by the heat. The next objection was that the Vacuum might not be formed by exhaustion sufficiently sudden or sufficiently perfect.

On tryall, tallow being employed to keep the piston tight answered perfectly well & did not appear to be changed in all the trials made, but on removing the tallow employed from above the piston it was found to be so tight as to Lose only 4 Strokes in ...[2] the want of it in Large engines this defect would be Lessened in proportion as the Squares of their ...[2]

In answer to the second Objection the vacuum was formed sufficiently sudden & very perfect that is within one sixth part for if this Reciprocating engine should not...[2] it must be some mechanical difficulty which I think we may certainly get the better of. If Doctor Small should chuse to be concerned with you in this I have reason to think it would be agreable to Doctor Roebuck & would be highly so to me. If you should not chuse to engage with this affair in its present state or at any rate you will lett this Letter remain a secret except to Doctor Small, I shall at any rate write you again soon & am Dear Sir

Your's James Watt

MB-James Watt 1/1

[1]John Craig, Watt's partner in his instrument making business.
[2]Letter torn and illegible.

DOCUMENT 3

To Matthew Boulton Kinneil May 17 1773

S^r.

Haveing seen your Letter to my Friend Mr James Watt of Glasgow dated the 29th of March 1773 makeing offers to me for my Property in the fire Engine as improved by him, I hereby accept of the same according to the tenour of your Letter. In the first place I assign to you my whole property in the patent obtained for the said invention being two third parts of the same and also the like share of the large model of the said Engine now at Kinneil, In consideration whereof you are to give me a discharge of the Account of Six Hundred and Thirty Pounds Sterling due by me to you and your Company and of all Debts I owe to you personally and you are also to refer to Dr William Small and Mr James Watt what part of the annual profits of the said invention you shall pay to me or my assigns during the course of the patent or of such agreements as may be made for the use of the said Engine and may subsist for a longer time.

I heartily wish you success in this Engine and hereby accept of your proposals in your Letter to Mr Watt as mentioned above and am S^r^ your Obed^t^ Ser^t^

John Roebuck

MB-Roebuck/23

DOCUMENT 4

Copy of Award between Dr. Roebuck and M. Boulton

29 May 1775

At the Request of Matthew Boulton Esq of Soho near Birmingham & Doctor John Roebuck of Boness Scotland, I have maturely considered the Allegation of both Parties Separately & also perused written Papers relative to a Steam Engine invented by Mr James Watt, by which it appears that Mr James Watt about the year 1768 agreed with Doctor John Roebuck to allow the Said Doctor two thirds of the Proffits arising from a Patent then obtained for the Sole making & vending thereof, on Condition of the Doctor finding a Capital for carrying it into Execution, but it not being convenient for Doctor Roebuck to advance more money than about one Thousand Pounds & it requiring as appears by the Expence Since, Fifteen Hundred or Two Thousand Pounds more to carry it into Execution, Said Doctor gave up his Share to Matthew Boulton, in Consideration of Matthew Boulton giving up all Claim on Said Doctor Roebuck as well on his own Account which appears to be Five Hundred & Eighty Four Pounds, as also that of Boulton & Fothergill,[1] amounting to about Six Hundred & Thirty Pounds, & further with this Proviso that if the Said Engine Should become an advantagious Concern, that Doctor Roebuck Should be allowd Such further Consideration as Mr James Watt & Doctor William Small Should award to him, but Said Doctor William Small being Since dead, I have at the Request aforesaid deliberately weighed the Allegations on both Sides, & do award that Mr Boulton do pay to Said Doctor Roebuck one Thousand Pounds out of the first Proffits arising from his Share of Said Engine to be ascertained by Mr Watt

& Mr Boulton, giving an Account of the Proffit thereof to Doctor Roebuck or his assigns on oath when required —

Witness my Hand this 29th May 1775

Wm Matthews[2]

MB-W. Matthews/15

[1]The toy making partnership.
[2]M. Boulton's London banker.

DOCUMENT 5

To James Watt Jnr

Thursday Noon
London 1810

Dear Sir

More engagements than my state of health is equal to must plead my apology for not having acknowledged before this your several favors.

Your request in regard to the accounts of the late M^{r} W Wilkinson I communicated to M^{r} J Weston, but I cannot promise myself that the communication has made much impression upon him.

A Weston[1] has withdrawn himself almost entirely from the Office, residing at Mill Hill ab^{t} 10 miles from Town, so that the increased bustle & hurry of James renders it more difficult than ever to call his attention to any affair of minor consideration — For this reason I could not expect to obtain any very satisfactory conference upon our intended partnership with Southern[2] &c.

It happens however somewhat fortunately for us that they have a measure of the same nature in contemplation in their own concern intending to form a connexion with two young men in their office of the name of Francis &—. The subject has therefore occupied their thoughts in some degree & the provision which I stated to him to have occurred to us as proper on the occasion seemed to coincide with their own suggestions. He much wishes to have the heads of our proposition stated in writing & promises to add in the same form what additional precautions he should recommend to be adopted — If you will have the goodness to furnish me with your ideas upon the subject I will add any of my own that may occur & lay the project before him for consideration — As I do not at present foresee anything to detain me in town longer than the commencement of the ensuing week I should be glad to receive your communication relative to this business as soon as your convenience permits —

I shall see Messrs Black Parry & Co tomorrow & if Mr Frost can be got at I shall endeavour to procure some information from him respecting the paper[3] — It might be well to make enquiries of Balstone[4] & if he cannot serve us he is a likely person to put us into a proper channel of procuring the paper from honest hands.

You will receive from Mr Woodward an enquiry as to the shortest time in which an Engine can be completed for Smith & Cooke — I have had at least half a dozen interviews with them & every one has served to confirm the impression I entertained of the first of the firm — Cooke is I think inclined to have one of our Engines & expresses without much reserve his opinion of him.

Believe me
truly yours
M. Rob^n Boulton

MIV Box 3

[1]For many years Boulton and Watt's legal adviser.
[2]Southern was admitted as a partner in the firm in 1810.
[3]For copying letters and drawings.
[4]A large paper manufacturer at Maidstone, Kent.

DOCUMENT 6

To James Watt Junr. Manchester 18 July 1814.

Dear Watt

Ewart[1] informs me in Confidence that Mr Greg and he are about to dissolve their partnership, the Causes of which are perhaps too complicated & delicate to explain. I think it proper to apprize you of it in consequence of some views which you once communicated to me — if the Arrangement which you have since made or other Motives have not chang'd them I should recommend to you to come here for a day or two in order to form your Determination, we can then explain to each other many reasons which it is difficult to convey upon paper and if you think it expedient you can then proceed to your Proposals & Negotiation.

The Talents and Integrity of our friend stand so high that many will have their Eyes upon him — I may tell you in Confidence that his Brother William has already made him proposals, but as I believe it was principally with a view to serve him they would not interfere with any better Prospects.[2]

For the reasons assign'd already I forbear to enter further upon the Subject and it is still less necessary to say any thing to you of the great Abilities or private Virtues of Ewart.

I am Ever Y^{rs}

Geo Aug: Lee.

M IV L

[1]Peter Ewart, millwright and former employee of B & W, was a partner of Samuel Greg, cotton spinner at Styal, Cheshire.

[2]Ewart did not become a partner with M.R. Boulton & Watt Jnr.

DOCUMENT 7

To James Watt Junr. Manchester 31 July 1815

Dear Watt

I wrote a few lines to you last night & have since receiv'd yrs of — Poor Southern will be a great Loss to his family in particular, & to the World for his peculiar & rare Talent — by you he will be much miss'd as a Companion and Assistant in Labour & Difficulties which will for a time fall heavy upon you — his Merits we shall all long recollect & his faults most probably arose from the constitutional & irritable Malady which is justly an Object of Pity rather than Resentment, and which ultimately was the Cause of his Dissolution.

You will now be unfettered in making your future arrangements and I trust you will be able to relieve yourself effectually in time. There is a considerable difficulty in incorporating a Stranger in any Establishment & particularly in yours but the greatest Obstacle consists in assimilating the new & the old Materials & that is now much diminish'd — by putting the two Creightons[1] upon their Mettle you may I think draw more from one of them at least than has hitherto been display'd and you will gain a Respite until you can find competent assistance — they know the Routine & will be humble.[2]

I have sent your Letter to Ewart & Henry — if you still think it probable you may make an Arrangement with the former let me see you — it will be necessary to consider maturely all the becoming Circumstances on both Sides & that each is now fitted & habituated to his respective Modes of thinking & acting.

Yrs mo truly

G A Lee

If you think you cannot come here without staying too long I will meet you at Hamiltons or Wedgwoods.

M IV L

[1]The brothers William and Henry were employed as engineers and agents. William was based at Soho at this time.

[2]Their letters to each other display entirely different sides to their characters.

DOCUMENT 8

To G.A. Lee　　　　Soho 12th Aug 1815.

Dear Lee

Ewart has, as you probably know, appointed to meet me on Wednesday at Newcastle. His and my father's communication hold out but little encouragement, yet something may result from our mutual explanations and at all events my mind will be satisfied to have made the attempt. I feel sensibly the kindness of your offer to meet us, but as nothing has been said upon that head perhaps it may be best for us to try first what we can do alone. Indeed my own feelings upon the subject of partnership are so delicate, that unless the arrangements appear not only mutually expedient, but even desirable, I think it better it should not take place as I would not for the sake of present convenience to myself wish any plan to be adopted that held out a prospect of occasioning regret to him at any future period. I shall however feel greatly obliged by the communication of any hints that may occur to you, if any should occur, which may tend to facilitate the object of our meeting, in which case you will please to address me at the P.O. Newcastle, or by Ewart.

(J. Watt jnr)

M IV L

DOCUMENT 9

To James Watt Junr. — Mr Lee's Cottage Augt 12th 1815

Dear James

We arrived here on Wednesday evening after a pleasant Journey. We found the family all well and M^{r} & Mrs Wallace & Mrs Hamilton still here, having stopt to see Newcastle races; for which reason we were obliged to set out immediately after Breakfast otherwise we could not have got horses — From the dryness of the weather & the Stony roads, the spokes of the wheels of my chaise got all so loose that upon a careful examination I thought it prudent to get a new set of wheels, rather than trust to expedients, which did not seem to me very solid, and Mr Lee's operator has accordingly accommodated me with a set fitted to Lancashire roads, upon which I mean to set out tomorrow.

On Thursday I went in to Manchester and called upon Mr Ewart with whom I had a conversation when he repeated his great inclination to join you & the arguments against it the same as you stated & that he had written or meant to write to you appointing a meeting at Newcastle on Wednesday if that suited you, and which I think you should attend if you can; but at the same time I think his arguments unanswerable against his having his present establishment and besides if he could do it, that could not be suddenly. I have also talked with Mr Lee who sees the same difficulties, and it appears that if Mr Ewart could get over his objections it should be a consideration with you how far it would be right to invite him to run so great a risk in the present circumstances. In short nothing new has occurred upon the subject & I must leave it to Mr Ewart and your own consideration, but I think you may both be gainers by the conference proposed.

I have heard of nobody else here who would be likely to suit you both Mr Lee & Mr Ewart promise to be upon the outlook.

I dined here on Thursday with a few friends, and yesterday at Mr Ewarts with Messrs McConnel & Kennedy[1] and their ladies some more ladies, a Mr King & M^{r} McCrae American Gent^{n} & Mr Dalton,[2] when you were kindly remembered by the whole company that knew you. The day proved a most unfortunate one it rained in torrents & a Stupid coachman, though well directed by Mr Lee drove me about all day without being able to find Messrs McConnel & Kennedy's factory. In the course of his wanderings however he brought me to Mr James Kennedy's[3] house, who greeted me kindly & directed the coachman to his Brothers house who was not at home, but in returning I met him half drowned with only an umbrella to defend him, & in consequence we deferred our meeting to Mr Ewarts house, with the latter I had some further conversation, but it was to the same purpose as the preceding day — In the morning part of yesterday I called upon Mr Henry[4] at his cottage about half a mile from hence I found him greatly changed since I had seen him, but retaining his faculties except his hearing, we are to drink tea with him to night.

The family here are all very well & exceedingly ... [5] Mr Ewarts little boy is thought to be out of danger & all the rest are well and thriving.

As to our health except slight headaches we have little to complain of — To day does not rain constantly & there is occasional sunshine but it does not promise to settle into fair weather yet.

If there is any fruit at Heathfield you have use for please send for it.

The ladies here join me in best wishes to you Mr Tuffin and Mr Boulton etc and I remain

Dear James

Yours affectionately

James Watt

MIV Watt 23

---

[1]Cotton spinners of Manchester.

[2]John Dalton FRS of Manchester.

[3]Cotton spinner of Manchester.

[4]William Henry, FRS, son of Thomas Henry FRS.

[5]Letter torn.

Section B

Engine Technology

Engine Technology

Documents 10-24

William Small's advice was sought by Watt on drawing up the first patent specification[10]. It is clear that Watt, at this stage, intended to convey as much information as was necessary for a person to be able to construct an engine from the specification. Boulton and Small thought otherwise and Watt was advised to be rather less precise in his description. Had Watt followed his own inclinations in this matter, the court proceedings of the mid to late 1790s might have taken a different path. Watt's letter of 1769 indicates that he was unclear about the operation of the steam wheel but wished to include it in the specification whilst at the same time he was continuing to modify the reciprocating engine.

From the beginning of their partnership, Boulton took a keen interest in both the minutiae and the major questions of engine technology. The subjects of the positioning of pipes, the design of condenser tops[11] and steam pressure[12] all drew comment and occasionally proposals for modification.

Boulton's engineering abilities coupled with his perception of market opportunity were nowhere better displayed than in his experiments on and promotion of rotative power[13 & 14]. His sense of urgency "let us be wise and take ye advantage", was impelling, encouraging the sometimes dispirited Watt to greater endeavours.

Boulton, moreover, took the lead in seeking orders amongst the London waterworks companies and breweries[11 & 19] as well as canal companies in the West Midlands.

The partners had, however, engaged an able team of assistants. The resident engineers at Soho, (Murdock, Southern, Lawson and later, W. Creighton), suggested important modifications and new developments of the basic engine like the governor, whilst other improvements were suggested in the course of visiting and corresponding with innovating entrepreneurs in different industrial sectors [20 & 22].

DOCUMENT 10

To Dr. W. Small Glasgow Jany 28th 1769

Dear Sir

I wrote you last Sunday with a coppy of the Intended Specification which I hope you have read. The plan I pursued in Drawing it out was in the first part to lay down my principles in as few words as possible so as to be Intelligible & Clear from Double Meanings, so that they alone should ascertain my Invention in so far as related to myself. This I have declared in the next sentence but as I have been Informed that some patents have been defeated, because the Specification was not Clear enough to enable other people to execute the schemes, I have added descriptions of the Machines with Drawings. However if something after the nature of the first part could be contrived to answer the whole purpose it would be better. The Description of the Wheel is very defective but I have so few Ideas at present that it is with Great pain I have been able to bring the whole the Length you see. I shall therefore be much obliged to you to sit in Judgement upon it & shorten it as much as you can to keep the meaning Intire.

I have not heard from Boulton yet; I fancy it will be best to defer the Bargaining till the Doctor[1] & I be in England which I expect will be this Spring.

N.B. There is a section of the wheel & a plan of the other engine to be added to the Drawing, When time allows. I have been trying experiments on the Reciprocating engine & have made some alterations for the better & some for the worse which Latter must return to their former form. I have improved the condenser by reducing what was before a number of small pipes to one single pipe surrounded at a little distance by another & making cold water run between them; this both answers the end better & will considerably lessen the expence in great. I have misimproved the Cylinder by making the bottom of cast Iron which being obliged to be thicker than the brass or copper ones used before, does not communicate heat fast enough to evaporate the water left in the Cylinder the first Stroke without being considerably cooled by it & consequently Condensing as much the next. I have also gone wrong in removing the outer cylinder which kept the inner one always surrounded by steam ready to supply any loss of heat that might happen by evaporation from the inside surface. This Cylinder should be of wood both for cheapness & to Transmit little heat to the external

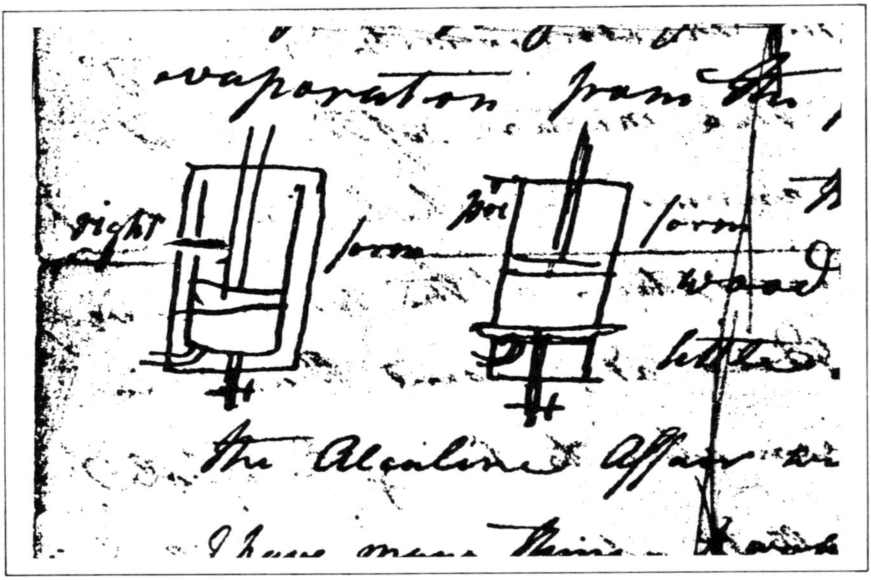

air. The Alcaline Affair will be put in Agitation immediately. I have many things I would talk with you about much contrived & little executed. How much would health & Spirits be worth to me! ...

Adieu

Watt

MB-Watt 1/4

[1] Dr Roebuck.

DOCUMENT 11

To L. Henderson,[1] Soho. London Augt 13 1777

Dear S^r

I have just wrote 2 Sheits of paper to Mr Watt & therefore must be short with you as its past 11 oClock — I thank you for your letter of the 11th Ins^t & for the attention you have shewn to sundry points but am sorry you have left my house. I would not impose any task upon you that is disagreeable to you & if you had done so by me I would have told you of it for I am disposed to more than meerly wishing you happy. I shall be obliged to you for news as often as any occurs. I had a long letter from Mr Watt today wth his Journal; he was at Chasewater on Saturday last & saw Wilson, Dudley & Hornblower. Wheal Busy[2] is in considerable forwardness. All Cornwall is agape to see its performance. Huel Virgin is to have another old Engine added wch will enable them to go about 8 fathoms deeper. They have bought a Boyler of a Monstrous size & have offered £50 to have it carried to ye place — Tingtang[3] things are exceedingly wanted as they have now come to Water pray write p^{r} first post to Turner at Bursham & tell him Tingtang things will be returnd upon our hands unless sent off immediately & I must also beg youl send to me p^{r} return of Coach all the plans & papers relative to Richmond[4] & Chelsea[5] also Mr Rothwells plan & Section of Shadwell[6] as they are now determind to alter the first intended Engine. I want Chelsea Letters & agreements as I may put them in the Lawyers hands. Pray has Mr Woodcock ye Lawyer sent Shadwell agreem^{t}. I calld there today & found him not, but his Clark said that Woodcock had sent it a week ago. I wish you would also let somebody assist you to write out a Copy of that description of our Engine wch Mr Watt wrote for prices Book as I want it for a special purpose — I am glad to hear that Belzibub[7] improves & think when the new pipe & top is put to ye Condenser it will still be better but not at best till the outward Steam Case is added, which I beg youl push forward. We must no more talk of out side Cylinders for I perceive by a Conversation which Mr Meason of Edinburg[8] hath had wth Smeaton, that names of things & external appearances bias the Judgment very much. Mr Smeaton speakes well of our Engine but always adds a bit & wishes it was more simple. The derbyshire folkes are erecting a larg Engine at Yate Stoop & are to pay 15/p^{r} Tun for their Coals aledging that our Engine is always out of order & that we charge 6^{d} p^{r} lb for our screws whereas they can get em at 4½ or 5^{d}.

Cummins is writing against Smeaton upon *Momentum*

do desire Mr Fothergill to send 3 or 4 very pretty setts of patterns of Buttons for Mr Des Cotes with whom Mr Carbarit lives and they must come very soon & desire Mr Egginton to send me the bit of papier mache which he lookd out some time ago. Mr Martines expects patterns of Buttons also make my Compliments to Mr Scale & present him with ye inclosed Button which are much wore; the Silver part is leaf silver if he hath none send to me for some[9]. I have not yet receivd Mr Simcoxs[10] Specification wch pray remind him of it pray cast your eyes around you & fav^{r} me with a line as often as occurances require pray tell my wife that I am very well but find London very hot adieu

Yours sincerly

M B

Box 20

---

[1]He offered Boulton an invention for rotative motion which was not proceeded with. He was a valued assistant at Soho until he emigrated to Italy.
[2]Chasewater no. 2, 63 in. cyl. 9 ft. stroke.
[3]52 in. cyl. 8 ft. stroke.
[4]Waterworks engine 18 in. cyl. 3 ft. stroke.
[5]Waterworks engine 30 in. cyl. 8 ft. stroke.
[6]Waterworks engine 27 in. cyl. 8 ft. stroke.
[7]The 33 in. cyl. 8 ft. stroke engine erected at Soho Manufactory in 1776.
[8]Proprietor of Wanlockhead mines.
[9]A reminder of Boulton's other business enterprises.
[10]Simcox did not patent his engine improvements but had some small financial reward from Boulton, in return for divulging his scheme.

DOCUMENT 12

To James Watt London Aug 18 — 1777

Dear S^r

I wrote to you about y^e 12th & rec^d yours of y^e 9th which is all I have yet $recv^d$ from you. I have this day seen Mr Edwards of Hales who is our staunch friend, he says Bouge[1] declines him any thing to do with y^e erection of y^e new 63 Inch Cylinder & therefore begs you will take all necessary information upon the premises. I hope youl not think Mr Bouge a man of consequence enough to be angry with, but treat him with a laugh or a smile of that sort which his sublime insignificance merits. Mr Edwards presses very much for expedition & used many arguments to show the importance & absolute necessity of it. I am to accompany him to Wilbies on Wednesday $morn^g$ with his friend Knill who is just returnd from Ireland & is now in London. It is only the Engine end of the Beam wch is wanted at Wheal union[2].

I have by this Coach sent down to Henderson the drawings of Chelsea & Shadwell to be Copyed & then I shall send the originals to you as you may give us your plan in words if not in Lines. In the last letter from Henderson he says Belzebub[3] goes well & as Violent as ever. I think y^e great fault is that tis only loaded to about $\overset{lb}{5.7}$ p^r Inch whereas if it was loaded to ab^t $\overset{lb}{7\frac{1}{2}}$ it would not be so very fierce. Chelsea Cylinder[4] is but $\overset{In}{27}$ & I think our expansive one should be no bigger however I submitt these matters to your better judgment & only wish you to issue your orders as soon as you can find time, & I will take care to advance them in their execution. Joseph Harrison hath tryed y^e 2 Engines at Bedworth[5] & says the new Engine will save ¾. M^r $Hend^n$ says he gave you a letter to M^r

Walter Grose who purchased all the Stores for Huil Virgin Mine & has almost y^{e} whole direction of it. I think you should see him. I am going to Mr Veres this afternoon & shall stop at Richmond to see y^{e} House set out. I expect to leave this Town on Wednesday night & stop at Coventry. There is a ship arived this hour from New York but its letters are not deliverd yet. They report that Burgoine hath taken Teconderogue[6] & advancing forward fast, that Lord Howe hath put a larg Army on board the Fleet to strike some unexpected Blow wch I hope he will do to ye tune of Britons Strike home.

I hope Mrs Watt hath recoverd the fateague of her journey & that she is as happy, as a strange Country, & the sight of so many old Fire Engines can make her. pray take care of your health & manage your mind like a philosopher. Vex at nothing, all's for the best, & contrive to turn all misfortunes into benefits.

Farewell remember me kindly to Mrs Watt & believe me

Yours sincerely

M.B.

I am impatient for news from you. Since the above I have recv^{d} yours of ye 14th & rejoyce to find you both well all ye rest is as I expected. The Building of a House w^{d} make ye engineers stare but I think it would be right so to do unless you can rent. I will write to Wilkinson this post & repeat Pauls words, do you think it necessary that I should go to Holland soon. If you work one Engineer against another we shall soon get a Majority. Edwards said that Bouge told 'em we worked wth a Slack piston or Buckett in our Pumps so that there was only a shew of working a long pump. Be wise as a Serpent &c: for you have devils to deal with. Do make ye table of Engines. The news from America is now believd & is important.

Box 20

[1]Budge erected several engines in Cornwall [129-139]
[2]A 63 in. cyl. 8 ft. stroke engine.
[3]The Soho engine.
[4]Chelsea Atmospheric engine.
[5]B & W's engine at Bedworth had a 58 in. cyl. and 8 ft. stroke. It was ordered in 1776.
[6]The fort at Ticonderoga was taken by Burgoyne in July 1777 and abandoned after his surrender at Saratoga.

DOCUMENT 13

To James Watt | Soho Sunday Jany 13th 1781.

Dear Sir

I wrote to you last post since which I have reflected so much upon your last Rotative Motion (or 5th Method)[1] that I could not refrain from making one this day instead of going to Church — I took two of our Reel Wheels[2] (wch have 35 teeth) & one of the crossing Machine Wheels for the fly & of them I made a working Modell about 1 Inch to y^{e} foot but instead of the groove in the fly wheel I was obliged to make a Link thus

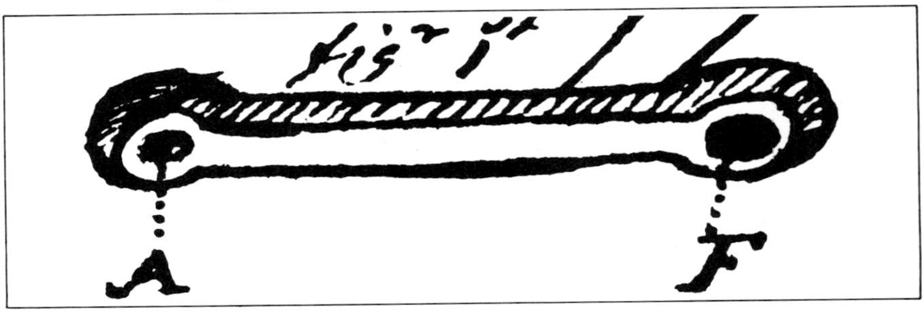

the back center of your Revolving Wheel (wch I think you call D) goes into the hole A of this link & the great axis moves in the hole F (fig^{e} 1). — for all that is wanting, is some frictionless means of keeping the Centers of the 2 Wheels at = distances in all parts of y^{e} Revolution. A weight being put on one side the fly serves (without y^{e} secondary wheel in y^{e} Crank scheme) to carry it over both Zenith & Nadir. It hath so many excellent properties & so few bad ones that I will now venter to pronounce it the very best of all the Rotative Motions. There is something so agreeable in a good Movement to ones Mechanical feelings, that I could not forebear playing with it all this afternoon as it is a desireable thing to have the pull of the Engine as near the bearing part of the Neck of ye Gudion or axis as possible, I was thinking of fixing a Center pin in a strong upright post so as to be Centeral to the great Shaft & instead of fixing the pin in the back part of the fixed Revolving Wheel I would fix it in the front: & there connect the 2 Centers or pins by means of a Link as above (fig^{e} 1st) thus you get the power of the Engine applyd nearer to a perpendicular direction to the bearing part of ye Axis & the fly thus lyes close on one side the neck & ye pull of the Engine Close on ye other side it, but these are circumstances that may be varied as experience shall direct. All that I possitivly know is, that it is the sweetest motion & most convenient for most sorts of Mills (particularly whirling motions) of any other yet produced.

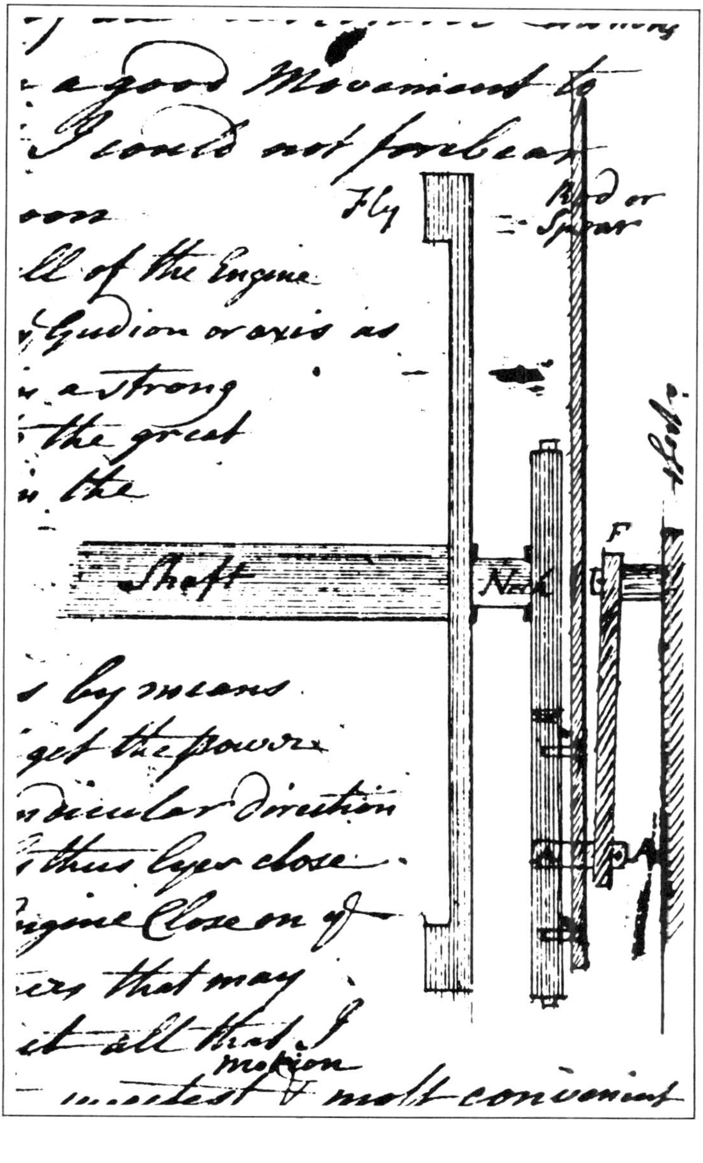

In all Rotative Motions derived from Reciprocating ones, the fly must be the equalizer But let me see if I understand how to calculate y^e effect of a Fly of any given w^t.
1st I will suppose a stout man will raise $\frac{\text{Hog}^d}{1}$ of ∇ alias $\frac{\text{lb}}{600}$ <u>ten</u> feet high in one Min^t or 6000 pounds <u>one</u> foot high p^r M^t w^{ch} ÷ by 60 = $\frac{\text{lb}}{100}$ one f^t high p^r second. thus 16 Men will raise $\frac{\text{lb}}{100}$ to y^e height of 16 feet p^r ″ from wch height it will fall in one second & will have acquired at y^e end of its fall a Velocity of $\frac{\text{feet}}{32}$. Hence $\frac{\text{lb}}{100}$ in y^e rim of a Fly moveing with a Velocity of $\frac{\text{feet}}{32}$ p^r ″ would require the force of 16 Men to stop it in a second but I will suppose y^e w^t in our fly to move only 16^{ft} p^r ″ & in that case will require only 8 Men to stop it. Hence a fly off one Ton say 2000 w^d require $\frac{\text{Men}}{8} \times 20 = \frac{\text{Men}}{160}$ but as a Fly ought to have a sufficient Momentum to prevent the smallest degree of retardation from being perceived I therefore suppose that where a Mill is required = in power to $\frac{\text{Men}}{160}$ that the fly should be $\frac{\text{lb}}{4000}$ at least. When you want to amuse your self with these subjects do tell me if I am right in my above reasoning — I am aware y^e Engine acts ⅓ of y^e time — Mr Dearman[3] was here 3 days ago to be p^d his acc^t when I had a good deal of talk with him about Engines in consequence of his telling me that a Mr Hornblower calld at his Foundry a few days ago but D: was out & did not see him. I shewd D: the old tube condenser & said only what I wish everybody knew — We talked also about Cranks & Crank Comp[4] in which I hinted peace & good will & just enough to shew ye Crank C^o that if they had a mind to make any offers of peace we should not decline any treaty they might bring upon ye tapis however I believe we had better lye quiet till we have exhibited all our Rotative Motions & I do hope you will send off soon the Specification to Handley that we may get to work.

The Birmingham Navigation C^o have wrote to us a letter requesting we would put their 2 Engines[5] in the best order possible directly & they likewise beg to see me at the Office as they want to erect another new Engine upon a new branch which they are going to Make to some new Works of Ld Dudleys near Wednesbury.

Your fav^r of ye 7th In^t I recv^d yesterday covering ye drawing & further directions respecting Walkers Engine[6] but shall directly exchange a few Letters with them before I put ye drawing into ye hands of a Founder & will endeavour in this engine to furnish them wth ye Cylinder & Nossell as ye surest & most expeditious plan for both parties.

You may be assured that neither H:[7] or any other person has, shall, or can, get sight of your letters. I shewd him a part of one wch after I had twice read I concluded it was proper, as it only related to ye Irish business. He has often come into my Room & did make an attempt to write in it (there being a good fire) but he soon perceived from my hints that it was not agreeable to me, he then retired to his old Bed Chamber late Playfairs & there he seems to take up his quarters night & day, except at Meal times at my house. Now I shall always be glad to accomodate him at my house wth Eating & drinking when ever he returns from a Journey or his residence at Soho is only temporary but when he comes to live there constantly I know what Madam will say — I believe I formerly told you that ye building of the new Engine shop w^{d} cost upwards of 130 pounds but now all ye Bills come to be settled & paid I find it has not cost £110 including Glass & everything.

Your discovery of the acid of Galls & sugar being ye same, is curious & is the only clue for discovering improvements in Ink. I shall call upon Galton in a few days & give him Bergmans new table of affinities which is printed by Mr Phillips' Son in London in English words. I bought 4 viz one for you, Withering, Galton[8], & Self. I shall take ye opportunity of asking Galton for a little of ye Tartarous Acid & try it. — Mr Keir told me that Dr Higgins is gone into partnership with ye Adams's to make F ⊕ v[9] & that Higgins & Fordice also have each of them taken out a new patent for decompounding Sea Salt & Liver of Sulpher by means of Iron or Lead or other Metals or their Calx's & also wth Calcarious Earth — Mr Moor of ye Adelphi told me that Dr Fordice told him that Scheale[10] had published in one of the late Volumes of ye Swedish Transactions a method of decompounding Salt by means of lime & I find it is true that he has pub^{d} such a paper — I dined a few days ago at Dr Witherings wth Parrott & Garbett where Garbett mentioned that Adams's have layd out 5000£ in building works at Battersey & at ye same time said he had seen, or got (I know not w^{ch}) a Copy of Higgins's Specification from whence I commence a soupsoni quelque chose. I should suppose that neither Higgins's or Fordices pattents are good, their methods being ye same & moreover I fancy M^{r} K^{r} was prior to eithers. If you should ever think of making ⊖ v Ireland is the place & tis a subject that Judge Foster of Ireland is calling up the attention of the Irish Parliam^{t} to encourage.

I will send you p^{r} next post the Birth, Life, & Death of phlogiston[11]. In ye interim I wish you easy health & good Spirits remembering me kindly to Mrs Watt

I am Dear Sir

Yours most sincerely

Matt^{hew} Boulton

Jim[12] desires I would ask you to give him leave to learn French. I wanted to set one of the Copying C^{o} Cabinet makers to make ye patt^{n} of ye Irish Nossell but J Buchanan says M^{r} K^{r}[13] says he would sooner pay him for standing still than he shoud do any work for Engine C^{o}. I dont understand it.

Box 20

---

[1]Commonly called Sun & Planet wheels, the method B & W and opted in practice.
[2]From the silk reels ordered for the East India Co in partnership with Rehe.
[3]Of the Eagle foundry, Birmingham.
[4]James Pickard of Snow Hill, Birmingham, had patented a crank rotative engine.
[5]Engines of 31 in. cyl. and 20 in. cyl.
[6]Probably the 40 in. cyl. 8 ft. stroke blowing engine for Holmes Ironworks.
[7]Logan Henderson.
[8]William Withering and Samuel Tertius Galton, Lunar Society members.
[9]Fixed alkali.
[10]Carl Wilhelm Scheele, the Swedish Chemist.
[11]Although members of the Lunar Society had believed in the Phlogiston Theory, & Priestley stubbornly tried to return it, the work of Lavoister ultimately overthrew it & established the true nature of combustion.
[12]James Watt Junr.
[13]James Keir who helped Boulton at Soho Manufactory for a time.

DOCUMENT 14

To James Watt

Soho the longest day
not long enough 21 June 1781

Dear Sir

I had the pleasure of writing to you p^{r} last post. This morn^{g} I receivd a letter from Jno Turner telling me W^{1} Crenver[1] & all the other goods were on board & that they were endeavouring to get a few Bricks to make up the loading so that with good luck & a good Wind I hope you will soon be in possession of the lovely Bettys Charms. If she arive soon I would have you take a lonely Walk & consider if you think it possible to get the Engines to work timenough to fork ye Water before the Wet season comes on, & if your own judgment (baring low spirits) tells you it can't be done, then set your mind at ease & dont vex about it, for what cant be cured must be endured. But if you think you can get them erected timenough; then do not suffer either Paul or Thomas or any other apostle or Cap^{tn} or Lieutenant to put you from your purpose by any of their perswations. (for two Mill is Money) however in either case don't vex yourself, it is not in the power of Mortals to command success. I think I understand all the Elements of the horozontal, Eliptical, Ecliptical, Conical, comical, Rotative, Mill; & a good Mill it is therefore I beg neither it nor any of the others which we have made (& nobody else) may be lost for want of Securing, as the people in London, Manchester, & Birmingham, are Steam Mill Mad, & therefore let us be wise & take ye advantage. I dont mean to hurry you into any determination, but I think in the course of a Month or two we should determine to take out a patent for certain methods of producing Rotative Motions from the vibrating or reciprocating Motion of the Fire Engine remembering that we have 4 Months to describe the particulars of the inventions —

At your leasure only please to fav^{r} me with your sentiments respecting a patent. I will take ye trouble &c of doing ye necessary.

Mr Hurd Junr & Mr Ruston had layd a plan to make some money of the Brass Comp^{y}[2] but ye latter being too impetuous & ye former sly, he (Hurd) hath bought a piece of Free Land near to M^{s} Farmers house for £900. The canal runs through the middle of it, & he has set about ⅓ or ¼ of it to the Brass Comp^{y} for 100 Years at 40£ a year. Hence the two Friends are at variance & quarrelld with each other at the last Committee.

Herewith I send you a Copy of Gore & Sheers acc^{t} pray was ye 5 p^{r} cent on ye last Ball^{e} paid in Copper agreed upon to allow them.

Henry Williams went to Ketley on Tuesday last, Robt Cameron is at work upon a working ∇[3] for ye little Engine, wch he will bring you an acc^{t} of.

I am afraid you will be incomoded with the paint pray live out of doors as much as you can till the smell is gone. I hope to hear by the next post of your arival & I hope in health with Mrs Watt & ye Children. Pray mention Greg in particular, Present my best Compliments to Mrs Watt also Mr Henderson & believe me

Your most sincerly

Matt^{w} Boulton

pray w^{d} you have us send some damping Boxes to Cornwall.

Box 36

---

[1]Wheal Crenver copper mine purchased a 48in. 9ft. engine.

[2]Birmingham Brass Co.

[3]Barrel.

DOCUMENT 15

To James Watt — Soho July 30 — 1781

Dear Sir

I wrote to you late on Saturday night in ye meridian of a bad head ache but by following D^r W:[1] advice I am better today though not lofty as ye tinners say. Yesterday morn^g I finished my modell of a forge which work'd 720 Strokes p^r M^t & I think is stronger & more compactly harnessed than any forge yet made & doth not depend upon any external fastenings such as frameing or timber above or under ground or Building or drone beam or any other appendages but like a press-frame or a rolling Mill frame, the whole power is exerted within it self & may be regulated to ye greatest nicety. The Visinertia of a forge hammer that works very quick is a great Tax upon the power & I yet dont see the means of lessening it without diminishing the effect. Wilkinson took ye drawing with him (for he w^d not go without it) for Casting 2 Cylinders 30 In Diam^r & 7 f^t Stroke. Mr Pickard[2] applyd 2 M^o ago to Whitmore[3] requesting that he w^d set up a Machine for dressing Rolls saying that he was constantly disapoint^d by ye other men, & upon a promise that Whitmore should dress all their Rolls, He accordingly set to work to make a 6^{ft} Lap & a formidable machine. Whilst it was under hand Pickard often visited him under ye pretence of assisting him in improving the Machine saying he had seen Mr Boultons & others. When ye machine was nearly finished Pickard applyd to dearman[4] & got another set of ye Castings from ye same patt^{ns} & has now nearly finished one of his own wch Whitmore proclaims for the honour of Pick^d. It is unnecessary to make any Comments on P as his Character is very uniform — Elias Wallen hath left Birm^{gm} on acc^t of a dishonorable amour. Our pious & Rev^d friend Cleveland of Worcester hath been detected in a similar case & he is forbid doing duty in his Church till the foul crime is done away by pennance. I was yesterday honourd with ye Company of the philosopher Walker[5], his Wife & Dr Withering — towards ye Even^g we talked of Engines when Walker said he had taken ye liberty of makeing a Modell upon your principle with some improvm^{ts}. He observd that he had seen Richmond, Chelsea & Shadwell & that ye power of ye Steam was much diminished at Chelsea by the length of y^e Steam pipe, that he had improved the Boiler by makeing y^e fire to asscend through tubes perpendicular through y^e Water, that he had placed his Cylinder within the Boyler & that his machinery was much

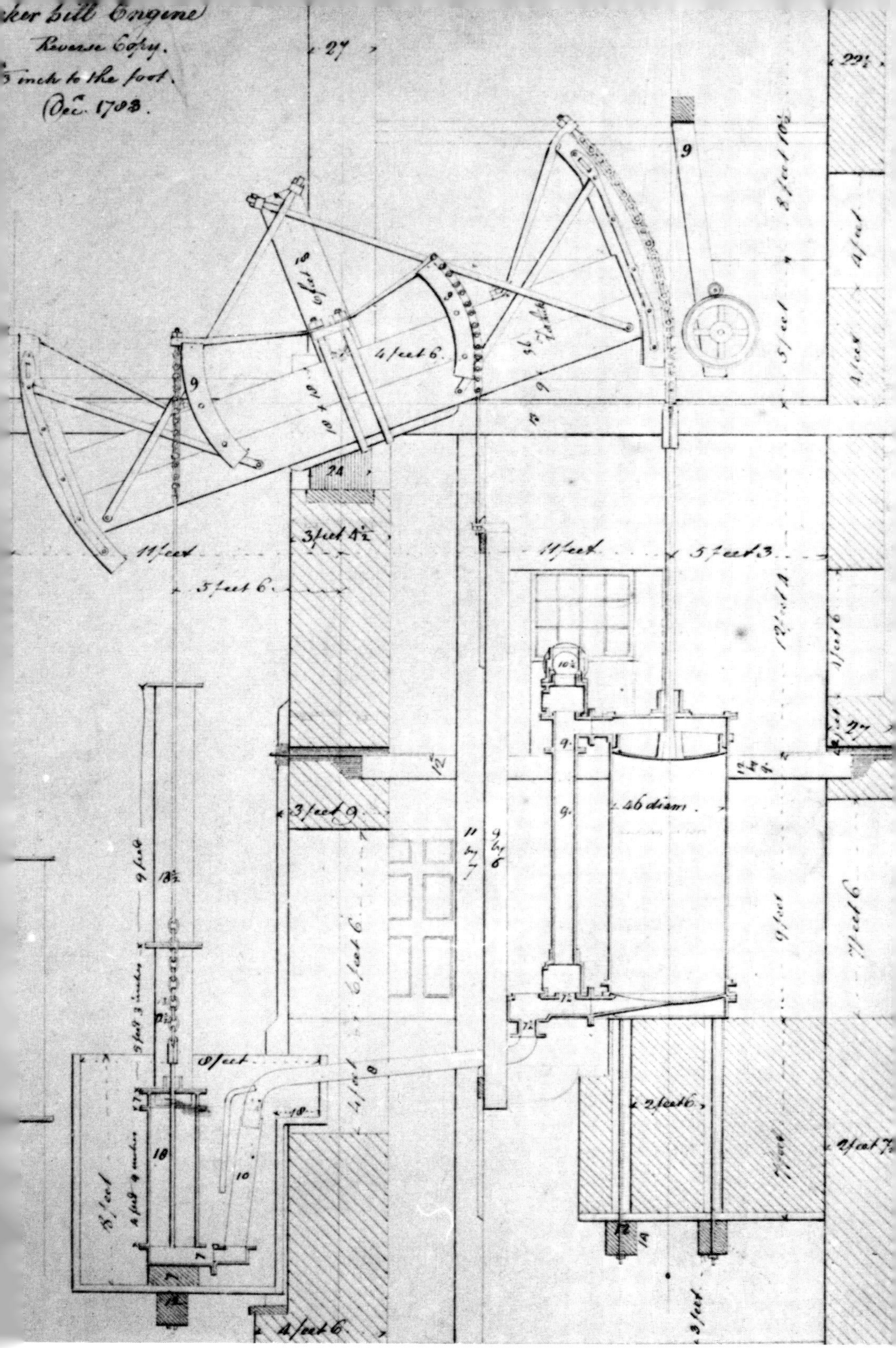

Plate [1]

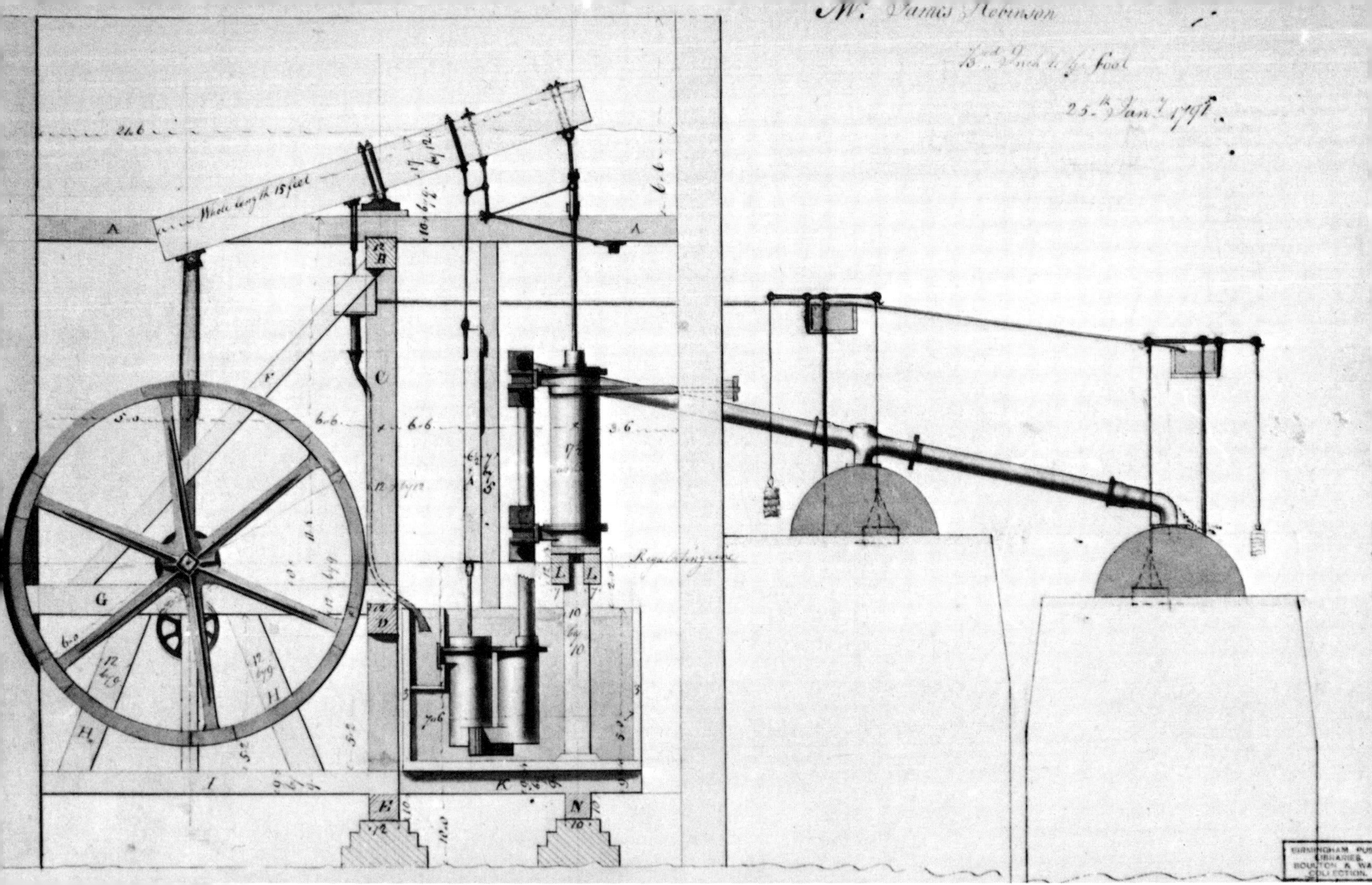
Mr. James Robinson
½ Inch to a Foot
25th Jany 1797
Whole length 18 feet
Regulating cock
A
B
C
D
E
G
H
I
K
L
N
BIRMINGHAM PUBLIC LIBRARIES. BOULTON & WATT COLLECTION.

simpler than yours. I shall not take up y^{r} time wth my remarks as they are obvious & yet these things go smoothly down with his auditors. I lowered his top sails by showing him we had executed all his inventions prior to himself, in a superior degree & after all considerd them as baubles. Hence you see a sort of reason for our executing every Idea that may seem plausable to Strangers such as putting the Cylinder in the Boiler, such as applying a sinking pipe 35 feet long — Such as using second hand Steam, such as working the Condenser by the least power possible & any other thing that may seem likely in order to shut the mouths of ignorant boasters. I wish you would make a drawing of your plan to work the Condenser — without Y^{e} Engine power, also of an Engine to work by the alternate expansion & Contraction of Steam sans Condensation. I think all the possabilities of a good Steam Engine lye within a little compass & therefore we should execute all, and I also think, so doth Dr Withering (who says he has thought much upon ye subject since he was acquainted with you) that all the possabilities are included in your patent.

Now my dear Sir I beg you will consider Hornb^{rs}[6] patent in the same light as I should do any mans pretentions who says M^{r} B: you have pleasant spot upon your freehold farm at Yoxal & I mean to build Several Houses upon it, & let them out to Tennants, to which I object, saying it is my own freehold Land & that I will not let any person build upn it without first obtaining my consent, but he perseveres & says he will build upon it in spite of me. I lay my title before Council who says it is as clear & as good as a title can be & better than titles in Common, because it has been confirmd by 2 acts of Parl^{t} — I am also satisfyed in my own mind that it is. The let him build, I will give him propor notice when he begins, & treat the conceited Blockhead with contempt, without suffering his folly to make me uneasy one moment. This is a similar case to Hornbl^{rs} & if we do suffer our selves to be pissed upon by that family we deserve to be sh-t upon. For Gods sake dont loose a moments peace upon that head but proceed, get the Engines erected & the Money in our pockets & not doubt whether we have a right solely to press down the piston by Steam or Condense in a Seperate Vessel.

doubt the Sun is fire
doubt all ye powers of sight
doubt truth to be a Lyer
But never doubt our right

I am sure you want assistance shall I send Harrison or will you send to Scotland for Wright to come directly.

As I was engaged yesterday wth Comp^{y} I was obliged to get Z. Walker[7] to write you a line & send you Hanleys Letter by wch youl see that Horn^{s} Pattent passed ye 23^{d} In. I wish we had sooner because I had orderd Hanley to Stop it but too late. However I have some notion that it will make difference between one executing any project before his Specification is enterd & after, & therefore if you wish any thing more to be executed let me know. As I shall set out early on Tuesday morn^{g} I can't wait till the post comes in so that I fear it will be Saturday before I can write again. May the giver of all good things give you easy health & Cheerfull Spirits is the fervent wish of Dear Sir

Your aff^{r} friend
M. Boulton

Q^{y} if we were to publish ye state of our Case & an explaination of our Specification & ye Spirit of it in ye Sherborn paper I will write one & send you & I will consult Council next week. Be assured ye Subject is uppermost. If we are firm we shall keep our Ground.

Box 20

[1]Dr. William Withering.
[2]James Pickard, crank engine patentee.
[3]William Whitmore, engineer, of Birmingham.
[4]Richard Dearman of the Eagle Foundry, Birmingham.
[5]Adam Walker, Junr., Lecturer in natural philosophy.
[6]Jonathan Hornblower obtained a patent for a steam engine in 1781.
[7]Zach Walker, Boulton's cousin, employed at Soho.

DOCUMENT 16

Rotative motions (in J. Watt's hand) Novr 10th 1808

From the time of the first invention of the improved Engines I was desirous of applying them to giving motion to Mills in some better way than by raising water to turn a water wheel. One method I described in the patent of 1769 the reasons why that method was not pursued it is now unnecessary to enter upon. Of another method I made a model soon after which still exists in an unfinished state & several more have since been devised by myself and others. All these were self acting rotative engines, not derived from the rectilinear motion of a piston in a Cylinder. They have their respective merits but instead of being more simple in their construction they are more complex than those derived from reciprocating motions & more difficult in execution.

The first I know of the latter kind was one I saw at Hartley colliery about 1768, which consisted of a toothed sector on the end of the working beam, working into a trundle which by means of two pinions with ratchet wheels, produced a rotative motion in the same direction, by both the ascending & desending Stroke of the arch & by Shifting the ratchets the motion could be reversed at pleasure. It was employed to draw coals out of a pit, had no Fly wheel, and went sluggishly & irregular, the name of the inventor I have long since forgot. One Stewart[1] had a patent for an engine which produced a rotative motion, by a chain going round a pulley and round two barrels furnished with ratchet wheels with a weight suspended to the free end of the chain which served to continue the motion during the return of the Engine. I have never seen this but believe there was no fly. Whether this was the same as Mr Clarks or not I know not.

The next thing I know of is Matthew Wasboroughs[2] which was virtualy the same as I saw at Hartley, but Matthew *had added a fly wheel* which as far as I know was the first time that had been employed for that purpose. I at present recollect nothing of Fitzgeralds rotative Engine but think it was something of the ratchet kind.

The true inventor of the crank rotative motion was the man (who unfortunately has not been deified) that first contrived the common foot lathe. My Share in the application I remember perfectly to be as follows. One of Matthew Wasboroughs rotative engines was erected at Birmingham, for a rolling Mill & was much talked off, this set me again to think upon the subject & brought to my remembrance my former meditations upon the crank, the date of which I cannot ascertain. I was however desirous to render the motion continuous and equable without a fly wheel, the regulating power of which I did not then fully appreciate, nor the advantage derived to the engine from the crank causing the motion of the engine to be slow at the beginning and end of the Stroke, which however a few experiments with other contrivances soon made me sensible of. I then resolved to adopt the crank & to equalize the power, I proposed to adopt two cylinders acting upon two cranks fixed upon the same axle at an angle of 120^{d} from each other & to place a weight on the circumference of the fly wheel at an angle of 120 from each of the cranks which weight was to be so adjusted as to act when neither of the cranks could do so, & consequently to render the power in itself nearly equable. Or to place the two cranks at right angles to each other, & to load the outer end of the working beams of the two cylinders with weights equal to half the powers of the respective cylinders, which might act during the time of the ascents of the respective pistons. Of this I caused a model be made which performed to satisfaction, but being thus very much engaged with other business I neglected to take a patent immediately, and having employed a blackguard of the name of Cartwright (who was afterwards hanged) about this model, when in company with some of the same sort who worked at Wasboroughs Mill & were complaining of its iregularities & frequent disasters, he told them he could put them in a way to make a rotative motion which would not go out of order nor Stun them with its noise, and accordingly explained to them what he had seen me do. Soon after which John Steed who was Engineer at Wasboroughs Mill, took a patent for a rotative motion with a crank & applied it to their Engine. Suspicion arising of Cartwrights treachery he was strictly questioned & confessed his part in the transaction when too late to be of service to us. I had afterwards a conversation with Steed upon the subject who said that he had made the invention before he was informed of Cartwrights communication & had taken his ideas from the common foot lathe & that Cartwrights information had no other effect than to accelerate his taking out the patent. How this was I know not, it is certainly possible that Steed might have invented it, but what I have related is the fact as far as concerns my invention.

Finding the door thus closed upon us & circumstances making us unwilling to go to law with the patentees which could only have had the effect of throwing the invention open to the publick if we had succeeded, we judged it better to let it remain with them who did not seem capable of doing us much harm, than to let it get into the hands of men more ingenious, being sensible at the same time that their rotative motion could not do much good without it were attached to our engines. We therefore thought it better to take a patent for several contrivances on the same principle, which the former pattentees never thought proper to call in question. The revolving wheels was the one which we principaly adopted but the advantages of the crank in point of simplicity have made it now be generally used, and the other to be laid aside perhaps for ever.

Double Engines

The idea of this construction occurred to me very early. A Drawing of it was produced to the committee of the house of Commons at the time we procured the act to prolong the patent (1774-5). It however lay dormant till about 1781 or —[2] when I inserted it in a patent with other things having first executed one on this principle at Soho.[3] I think the Albion Mill was the first publick exhibition of it. Mr De Betancourt came recommended to us as an ingenious man & as such we showed him every civility, among other things he was shown the Albion Mill, which he was told acted by the power of steam both in the ascent & descent of the piston. How

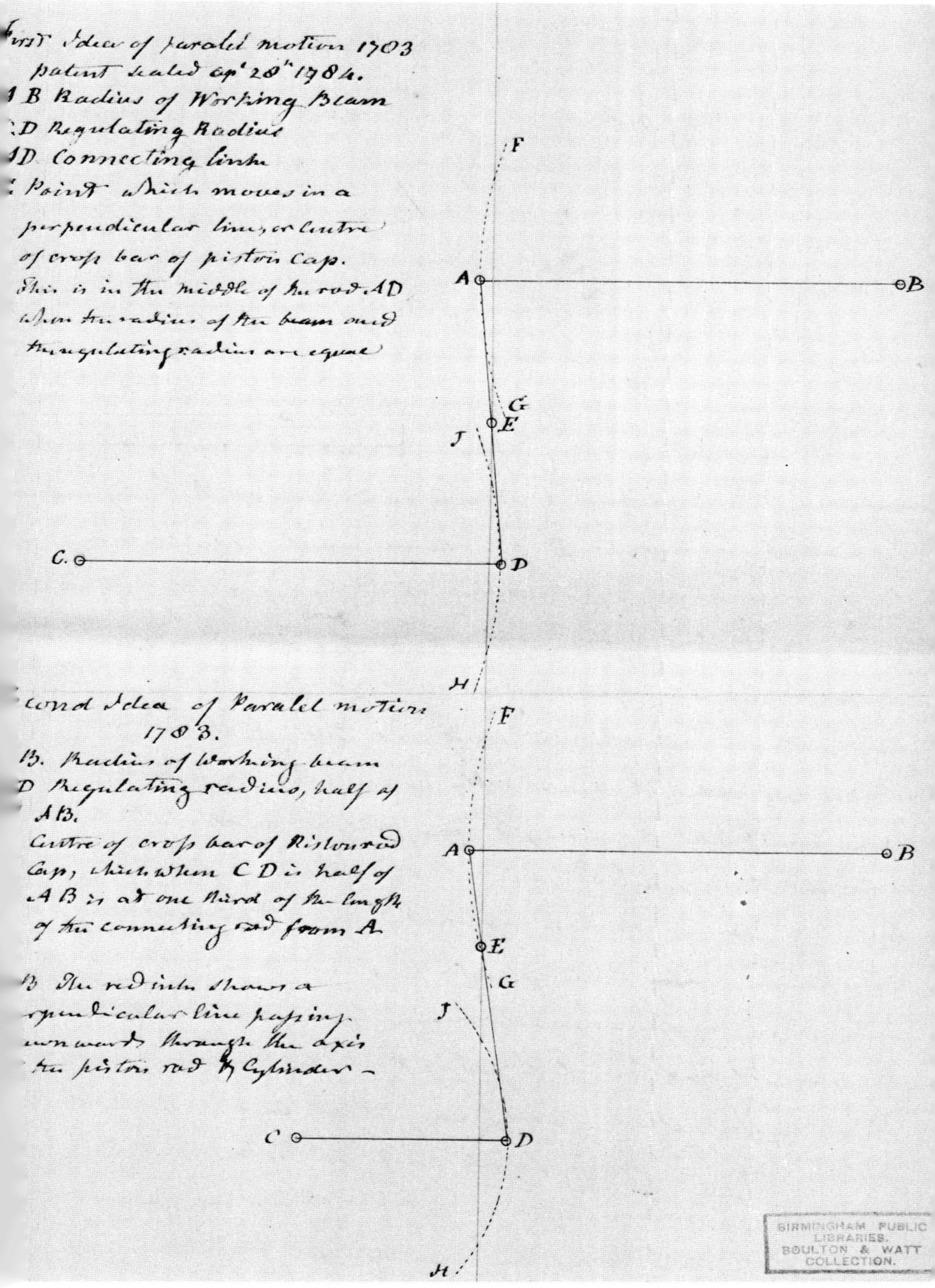
irst Idea of paralel motion 1703
patent sealed Apl 28th 1784.
AB Radius of Working Beam
CD Regulating Radius
AD Connecting link
E Point which moves in a
perpendicular line, or Centre
of cross bar of piston cap.
This is in the middle of the rod AD
when the radius of the beam and
the regulating radius are equal
F
A
B
G
E
J
C.
D
H
cond Idea of Paralel motion
1783.
B. Radius of Working beam
D Regulating radius, half of
AB.
Centre of cross bar of Pistonrod
Cap, which when CD is half of
AB is at one third of the length
of the connecting rod from A
B The red ink shows a
rpendicular line passing
wnwards through the axis
the piston rod & Cylinder –
F
A
B
E
G
J
C
D
H

Plate [3]

O Whatever may be the position of the working beam within the angle dra

d idea of parallel motion
1783
being the radius of
working beam, and
ar a.b. of a given length
g required to slide thereon
as the end a a shall always
n the line a a, ^(the opposite O) required
lengths of a regulating
ius C D which being
ted to end D shall
e it to comply with the
condition
ide the length of the
he a a into any number
qual parts & draw lines
n each of them to the
e B. upon ^each of these lines
off a b which will form
series of points b b
anged nearly in an
h of a circle, the centre
that arc C is the centre
otion of the required radius.
d as sliding motions
e not desirable in
tice, the parallel bar
was suspended by the
ecting links A E. F D
the regulating radius
being jointed to it
D & moving on a pin
e C obliged the point

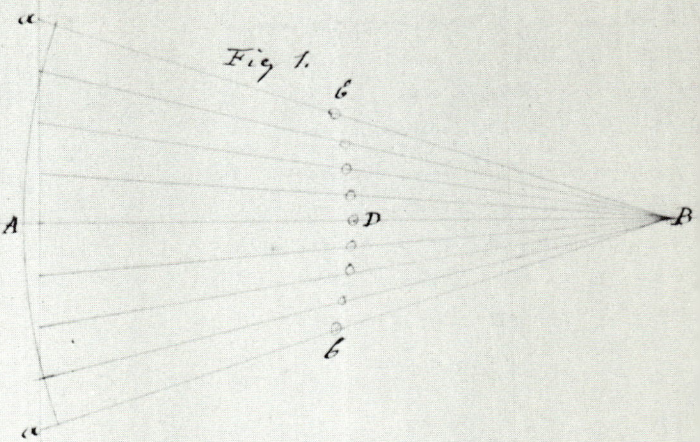

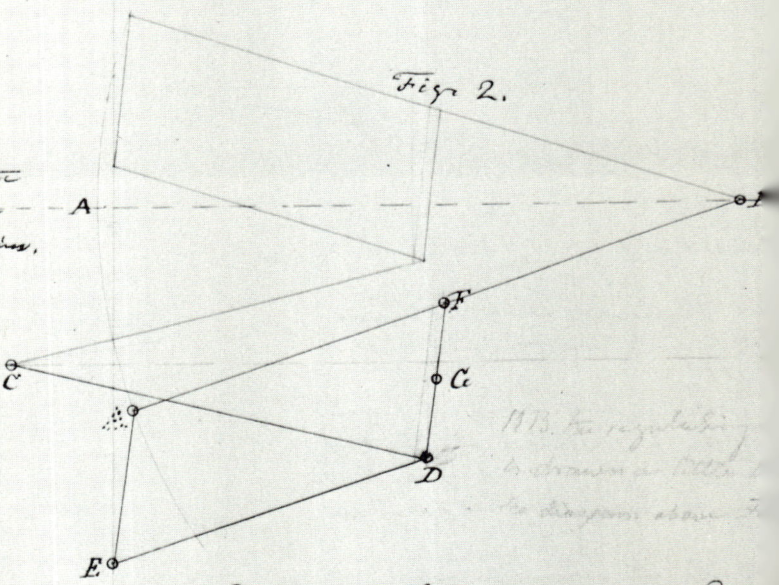

which the piston rod was suspended to move up & down in a perpendicu
& the top of the air pump rod being suspended from the centre G would
e up & down perpendicularly, with half the velocity & through he
that E does. The black lines shew the position of the parts whe
beam is down & the red ink lines the same parts when the beam
ly up. ——— Jan 1814 Not being aware that the ideas of
on this subject as they arose in my mind in 1783 is any where
d I have now put them down. Mr Southern soon after contrive
metrical construction & an analytical formula which gives the lengths
rts more accurately & more expeditiously

Plate [4]

far the particular mechanism was explained to him I do not now remember; but he had the modesty to ask to have the engine stopt, on which 50 or 100 men were attendant, that he might see the inside of it which was refused. But the outside he was at liberty to examine & the principle of action was explained to him.

Paralel (Sic) motion.

I have been accused by Geo. Adams of Stealing that from Luardi, in which it is not contained, & that I should have owned it to some person or other.

The idea however originated in this manner. On finding double chains, or racks & sectors very inconvenient for communicating the motion of the piston rod to the angular motion of the working beam I set to work to try if I could not contrive some means of performing the same from motions turning upon centres, and after some time it occurred to me that A.B, C.D, being two equal radii revolving on the centres B & C & connected together by a rod A.D in moving through arches of certain lengths the variations from the straight line would be nearly equal & opposite, and that the point E would describe a line nearly straight, and that if for convenience the radius C.D was only half of A.B by moving the point E nearer to A the same would take place, and from this the construction afterwards called the paralel motion was derived. Luardi shows how an equilateral triangle may be described by his Geometrical compasses but it is by a construction widely different, so that If I had availed myself of him at all it could only have been by taking the idea from him that a straight line might be derived from circular motions compounded, and it is possible I might say that; But the paralel motion originated from the reasoning I have mentioned. Even granting I had taken the idea from him, there remained much room for invention to produce the paralel motion & to adapt it to the engines as has been done. And I can safely say that in all such cases first ideas are a very small part of the work. Though I am not over anxious after fame yet I am more proud of the paralel motion than any mechanical invention I have ever made. As to Mr Prony[4] I acquit him of all blame or envious intention, he was merely the chronicler of what was related to him & with such relaters as Perrier & Betancourt at his ear, what better could be expected, he knew nothing of me or my works but what they pleased to relate. He is himself a most ingenious modest & candid man & regrets much his having published what he has done, and offered to insert in his next publication whatever I pleased to communicate on the Subject. I had begun a Short history of the invention for him but the war breaking out it was never completed.[5] I send herewith what I had seen of it which as I have no copy I beg may be returned with the other papers. I cannot at present find his second vol so am unable to reply to some other of his imputations he has published. To answer all the accusations that have been brought against me by the envious & misinformed would require a larger volume that I am now able to write, many of the circumstances are lost to my memory & the labour is more than I can undertake I must trust to the Equa posteritas.

Arthur Wolfe

I flatly deny having ever said or thought, as he states me to have done, and moreover do not believe a word of his theory which it would be a waste of time to try to refute being in direct opposition to facts; for Steam to expand into twice the space, with the elastic power of the atmosphere, it must be of 251^{d} of heat & its elastic force equal to 2 atmospheres when confined, and unless it were maintained of the heat of 251 it would not expand to twice the bulk upon being admitted into a vessel of double contents, but if left to itself would immediately fall down to 212, or lower, because the latent heat would be answered at the expense of the sensible. But this is an intricate subject which I cannot at present clear up my ideas upon.

MIV Misc (3)

---

[1]Patent no. 859.
[2]Patent no. 1213.
[3]Watt patented the double acting principle in 1782 (no. 1321) but it was, in fact, already covered by Jonathan Hornblower Jnr's patent of the previous year (no. 1298).
[4]G.C.F.M.R. de Prony published the first description and drawings of Watt's engine in book form in his *Nouvelle Architecture Hydraulique* (1796).
[5]The draft is in MIV.

DOCUMENT 17

To James Watt — Papplewick, Notts. Feby 21st 1786

Sir,

On saturday last, we compleated our Engine, & on Monday we made a trial, & to day we set all the Shafts, on, without any other of the Machinery, we find no Ill conveniency at all attend it, any more than the fly Wheel which is very liable to go the contrary Way, in setting on & stopping the Engine, which is very hurtful to the Machinery in the Mill[1] — But we intend to Erect a Catch for the Wheel, so that it cant go back

I am Sir Yrs
&c at Command
Rich. Dayus.

P.S. Mr Robinson seems extremely well pleasd.

Box 4 D.

[1]Robinson's Papplewick Mill was the first cotton spinning mill to be driven by a B & W engine. The episode with the fly wheel also occurred in early attempts to use rotative atmospheric engines in textile spinning.

DOCUMENT 18

Donington furnace Engine[1] Augt 26th 1783

Engine for drawing Coals at Donington Wood — It consists of a vertical wheal whose axis is placed about 10 or 12 feet above the ground, the wh^l is 12ft 4inches di^{ar}, its Gudgeon ab^t 3 inches dia^r of cast Iron fixed in by a cross — from one side of the wheal is suspended the coal basket by a rope about 2 inches dia^r, and on the other side of the wheel is a large tub. which being filled with ∇ over ballances the coal basket and raises it. It is regulated by a brake applied to the circumference of the large wheel — The rope is = ized by a chain with large circular heavy links, which suspended by a rope from a wheel about 4 feet dia^r, and as the rope winds off are gradually deposited on the bottom of a bye pit, untill the meeting of the water tub & coal basket when they are gradualy raised again by the ropes winding on in the contrary direction on the small wheel — 3 shafts are necessary in this way of working — 1 for the Coal basket, 1 for the water bucket, and one shallow one for the = chain — they wind 7 or 8 cwt at a time, from 75 y^{ds} deep but it is a slow operation as the coal basket returns empty & the tub is some time in filling it requires 2 men one to mind the brake and one to land the coals.

J. Watt Blotting Book, p. 53

Box 53.

[1]in Shropshire worked by William Reynolds.

DOCUMENT 19

To James Watt London May 28 — 1788

Dear Sir

I have your fav^r of ye 20th In^t. I will wait upon ye Lead Hill C^o[1] & pursue your directions.

Mr Matt^{ws} says he believes he hath Bell & Goss's[2] agreem^t but he is so hurryd wth ye business of others that I have scarcely have had any opportunity to speak to him I will therefore go this even^g with him to Gypsey & return to breakfast — I shall see Mr Foreman soon & consult him about Trumans.[3] I will afterwards call on Trumans, Liptraps[4] &c.

I have no Choice about which side of ye Mint Engine ye Boiler stands & therefore submitt it to you. I am glad you have wrote to Hanley whom I will also see.

Mr Motteux hath appointed to accompany me to Count Lucchesa in the morn^g but upon further reflection I think it better to wave our visit untill I receive the drawings, with their own plans, wch I must request you to send directly.

Renne[5] says the piston Rod for Polis[6] Engine is ready but there is no Cap to it, as he says non was orderd — Q^y is there one made or must there be one made here? Q^y if the other piston rods are made for air pump & ∇ pump & if not say what must be done.

Mr Van linder[7] with his Sister are here. I have accompnyd him to ye mill[8] this morn^g where I find things go on Sluggishly — Mr Freer Complains but Wyatt is good humourd.

Parkers Scheme is to make a Rotative Mill without a Beam by means of a great fork with a wheel between y^e two tangs of it

with teeth half way round. I also find many new mechanical Schemes executed at the Mill. Those by Mr Parker are all as innocent as this fork.

There is one now at work (not Parkers) for regulating the presure or distance of the top Mill stone from the Bed Stone in such manner that the faster the Engine goes the lower or closer it grinds & when the Engine Stops the top stone rises up & I think the principal advantage of this invention is in makeing it easy to set the Engine to work because the top Stone cannot press upon the lower untill the Mill is in full motion. This is produced by the centrifugal force of 2 lead weights which rise up horozontal when in motion & fall down when ye motion is decreased by which means they act on a lever[9] that is divided as 30 to 1 but to explain it requires a drawing.

M^r V^s[10] will turn out a very bad affair but the House will stand good, S^r C R^d will bear a share in the loss & V. will have as much remaining as will enable him to live in y^e same stile he usualy hath done provided he will consent to go out of business. He hath given full powers to Mr Matthews & Mr Foreman to act for him in all things for he is incapable of acting for himself & besides nobody will attempt to act with him.

I have not yet seen $Mess^{rs}$ Garbetts or any of the Ministers.

Remember me kindly to Mrs Watt & believe me

Dr Sr Yours $affect^{ly}$

Matt: Boulton

Bouboulon yet sticks as Close to $parl^t$ as Irwin did

Box 20

[1]Leadhills Co., Lanarkshire, ordered a 30 in. Cyl. 8 ft. stroke engine.

[2]Bell & Gosse, London, distillers, purchased a 14 h.p. engine in 1786.

[3]London brewers.

[4]London distillers.

[5]John Rennie was employed by B & W for a year before he set up his own business in London.

[6]This was the 28 h.p. engine ordered for the King of Naples.

[7]J.H. van Liender obtained the Dutch patent for B & W [31-34] besides being an important customer and intermediary in the Netherlands.

[8]Albion Mill, Blackfriars, London.

[9]The Governor.

[10]Vere of Lowe, Vere, Williams & Jennings, London from whom Boulton borrowed money.

DOCUMENT 20

To John Southern — Walker Colliery Feb^{ry} 23^{d} 1798.

Dear Sir,

Permit me to remind you that you are in possession of a Letter of mine (at least I hope you are) under date of the 18th of December 1797 to which I flatter myself with the honour of a reply when it is perfectly convenient to you but not before — Need I observe that the worst payers are generally the best cravers?

In the next place as you & I have conversed a good deal upon the subject of Ropes permit me to inclose you a Table drawn from a good deal of observation, shew^{g} the Wear & tear of that Article in different places, in which you will perceive that Messrs Boulton & Watts Engines cut a good figure as consumers of the Article, & on this Account I wish to draw their attention seriously to the business, for it is the only thing now that the would-be Engineers & Mechanics of this Country have to say against the Engine, & I am sorry to add that Buddle seems to join that herd cordially, & has very lately declared himself a great advocate for the Old Water Wheel with all its pariphernalia of Lumber! — I have therefore to request you will lay the inclosed before Messrs Boulton & Watt, & as I know their Engine capable of excelling in the regularity of its motion, any other contrivance of the kind, lend us your aid to do away this objection.

I have studied the Case a good deal & am very decidedly of opinion, that the Wear of Rope in all cases is more brought about by rapid motion, irregularity in the velocity, & change of Load, than by the mere weight, for nothing wears out fibrous bodies so fast as alternate strain & ease — Now the proprietors of your Engines here, will, notwithstanding all that can be said to them drive the Engines with considerable velocity (more by far than in general there is occasion for) start them suddenly, & stop them again as suddenly, I must not include under this head the King Pit & Benwell Engines for they are under good regulation, and the wear of Rope in the former is chiefly occasioned by the salt Water springs of the Shaft. The Benwell Engine lies under the disadvantage of having to draw a light Corf which I need not tell you will always cost more rope in proportion to its weight than a heavy one — besides you will see that the Ropers have made the Cordage too heavy, as indeed they have done at the King Pit, which I shall take proper care to prevent for the future.

Having promised this, I have next to propose a remedy which whether it answers or not is on the point of trial—I know you will anticipate the Elastic connector between the Rope & the Corf, but I do not rely upon this alone — I really think that as much good may be done by making the Fly Wheels heavier, which will not only make the Engine start much more softly but in every vicissitude of motion will make the transitions more insensible — I can feel very sensibly for instance in going down the Pit or coming up every stroke of the Engine while the Men are handing the Gear, & as the Engine has a great deal to do handid I think this alone a great enemy to Rope, but this would be considerably diminished if the Fly Wheels were more loaded, and I have really felt amazing alteration already by adding a Rim of Lead 6 Inches broad & $1^{1}{}_{8}$ thick to the King Pit Fly Wheel which makes the motion much smoother — It is sometime since I learnt our Enginemen to draw up the Men entirely with the Throttle Valve which makes the Engine work very smoothly, but they cannot let down weights in this manner — Since I begun to write this letter however it has occurred to me that they may do it with all Valves shut, & a slight touch of the Brake, which I shall try immediately, then I hope that except in blowing to start in the morning, I shall get their clumsy fingers entirely removed from the hand gear.

You will perceive that Camerons[1] Engines have worn but little Rope — I attribute this to the light load & short stroke of the Engine, the former compels the Engine to go very quick in proportion to the Corf, & the latter makes so many more revolutions of the Fly Wheel, which being heavy gets a most dreadfull momentum. Indeed the small Engines at Mount Moor use less Rope more in consequence of the slow motion of the Corf than any other Cause.— As to the Water Wheel its excellence lies in being the best Fly Wheel imaginable.

I beg you will as soon as convenient send me a little of the Science of Soho on this subject,[2] & say whether you think it would be a good speculation for me to make a bargain with Buddle to find Walls End with Rope. I think there are ways & means to silence him.

Newcastle is going mad for little 3 horse Engines of B & W^{s} construction in consequence of the little $\overset{In}{5\frac{1}{4}}$ Cylinder I have put up for my Brother's shop which goes like the Devil — I hope to Write to them on this & other subjects on Sunday or Monday next. In the meantime please to present my very best wishes for their success & believe me to be Dr Sir,

Yours most sincerely

Thos Barnes.

PS M^c^Murdo waits for Orders
So does Taylor, having finished
at Lawson Main a Fortnight ago.

Box 6 Barnes (4)

[1]Robert Cameron patented machines and engines for raising coals, ores and water from mines in 1786 (pat. 1525) and 1789 (pat. 1702).
[2]Watt made various suggestions for improving colliery winding engines and a number were made at Soho.

DOCUMENT 21

To James Watt Junr. — Manchester 9 Nov. 1798.

Dear Watt,

I wrote you on the 3^d^ and am without your Reply.

I felt so much Remorse at the Idea of relinquishing the Expansive principle in the new Engine principally upon account of the Difficulty of regulating the Speed as to encroach very much upon my mechanical Equanimity[1] — Our friend Peter[2] and I have been in deep consultation and inclosed you will find his very ingenious Mode of Obviating the Objections — I should wish to know the opinions of the Conclave, having had no Experience of the Expansive mode of employing Steam, I feel extremely diffident in forming an Opinion.

If this or any other mode should appear elegible, we might perhaps squeeze in a 30 horse Engine — if it would be in other respects advisable both might work expansively in power equal to 45 horses and by connecting them at will by the Main Tumbling Shaft together, at proper parts of their respective Strokes correct each others Inequality.

A Neighbour Mr. Joule a Brewer who has one of Sherats[3] Engines wishes to pay the premium (ab^t^ 8 horse) and have an Air Pump made by you or him — I said Mr. Lawson would probably be here in a fortnight — if I am mistaken can I communicate anything from you, as he is stimulated by the Magistrates who threaten an Indictment for a pestiferous smoke upon their Sanctuary the New Bayley.

I am
Y^rs^ truly
G. Lee

If the whole be so fixt that the Catches 1 & 2 are, by means of the Gover^r alternately presented opposite the Xs G & H as the Engine goes too fast or too slow, the Pinion will be turned a quarter round every Stroke & the Slides be moved so as to shut the Steam Valves sooner or later as required.

I suppose sufficient friction may, without detriment, be given to the Slides & the Spindle of the pinion to Shut the <u>Steam</u> Valves without being push'd back — But if that will not do a wheel with notches in it such as M may be fixt on the Spindle of the Pinion — & a spring N with a tooth on it be fixt to the Plug having a Scapement O, P a Peg fixt to the Slide K may be presented to the Scapement just before the X meets the Catch by which means the tooth may be removed whilst the Pinion is turning & dropt into the Notch when it should be held fast.

P. Ewart, with great deference submits the annexed crude Sketch to the consideration of his friends at Soho — & shall be proud if any part of it meet their approbation — He is sensible that it is too much in the gimcrack way, but thinks it might answer under the judicious management of Mr. Lee who is laudably inclined to make the experiment of an expansive rotative Engine.

(In hand of Peter Ewart)

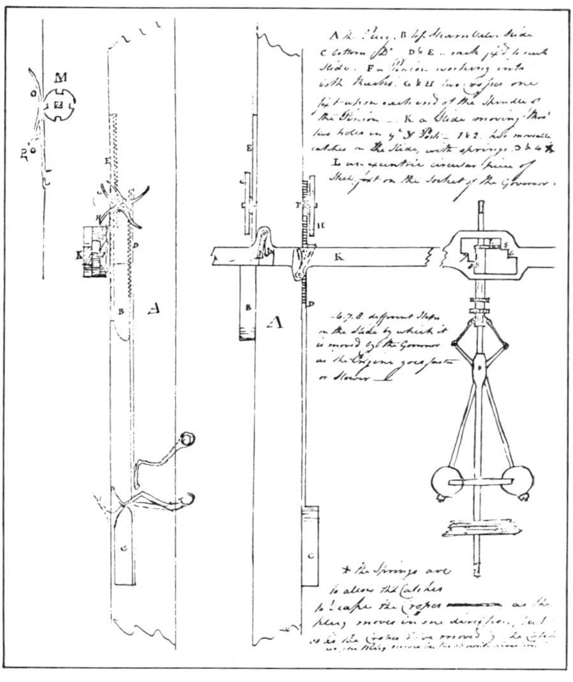

Box 3 L

[1]Lee's interest in things mechanical overruled his business sense on occasion. His mill was one of the early iron-framed ones. It was steam heated and gas lit besides being powered by several B & W engines.
[2]Peter Ewart, former employee of B & W, then a millwright in Lanchashire.
[3]Bateman & Sherratt, ironfounders and engine makers of Manchester.

DOCUMENT 22

To James Lawson — Manchester November 9 1798.

Dear Lawson,

Since your Departure nothing material has occurr'd here requiring your attention except two Applications about the price of Engines for wch I have referr'd them to Soho, informing the parties you were shortly expected here.

Mr. Ewart & the Engineers at Soho are scheming a mode of working expansively[1] and to regulate the Steam Valves by a Governor — if this can be brought to bear we shall enlarge the new Engine to 30 horses and work them both expansively & possibly connect them together at the main tumbling Shaft.

If you are not coming soon, let me hear your opinion if you are going soon to Soho, Peter will accompany you and we may make a Trio and settle this knotty point together.

Yrs truly
G. Lee.

Our neighbours the Goulds are return'd.

Box 5 X

[1]Watt found that, in practice, the expansive engine saved very little fuel and few of these engines were made.

DOCUMENT 23

To J. Watt Junr., London. Soho, 22 Apl. 1799.

Dr Sir,

I had the pleasure of writing by post which has just left. Now for Simonds,[1] I cannot say that I think the proposed construction of small engines with sliding valves is likely to be found in use. I am much afraid that the rubbing planes will get shited, and should that happen the repairs will be so difficult and expensive, that you will find the loudest complaints will be made against you. I cannot but observe that I think you are attempting too many and too great changes in the construction of these engines at once; & while I feel the difficulty of resisting the torrent of ingenuity which Murdock's genius pours forth, I beg to be understood as neither wishing to repress improvements, nor as approving all he proposes, which I am fully convinced is always with the purest views.

If therefore a small engine for Mr. S or any other gentleman must be made before you have had sufficient experience of the new sliding valves, which I should not think less than 6 months use could be called, I am of opinion that the Valves may be made the same way as the new mint engine ones, with a working gear nearly on the old principle — say the centres of the parts in a straight line when the valves are shut, to make the opening easier. With respect to the beam, that it seems must be given up as opinions, if not prejudices, are against it — nevertheless I hereby declare that I think without them the engines will not be so convenient, though, perhaps, a little simpler and requiring less room.

Issue your fiat as to the construction with what sort of valves, and as to time please to recollect that nobody but W. Harrison can fit them up at present — that patterns are to be made if the valves are as I propose — that Murdock may want some of his assistance in turning the great rod — and you will be induced to take the longest time possible — if you can get 4 months it is very desirable you should. If not, pray take 3 at least. You may tell Mr Rennie that as we understood he had all the machinery to make, we thought the hurry was not so great, or calm him in any other way that your resources offer.

Let me add, that I think the construction of the sliding valves as now making is quite unexceptionable, and that there is nothing to be hoped from improvement in that way if they should be found to fail.

And also let me add, lest you have misconceived me, that I have the highest opinion of Murdock, both as to his ingenuity & his heart, that I am far from thinking him obstinate in his opinions* but I do think he is too sanguine in some of his expectations.

I remain Dr S^{r}

Yours Truly
John Southern.

*Dont you think that this character, if not to him applies to somebody else?

Foundry Letter Book 68, p. 127.

23[1]The Reading brewer.

DOCUMENT 24

To Erasmus Darwin — Soho Jany. 4th. 1790.

Dear Doctor

I should have answerd your first letter before this time had I not been obliged to go to London Bath &c. Your last of ye 31 Decem^{r} I will now answer but I have not yet considered of any arangement or what part of the subject to begin it I will therefore give you only a Catalogue of Facts & leave you to make any use of them you please.

Steam Engines of B & W construction differ from Newcomins in 6 principles

st

1. The acting power is Steam on ye piston, & Not the Atmosphere, as in ye Com^{n} Eng^{ne}
2. The Steam is condensed in a separate Vessell, & Not in the Cylinder as in D^{o}
3. The Cylinder is kept as hot as Steam, & Not alternately heated & coold as in D^{o}
4. The air is extracted by an air pump at each Stroke & not blown out at Snifting Clak
5. The packing of the piston is preserved air & steam tight by oyls ... & not by Water
6. The Vacuum is alternately made above & below the piston & thereby obtain double power

Conveniences & properties of B & W improvd Engines
a — It is the most powerfull Machine in the World &
b — It is the most tractable
c — It is the most regular & hath the property in itself of conforming to ye least work
d — It shews by a Mercurial Vacuum the degree of Vacuum — in ye Cylinder
e — It shews the Strength of the Steam by another Mercurial Gage
f — It registers (sans error) the number of Strokes it makes in $\overset{\text{Years}}{100}$
g — It is applicable to every purpose that requires either Rotative or Reciprocating motion

(a) It is the most powerfull because such an Engine as we have erected at Wheal Virgin Mine in Cornwall is = to all the work a Thousand horses can do Pr 24 Hours ———— The Cylinder is 63 In^s diam^r, its area is 3117 Sq^r In^s works $\overset{\text{lb}}{10}$ down & $\overset{\text{lb}}{10}$ up Pr Sq^r Inch = $\overset{\text{lb}}{20}$ Pr Sq^r In but can work 21 or 22 $\overset{\text{lb}}{22}$
$\overset{\text{sq In}}{3117} \times \overset{\text{lb}}{20} = \overset{\text{lb}}{62340} \times \overset{\text{feet}}{9}$ Stroke = $\overset{\text{lb}}{561060}$ one ft. high Pr. Stroke × $\overset{\text{Strokes}}{14\ \text{Pr.}}$ Min.$'$ = 7854840 lb. Pr. Minute one foot high ÷ $\overset{\text{lb in a Ton}}{2240} = \overset{\text{Tons}}{3506,6}$ Pr. M$'$. 1 ft. high × $\overset{\text{Minutes in 24 Hours.}}{1440}$ is = to $\overset{\text{Tons}}{5049504}$ one foot high in 24 Hours

Horses will only raise 33000 lb 1 ft. high Pr. Min$'$. × $\overset{\text{Mt.}}{60}$ = $\overset{\text{lb}}{198000} \div \overset{\text{lb}}{2240}$ in one Ton = $\overset{\text{Ton}}{883,9} \times \overset{\text{Hours}}{6} = \overset{\text{Ton}}{5303}$ NB. Cant work at that rate more than 6 Hours in ye 24 & keep up their flesh

Hence $\overset{\text{Tons}}{5049504} \div \overset{\text{Tons}}{5303}$ is nearly equal to 1000 Horses But we have a double Engine (viz. up & down) now orderd of $\overset{\text{In}}{72}$ Diam^r & for a Rotative motion to work a Mill but we shall endeavour to diswade the orderer from so great a power as the Sqr. of 63 is to 1000 Horses so is the Sqr. of 72 to 1306 Horses & this Machine can work at that rate every Minute from Sunday evening untill the Bell tolls for Church on the next Sunday Morng. and thus I hope I have provd to your satisfaction That it is the most powerfull in ye World

(b) It is the most tractable because we have applyd it to the Weaving of Ladies Garters, fine Muslin, spining silk, spining Cotton, drawing Coals, Copper, Salt & Men from deep Mines — It in some cases discharges its own Load & stops of it self at any part of the depth of ye mine it is required to do. It is applyd to the Draining & Watering of Land & to make Canals navigable that wd. not otherwise be so. It presses Oyl it grinds Mustard, Malt, Dying Woods, & killing Drugs, it makes paper, presses Sugar & Grinds Corn &c. &c. &c.

(c) It is more regular & uniform in its motions than Water Mills because if the power is adopted to turn Mill stones or ten Thousand Spining Spindles & if one be detatched from the general movements the primary motion will not be accelerated or if 2 or 3 or 9 be detached it will not be accelerated in its motion as Water Wheels would be for it has a power within itself of adopting itself to the power wanted & of raising & consuming steam in a certain degree proportiond to the power used or wanted & this it will adapt itself too even if the Fire man is fast asleep & no Notice given of the detatching or stoping 9 out of the Ten. The next explains itself & you understand perfectly.
The absolute power or Load of our Engine, is equal to the weight of a Column of Mercury whose Diam[tr] is = to the piston of ye Cylindr. & whose height is = to the height of the ☿ in the barometer connected with the Engine to shoew its Vacuum + the height of the ☿ in the Steam Gage minus all the frictions & Visinertia of the whole machine & its appendages & if the Engine works up & down or what we call a double Engine it is equal to twice the weight of the aforesd ☿ raised as many feet high at each Stroke of the Engine as is = to ye length of the Stroke. In which Engines we no longer use arch heads at the extreamities of the great beam but by a sort of parralell Motion annexed to the beam a Straight lined Motion is derived from a Circular one & by this means the assending power acts upon the beam which could not be accomplished with Chains as in the Com[n] Engine.

(M. Boulton)

MB — Darwin/38

Section C

Boulton and Watt's patents and privileges at home and abroad

Boulton and Watt's patents and privileges at home and abroad

Documents 25-40

Besides his first patent of 1769, Watt held others taken out in 1781, 1782, 1784 and 1785. In the months leading up to the granting of these subsequent patents, Boulton played a vital role as adviser, encourager and friend whilst actively seeking information on competitors and lobbying for such external support as was necessary. He constantly reminded Watt of the superiority of his invention over all others[26] yet experimented to ascertain the best practical mode of deriving improvements to the engine whilst seeking reinforcement for his own practical ideas[28]. It was more often Boulton who would suggest bringing certain extra practical applications of steam power under the umbrella of an impending patent application as in his letter of 1784 discussing steam road carriages[28].

As is illustrated in Section D, the partners had persistent trouble with patent infringements in the home market. Nevertheless they carried out a deliberate policy of seeking overseas patents and privileges, paying particular attention to France, the destination of their first overseas engines. Considerable lobbying was required[29 & 30] and it is significant that the drafts of letters to the French government should be in Matthew Boulton's hand.

The Dutch patent was promoted by the Batavian Society of Rotterdam through J.D.H. van Liender who appealed to Watt as a scientist of international repute. Insisting on the fact that his promotion of Watt's interests was purely a matter of scientific interest and patriotic zeal, van Liender both obtained the Dutch patent and channelled the subsequent engine orders to Boulton and Watt without pecuniary reward[33 & 34].

The promoters of the Spanish patent, on the other hand, were rewarded with the grant of a sole engine agency by Boulton and Watt[37] although, in fact, few Spanish orders were obtained.

DOCUMENT 25

To James Watt. 12. Aug. 1768.

Dear Sir

I am exceedingly glad to find that you are in London, and taking the steps necessary to secure proper advantages from your excellent invention. Mr Boulton had hopes of being in town about this time, but is prevented. A circumstance also has arisen very lately that will engage him to delay the execution of his Steam engine for some time. He, Dr Darwin, & I will rejoyce to see you, & to tell our neighbors the services you can do them. Mr Edgeworth[1] is a gentleman of fortune, young & mechanical, & indefatigable. He is not acquainted with Heatly,[2] but has taken a resolution of moving land and water carriages by steam, and has made considerable progress for the short space of time he has employed himself in that study. He knows nothing of your peculiar improvements, but seems to be in a fair way of knowing whatever can be known on such subjects. At present he is in Ireland. My respects if you please to Mr Cummings. Get your patent and come to Birmingham with as much time to spend there as you can. I am with great attention

Dear Sir

Your most obedient Servt

W. Small

MB-Small/2

[1]Richard Lovell Edgeworth, member of the Lunar Society.

[2]Joseph Hateley held pat.no. 895 for 'A Fire Engine with Boiler' dated March 1768. He subsequently took out two other engine patents.

DOCUMENT 26

To James Watt Soho 28th July 1781

Dear Sir

My last was 26th In.^t I have now to acknowledge the reception of your two letters dated ye 21 & 23^d In^t—

Be assured I am upon good terms wth Snow Hill Engineers.[1] I meet Sampson[2] every Monday & all is well & likely to continue so.

As We shall neither be ye makers users or Vendors of the intended Forge & as Wilkinson is fully apprised of all the circumstances & has agreed to take all disputes upon himself We thereby stand clear. Moreover as Wilk^{n} spent an afternoon & even^{g} wth Jno Steed at ye Castle when last in Town he is perswaded that he not only can make peace betwen them & us but can make them usefull allyes. He thinks he can do the same thing between them & us as he did between Jesson & Wrights C^{o}[3] & three or four other Iron masters who now make bar Iron wth pit Coal under their patent. At least Wilkinson will bring the matter to a friendly issue which is better for all parties than remaining in suspence and Jealousy. However it is nevertheless adviseable to include the double devil in the patent which I hope you will not delay as delays are dangerous. I have wrote to Hanley some days ago to send me all the particulars he can respecting the Horners[4] patent — Dr P.[5] & my self are now persuaded that non of his airs can produce Mechanical power so cheap as Steam but he could not finish his Exp^{ts} so as to give the results Mathematically on acc^{t} of an accident befalling his Glass tube, but resume ye exp^{t} for my satisfaction although he is now possitive that nothing can be done in that line.

I have within these ten years past conversed with numbers of Enginers, philosophers & Mechanicks & I never heard of any man that ever pretended to have thought of, or practised pressing down a piston with Steam or Condensed Steam in a Separate Vessell. It was quite new to Smeaton & so new to ye Society of Arts that when our Engine was first talked of they would not believe that Steam w^{d} do instead of ye atmosp^{e} — I believe you may easily prove (as far as a negative can be proved) by all the first men in the Kingdom in that line, that no man ever did erect an Engine upon your principle or propose such a thing — nay I can possitively prove by Mr Edgworths evidence that in ye year 1779 he mentiond his intention of driveing a Wheel Carriage by Steam to Gainsborough[6] & that he found him totaly ignorant of the principles of the Com^{n} Engine haveing at that time never seen one or even made any Exp^{ts} upon Steam or Steam Engines. You have not done as Arkwright[7] did but you have deliverd a clear specification of the principles & you have illustrated those principles by the erection of 50 Engines which are all exposed to publick inspection & not kept secret like the Cotton Mills — Therefore I & all men must conclude that you have the sole right to the invention of making & useing Engines where Steam is y^{e} acting power upon y^{e} $\overset{1}{\text{piston}}$; Condensing y^{e} Steam in a $\overset{2}{\text{seperate}}$ Vessell & Takeing out y^{e} air by a $\overset{3}{\text{pump}}$, of keeping y^{e} Cylinder $\overset{4}{\text{hot}}$ as Steam & of useing $\overset{5}{\text{unctuous}}$ matter to keep y^{e} piston air tight in lieu of Water[8] — No man shall use any of these principles in Fire Engines to our prejudice by God.—

Their Condensing in a sinking pipe, Eduction pipe, or by whatever name they may give it is point blank against the Law, & I think a small recepticle within ye Cylinder doth also come within the meaning of it. Yet I have no fears from that as Smeaton told me he had applyed our Condenser to his Engine & it made it no better than before owing to the Skin of ∇ left on the inside Surface of ye Cylinder from y^{e} ∇ on ye piston & if they use grease in lieu of Water they come within ye Act.

When I reflect on the obstinacy of Old Horn[9] & when I remember Jethro's speach upon seeing Wh^{1} Spirit first going its a bauble I would not give 2½^{d} for it & when I consider your generous Conduct towards that family in being Engineer for them at Poldice & giving them the reward of that office as well as the battles we have fought for them on every occasion, I say when I reflect upon these things I am fired with resentment & therefore I would expell them root & branch from ye Engine trade in Cornwall. We have dealt generously by them & by all our employers in Cornwall & we have got more in store for their interest — We found their Mines in a ruinous State, we shall put them in a prosperous state, & when ye present Engines are got to the extent of their powers, I believe you can enable them to go perhaps 20 fathom deeper. I never mentioned that circumstance to anybody except Mr Wilson. I have also something in view that will preserve them from the oppression of Copper Smelters & keep up their ores to a reasonable price. We have truth, justice, honour, power, & Law of our side; therefore Cheer up dont be low spirited but support your self as a Man against the folly & Villany of fools & raskels. If we permitt the Horn^{s} to smite us on the right Cheek & thereby intimidats us to turn ye left also I think we ought to wear horns as long as we live.

Let us dismiss this disagreeable subject untill we have new matter to go upon, dont spare your threats. I think we should prevent any of the Shropshire founders from Casting their Cylinders — I intended seting out to London on Monday but I have this day receivd 2 letters from Gregory Mine telling me the Engine[10] as set to work last Wednesday & requesting me to meet the principal adventurers there on Thursday next Augt 2^{d} & from various circumstances I think it very necessary that I should do so in order to guard our profits & our Character. I think our first Condenser by transmitting Cold through the metal & a pipe more than 30 f^{t} long in many Cases may be better than ye present as I think you make a better vacuum. Q^{y} what would a thin Diaphragm of Copper or brass do near the bottom of ye Cylinder & apply Cold water under it when you want to Condense — It cant be so good as ours — The philosopher Walker is come to Town & begins a Course of Lectures on Monday I expect from his Sylibus he will set people a talking about Engines. I have Subscribed but cant attend. He dines with me tomorrow. I think it better to make him a friend than an Enemy. I have wrote this post again to Spedding & to Jn° Turner. I had a letter today from Henderson he seems offended I dont write to him. Tell him I will write in a few days but I am overworked – Tell me when you find it necessary that I must come to Cornwall & I will come. Fothergill after talking as foolish at Bath as to Vere is now gone to London for ye benifit of ye air. Present my best respects to Mrs Watt & my Wifes who is but poorly. Adieu. God grant you health & Spirits is the sincere wish of your aff^t friend

M Boulton

Box 36

[1]James Pickard, patentee of the crank engine (Pat. 1263). This prevented Watt from using the crank and he avoided it by designing the Sun and Planet wheels.
[2]Probably Sampson Freeth.
[3]Ironfounders of Wren's Nest, Salop and West Bromwich. Their 'stamping & potting' method of making bar iron was patented in 1773.
[4]Jonathan Hornblower, patentee of a steam engine (Pat. 1298) in 1781.
[5]Dr. Joseph Priestley, member of Lunar Society.
[6]Brother of the painter, supposed inventor of a steam engine.
[7]Arkwright's carding patent had recently been invalidated in the courts. One of the points at issue was the vagueness of the specification.
[8]The air pump was not invented by Watt – it was used on atmospheric engines.
[9]Hornblower.
[10]A 46in. cyl. engine.

DOCUMENT 27

To M. Boulton Cosgarne Jany 23rd 1782

Dear Sir,

I received yours of the 18th today. You are right in the double tooth method, the rotative wheel will always make one revolution + the number of teeth in the Spear Wheel – I did not want to try the rotatives with pumps, but to load them with Lapp's Rolls or grindstones and to see which had the fewest inconveniences – I agree with you that it will be proper to employ Mr H. about the engines in your neighbourhood but I think the scheme of building a house is of a piece with some other of his Schemes, I hope you do not mean to advance the money for that purpose at a time when we want it so much. I beg you would take care how cash is advanced to him as if he continues to keep Miss P. he can scarcely avoid running into our debt.

I have since my last considered of Hoopers mill and cannot satisfy my own mind about the propriety of erecting it. The Engine, if made to draw water, will cost about £800 or £900 & if he gives us a proper premium that must be added to & N.B. It ought to be ready money otherwise it will come to so little by the year as never to pay our trouble. As to any other of the ways the Engine will certainly come Cheaper & do more work therefore will pay better but again he is no proper man to try experiments upon — I have therefore taken my resolution which is to lett his Engine Stand untill I can clear my hands of those who are before him in my books, which will not be this long time, as both my health and spirits are very bad, and I am particularly molested with a pain in my back — It is impossible for me while here to keep up any regularity in our business or to attend to the many and different things which pour in upon me. Indeed my head grows every day less fitt for it, and drawing is attended with so much increase of the Pain in my back, that though I cannot approve of Playfairs Specification drawings I believe I must use them from inability to make others, (although I have got one copy partly done in a much better manner). I therefore beg and pray that I may not be urged nor more new things laid before me untill I am delivered from Wheal Virgin, besides it will be necessary soon that I sett about the Irish and Expansive specification in the drawing of which my attention should be as little distracted as possible, and therefore every order which is not worth £500 ought to be deferred (at least) — My inclination & feelings would lead me to abandon both Cornwall & Wheal Virgin forthwith & to attend to and amuse myself with these rotative machines &c but it would be dropping the substance to catch at the shadow. I have a very mean opinion of the rotatives' profit, and the trouble with each of them must be at least double that of an Engine which raises water — Peace of mind & delivery from C — L[1] is my prayer.

This morning I rose confused after a very bad night's rest and to breakfast received the news that Woodmasons house was burnt by which we shall be considerable sufferers as besides the stock of Machines we had there, I fear that he is much in our debt and ten to one the burning the house will not be the only misfortune — I must trust to your prudence and Mr Keirs (with whom I beg you would consult) to have proper care taken of any Machines which may have escaped the fire half burnt, and to take proper steps towards recovery of the debt in case of the worst. This copying scheme[2] has been a most unlucky one in many respects. I shall not write Mr Keir this post as I have nothing to say except that above, and that I shall chearfully concurr in whatever and he thinks is for the general interest.

Jany 24th

Since the above was written I have done a good deal towards the fair copy of the drawings which I hope I shall be able to finish in time — Wheal Crenver is gone to work with elat & to the adv^{rs} satisfaction — my back & spirits are better, and I hope for better times — I pity poor Wood^{n}. very much and would not have any thing done to increase his present distress only to guard against contingencys as well as you can — Adieu

Yours most sincerely
James Watt

Parson Roblyon having dined here today I have not time to add several things which will be the subject of my next letter —

MB-J.Watt 3/10

[1] Cornwall. In fact Cornwall continued to be a lucrative market for B & W well into the 1790s.
[2] Watt patented his copying press (pat. no. 1244) in 1780 and in the same year a new partnership, James Watt & Co, in which Watt, Boulton & Keir were partners, was formed to exploit the invention.

DOCUMENT 28

To James Watt — Cosgarne 6th Augt 1784

Dear Sir

I have had a little conversation wth Brown about Wheel car^{gs}.[1] He proposes to catch most of the condensed Steam by makeing it strike against broad thin Copper plates & the condensed part trickeling down may be caught & returnd into its Boiler or other reservoir. This may do some good in rain or frosty weather & he proposes to have different sized revolvers to apply at every hill & every vale according to their angle wth ye Horizon. what he wishes to do may be right but ye means he proposes will not do. — I find his Father hath a w^{1} Carriage to go without horses which I believe Logan Malcomb lately saw in Scotland & as this was one of the first mechanical amusement wch Brown ever turnd his attention to in his youth it serves to account for the furor that he is seized with: I verely believe he would sooner give up all his cornish business & interest than he deprived of carrying the thing into execution. When a Man is mad in any way it is in vain to reason with him about his disorder. He doth not directly say that he hath a right to work pistons by Steam but he says he thinks or fears that the Trumpeters[2] or somebody else will & then take out a patent for W^{1} Carr^{gs} wch is the thing he says that makes him so sollicitous to have it secured. Now as you are going to specify sundry new applications of Steam Engines Q^{y} if it may not be prudent to specify the application of it to W^{1} Carg^{s} without makeing any drawing & only discribe ye application of the elastic force of Steam to act upon a piston or pistons in Cylinders & that force applyd to the turning of the Wheels by means either of Cranks or by one W^{1} revolving round another or within another or by any other of the methods now used to convert the reciprocating Motion of a piston moveing in a Cylinder (or of a reciprocating beam moved thereby) into a rotative motion or by any other of the rotative motions invented by you & discribed in your Specification dated — 1782. I propose this by way taking possession & saveing the expence of a patent.

I believe a Steam Engine Car^{g} may be made lighter than a common post Chaise if you include the weight of the Horses for in going up hill animal power must be furnished by them to raise the weight of their own bodys wch I consider only as other kinds of Steam Engines.

I think a very light & cheap modell might be made takeing one of our drawn tubes of an Inch diam^{r} to make the Cylinder of. I am sensible 'tis wrong to crowd you with new schemes at this time but my only reasons were the two following viz — 1st to seize the present moment of putting it into your specification & 2^{c} to ask you whether I shall make any agreement with Brown or not or acquiesse in any thing wch tends to keep peace & that doth not interrupt or retard our present real business.

We must have some more assistance here for Robert Moor is got into debt & into a Young Woman who hath or will swear a Basterd to him & I believe he is off.

Augt 7th I wrote you the last post since which I have recv^{d} yours of ye 31st Ultm^{o} previous to which I had wrote the foregoing otherwise I w^{d} have waved the Brown subject at present. I am sorry Wilson told you even the truth of him as it only serves to vex however I hope will not pay your self so bad a Compliment as to put yourself upon a footing with him. He may be a usefall tool in our hands if we make a proper use of him.

I yesterday met Mr Jenkins Danbos & Nankivel at Truro they seem'd desireous of taking a quarter in Chase ∇[3] but require a week to settle amongst their own Comp^{y} before they give a final answer. I have been at Camborn today & to Teheddy taking Kevil but to little purpose. Pool is on Monday but I cant attend it, as I have promised to attend polgooth acc^{t} on the Mine on Monday.

Jno Williams & Co want to take up Scorior & continue the Working of that Mine (wch is now stoped) & if pool stops next M^{o} Scorier Engine may remain & pool go to Chase ∇ — We are all tolorable well & joyn in best respects to Mrs Watt adieu I am Dr Sir affectionately

Yours M Boulton

dont vex but be firm. I'll stand by you as long as I live

Box 20

[1]Wheel Carriages. Whilst steam road vehicles did not catch the public imagination to the extent that steam ships did there were a number of enthusiasts — and possible B & W competitors.
[2]A reference to the Hornblower family.
[3]Chasewater copper mine.

DOCUMENT 29

The French Arrêt

Copy of answer to Mr. Magellan relating to Count d'Heronville. 1778

I have read the several papers you delivered to me relating to our obtaining an exclusive privilege in France for the Fire Engine of our new invention, and I beg you to assure the Count d'Heronville that we are fully sensible that he has done all that is possible towards expediting the affair. It remains for us to make the required trial in France. The superiority of our Engine, over those of the old construction, can be ascertained only by a comparison. But as there is no old Engine at Paris with which the comparison can be made, the erection of an Engine there would not be effectual towards obtaining a second Arrêt de Conseil granting an exclusive privilege. As the Count d'Heronville is going to sell the Moeres (Meres) we cannot make that the place of trial, unless the new proprietors apply to us. M. Jary, Concessionaires, &c. has applied to us to erect one of our Engines at or near Nantes in Bretagne in place of an old one, & he has come for this purpose to England. This would give us a good opportunity to prove the superiority of our Engine, but from the copy of the Arrêt de Conseil we have received from you, it appears the trial must be made at Paris or the Moeres. If the Count could obtain permission for us to make the trial in Britain & if the trial as such would be effectual towards the obtaining the second Arrêt, we could then proceed to erect an Engine for M. Jary.[1] We have not thought proper to make M. Jary acquainted with the Measures taken towards obtaining an exclusive privilege, but have delayed giving a decisive answer to his propositions till we know whether his Engine can be made the subject of the trial required by the Arrêt.

As we have not had leisure to carry the Circular Engine into execution, farther than was sufficient to convince us of the truth of our principle, we cannot speak with the same certainty and precision we can do of the Reciprocating Engine: but we have not the least doubt of its being applicable to any purpose in which a circular motion is required, viz. Mills for Grinding Corn or rolling Metals. A power equal to the strength of two Horses may be obtained by means of 112 pounds of Newcastle Coals per diem. I must defer answering the questions relating to the articles and prices of my manufacture But I beg you to acquaint the Count that it consists of whatever the Fashion of the Time requires to be made of Copper, Steel, Silver, plated or gilded, with other articles such as Paintings. I have sent a specimen of these paintings (Penelope) to the Count, which I beg him to accept.

(M. Boulton)

MB — M1/58

[1]Permission was given for the trial to be held in Brittany at Jary's mine. Jary was rewarded by being given the sole selling agency for B & W engines in France, a contract that B & W subsequently ignored.

DOCUMENT 30

Copy of the Memorial delivered to Mr. Genet at Versailles Nov. 27 1786. (in hand of J. Watt)

Showeth

That, the General theory & Principles on which the perfection of Steam Engines depends, were first discovered by J. Watt at Glasgow in Scotland in the year 1763 & were the consequence of a laborious course of experiments he made for that purpose.

That, although the theory & Principles then established have suffered no alteration since, nor are susceptible of any, yet in reducing them into practice many mechanical difficulties arose which prevented him from taking a Patent for his invention before the year 1769.

That, from various circumstances the public execution of these machines was unavoidably delayed till the year 1774, when JW removed to Birmingham, executed an Engine in ye manufactury of M.B & it then appearing that new tools & machines of various kinds must be made, Workmen must be trained up & the prejudices of the public must be removed by a long & expensive course of experiments on Engines in great, before the invention could be made profitable; It was judged necessary to apply to the British Parl^t for a confirmation & prolongation of the privilege which they granted in the year 1775 for 25 years & without which the invention must have fallen to the ground & the world might have been deprived of it to this hour.

That immediately after the passing the act of parliament a partnership commenced between M.B & J.W.

That the Expences attending the Establishment of the Business swallowed up very great sums of Money & Memorialists pursued the Business for many years with much labour but no profit to themselves & it is not very long since it has repaid those expences.

That, In the year 1777 a French nobleman[1] applied to the Mem.^sts for the erection of some engines at Dunquerque which they agreed to upon condition that he could obtain for them an exclusive privilege in France.

That, in the year 1778 the Mem.^sts were informed by the said nobleman that in consequence of a application made in their name an Arret du Conseil had been expedited granting to them an exclusive privilege in France upon condition that their Engines should be found by les Comissairs de l'Academie to be superior to the common ones, upon trials to be made upon Engines to be erected either at Paris or Dunquerque.

That soon after the said Arret M. Jary concessionaire proprietaire des mines du Nord in Bretagne applied to the memorialists for an Engine & procured a fresh Arret granting leave to make the trial on the said Engine, if necessary, which Engine the Mem.^sts furnished M. Jary with & also with full instructions for the erection of the same, for which services they have received no profit or reward and this was the first Engine of this species erected in France.

That, soon after Mr Jary, M. Perier[2] came to England with an intention to procure Engines of the invention of memorialists & finding that could not be done without their consent on account of the British Act of parliament he applied to them & on certain conditions which they made more favourable than in the like cases in England on account of its being of publick utility to his Majestie's

Capital, they consented & furnished him with Materials & also with all the instructions required for erecting the engines, which were accordingly erected & approved by the academie. It was also by the intercession of the Mem.^{sts} that passports were procured from the British Courtt for sending not only these Engines but many other useful materials for the Forges in Burgundy to France during the time when war unfortunately raged between the two nations.

That, after the erection of the Engines in Bretagne & at Chaillot the Mem.^{sts} were discouraged from coming to france by their being informed that the Arret du Conseil could not be put in force on account of the want of some of the necessary forms.

That, notwithstanding this discouragement & the delay of the conventionary payments from the Compagnie Des Eaux de Paris on M. Wilkinsons representing to the Memorialists that it would be a means of conciliating the favour of the government of France, they furnished Drawings, & directions for the erecting of an Engine for Blowing the Great Iron Furnace at Mont Cenis, and they have disinterestedly permitted Mr Wilkinson to furnish France with several Engines which they might have prevented if they had chosen to act with a strict attention to their exclusive privilege, by which they can prevent any person in Britain from making, vending or using their Engines.

That the Mem^{sts} from their very extensive experience have acquired more knowledge in the execution of their Engines & in the machinery applicable to them, than has been in the power of any other persons, and that in consequence they conceive themselves to be more capable to further the views of the Court of France in the extension of manufactures & public works.

The Memorialists therefore Being the first & real Inventors & Improvers of these Engines, Having, by the erection of about 200[3], acquired more experience & knowledge in the proper construction of them than any other persons, having In consequence of the confidence they reposed in the Arret du Roi en son conseil with wch they were honoured been virtually the first introducers of them into This Kingdom, where they presume these Engines have been & will be of essential use & whereby they have as far as lay in their power complied with the Terms of ye Arret and having hitherto received no pecuniary reward for the same except what was stipulated by la Compagnie des Eaux de Paris they most humbly hope that His Majesty will reward them by confirming the Arret by his Royal lettres patent or in such other way as in his Royal Wisdom may seem proper.[4]

MB — G1

[1]Count d'Heronville.
[2]Of the Paris waterworks at Chaillot.
[3]They had erected no more than about 120 engines, in fact.
[4]After B & W's visit to France in 1786 the matter of the French patent was dropped.

DOCUMENT 31

Dutch Patent

To James Watt, Glasgow. Rotterdam 11th Mai 1775

Sir

Having seen in one of your periodical papers that you have obtained an exclusive right for a new invented fire Engine, I take the Liberty of addressing myself to you by this letter begging you will be so kind and mention to me what sort of invention or approvement yours is, because I am busy at present in company with some few other Gentlemen of erecting at the rampart of this Town a fire Engine which will be a common lever Engine, for which we have received from New Whilly[1] in Shropshire the Cylinder of cast Iron of 52 Inches diameter and 9 feet long and the boiler of Iron plates nailed together, The diameter of the flange 18 feet, and at the bottom 16 feet, the height from the bottom to the flange four feet; and from the flange to the top of brim 7½ feet. Th' intended height of the water ½ foot above the flange; this boiler will be with an inside flue of the height of three foot and of the breadth of 2 feet and 18 Inches, which you will conceive better by th' annexed draught of its profile. This our engine will serve for bringing up the water from 6 Inches to 5 feet, never higher, with wooden pumps some round of 6 feet diameter, and some square of 6 feet Square, of which Eight are already fixed, and within a few days we think to put up our boiler and within three months we hope to see th' Engine work, which will be the first ever erected in this Country; and when its effect answers but tolerably our expectations must be a certain forerunner of a great many others.

One Mr Chrÿsell[2] from Bristol has obtained a patent for his boiler of a new construction; which according his brief account must be a very shallow one; me thinks his notions about that point are not at all absurd, and I believe that in common the depth of water in the boilers is to great for to have a quick boiling, and an economical expense of fuel. I am very curious to know Sir where in your improvement consist and in expectation of your desired answer I remain with Esteem

Sir

Your most ob^t Humble Servt

J.D. Huichelbos van Liender.

Box 36

[1]New Willey Works belonging to John Wilkinson.

[2]Christopher Chriscel's patent no. 1056, dated March 1773.

DOCUMENT 32

To M. Boulton — Rotterdam 25th Sepr 1778

Sirs

I was favoured in due time by your Agreable answer upon my letter to Mr James Watt, and saw with pleasure that gentleman constantly employed in erecting new Engines according his improved method, but it gave me some concern, that neither he nor you were in no manner Sollicitous for honorary rewards, the more as the desired Solution and answer upon the question could have been composed by Mr Watt for diversion's sake, and in his leisure hours in th' approaching winter evenings, without doing him any harm, and where there was â prospect, that by obtaining the promised reward, his person and method would be more effectually recommended in his Country, than by all th' exclusive priviledges you may have, besides this would never impede to obtain in this Country th' exclusive privildge, you seems to desire. In the contrary It would greatly promote its obtaining, and must certainly be the best recommendation of any one's abilities in that Sphere, and the preference he merites above all his competitors. These considerations I hope may alter yours and his resolution, and induce him to try his abilities in the discussion of this Scheme. My friends and I[1] are quite unable to employ your Mr Watt, we have no lands which want to be drained, all whatever we proposed ourselves to do had been nothing else than to demonstrate to people in general, that fire Engines could be used in this Country to better purpose, than the common wind mills, whose defects and insufficiency were known and seen by us since many years. We never hath any other or special view of self interest, we exspected not the least reward nor profit, not even the reimburse of our expences. The glory of doing â Signal service, to our Country, was the only reward we aimed at, and those are the Sentiments we stand in till this very moment, and as I hope we will never alter. We were offered an exclusive priviledge for the common fire Engines, but we have declined it, and In the same time we have served other people. Mr Blakey[2] should never have obtained th' exclusive priviledge for his invention in this Country, If I hath not given my formal consent to it, and I am ready to serve you and Mr Watt, where I can, and assist you in procuring th' exclusive priviledge, if you think fit to obtain it; but to tell you the truth I can see no great advantage for you in it because the Dutch have no faith at all, they must see the thing with both their eyes, and then (when it is a new thing) they are with great difficulty persuaded to adopt it, and this is the Case with most nations.

I am very respectfully,
Sir
Your M^t ob^t H^{ble} Servant
J.D. Huichelbos van Liender.

MB — L/6

[1]Of the Batavian Society.
[2]William Blakey obtained a patent for a steam engine in England (pat. 848) in 1766.

DOCUMENT 33

To J.D. Huichelbos van Liender. Birmingham March 29 1785.

Sir,

Your letter of the 30th Jany wch Mr. de Luc was kind as to put into my hands should have been answered sooner but I was obliged to make a long stay in London, where I was too much agitated with various matters to give your proposition the due consideration. Since my return home I have been much indisposed with a bad cold & a feverish disorder, from which latter I am just recovered, and my partner Mr. Boulton remains still in London confined by a sprained leg.

We approve in general of the proposition you have made us and if you understand it in the same light as we do, have no doubt but we shall agree with you. You propose that we should furnish you with Drawings and directions for erecting a steam Engine capable of raising 130 cubic feet of water 6 feet high pr stroke and of making 10 strokes pr minute, and that we should also provide, at your expense, the cast iron work & such parts of the hammered iron work as we shall judge necessary to send from England, and this without our charging any other profit than what may arise to us as manufacturers at the prices we commonly charge these goods to our customers here. That is that we should receive no premium for the use of the invention or for our skill & trouble.

On your part you promise to erect such an Engine at your own cost on a Polder near the City of Rotterdam, and to pay all the expenses incurred thereby and also that you will oblige yourselves to procure us an exclusive priviledge for the use of our invention.

As we are ignorant of the nature of such exclusive priviledges in your country, we wish to have an answer to the following queries.

1. Whether the patent or priviledge is meant to be for all the United Provinces or only for the Province of Holland.
2. In what manner must the invention be explained in the priviledge to secure against evasions, I suppose you know that in England we give in a separate writing called a Specification wherein the Invention is explained & the principles on which it proceeds and in which its novelty consists are set forth, so that any person making another machine which acts on the same principles however different it may be in form; is precluded. For instance suppose a man was the first inventor of spinning flax but the machine he used was a distaff (quenoûille) and he described the nature or principles of his invention to consist in the pulling out gradually a few fibres of flax and twisting them together by the whirling or rotative motion of a spindle which motion was to be communicated to it by twirling the spindle between the fingers, or otherwise as might be found convenient. Such patent would be exclusive against spinning by a wheel because the wheel in so far as regards the twisting acts on the same principle, yet the spinning wheel & the distaff differ toto caelo in form. Now our invention consisting entirely in the application of certain philosophical principles, which may be applied to Engines of 100 different forms, if the principles cannot be secured the patent is of no avail.
3. Is it understood that the priviledge is to be granted us free of all expense to us?
4. Whether drainages will not be required in other provinces besides Holland, & if so whether to any great extent.

We have had so much opposition to our Engines, to combat in this Country, & still have, whenever we attempt to open a new field, that it is with a considerable degree of reluctance, that we can bring ourselves again to lye under the disagreeable necessity of struggling with the prejudices of mankind, particularly as your countrymen have shewn a rooted aversion to Fire Engines in general. Yet if our doubts receive satisfactory answers, our confidence in your & other patriotic & enlightened Gentlemens exertions together with the known merit of our Engines (which we expect will speak loudly for themselves) will probably determine us to accept your proposals, though the load of business which presses upon us from every quarter of this country, and which increases daily, the bestowing the labour time and attention which will be necessary to erect such an Engine as you want, without any prospect of immediate reward, is perhaps not perfectly consistent with prudence, but tends more to indulge the love of fame and ambition of doing all the Good we can in the world, to which we are not a little prompted by the example of your truly patriotic society.

If the quantity of water you require is to be raised by a pump it must be 63 inches in diameter, and the cylinder of 6 feet stroke must be, [or ought to be to work in the most advantageous manner] about 37 inches dia^{r} or if 8 feet stroke 32 inches dia^{r}, and the consumption of coals by such Engine should be from 1¼ bushel of good Newcastle coals measured by the heaped Winchester Coal bushel which weighs commonly about 84lb avoir dupoise, but if scotch or free burning coals the consumption would probably be from 140 to 160 pounds weight per hour at 10 strokes pr minute, & this supposing the Engine to be kept in good order and the manner of firing well understood otherwise an ignorant fireman may easily consume more. I have in the above calculated upon the supposition of your using a pump, but I am not certain that we cannot serve you better by a rotative Engine applied to a water wheel but that must be subject of future deliberation, and I even think that I have found a better method than a water wheel which I mean soon to submit to experiment. These considerations will come best when we have settled the first points, but in the mean time we will be obliged to you for your opinion of the quantity of water lost by water wheels, that is the quantity they raise less than they ought to do by calculation.

We hold ourselves much obliged to you for the reference which you have so obligingly given us in this business and the attention which you have always paid to us.

Mr. Boulton desires to join in compliments to you and I will be obliged to you to remember me to my old friend Mr. J. Ainslie.

I remain, with much Esteem,

Dear Sir

Your most Obed^{t}

& most humble Ser^{t}

James Watt.

Office Letter Book 9 pp. 106-107.

DOCUMENT 34

To Boulton & Watt — Rotterdam 13th January 1786.

Dear Sirs

The 30th of last month, I have wrote you, desiring to return me the drawing made by our Architect here, to which I further refer these serving principally to acquaint you, that I have received this morning a letter from the Pensionary of Delft mentioning that yesterday afternoon in th' Assembly of their Noble & High Mightins our demand for an exclusive privilege in your favour is allowed, and that a penalty of ten thousand guilders is inflicted upon any person or persons who dare to infringe it. The Act of it dressed in due form will be delivered us, as soon as we are able to give in a drawing and description of the Invention, so that you will be so kind as to let me have these desideratas as soon as you possibly can, that we may finish this transaction, and give you an acknowledgement, that we never intend to make use of this Privilege, as to your advantage only; of which you will send me a Scheme, how you would like to have it; I wish you heartily Joy with this speedy success. The Pensionary has been so kind as to consult me before upon the principal terms, how I wished to have them stipulated; I beg the favour that now our undertaking may be furthered on your Side that it may be the means of giving you profit in its consequences and to us glory, which and the real advantages of our Country, is all what we are aiming at. I remain with every regard

Dear Sir

Your Mt Ob. He Servt

J.D. Huichelbos van Liender

Box 36

DOCUMENT 35

Proposed American Patent

To R. Barclay, London. Birmingham 22 July 1788.

Dear Sir,

We send this by Mr. Rumsey[1] of whom we have conceived a very favourable opinion & have in consequence made him the following proposals.

1st To give up our claims on the method of propelling boats by steam Engines, provided the following articles are accepted by him.

2nd To form a connection with him for supplying him with our steam Engines for that purpose, provided he shall employ no other steam Engines to move boats in any Country where he may obtain an exclusive priviledge. And upon this condition we also agree to pay half the expense of the patent he is now soliciting in England.

3rd The Engines to be furnished by us at the customary prices for Materials & premiums.

4th To form a connection with him to take out a Patent for our Engines in America in the Joint names of B & W & JR (say a patent for the Engines & all their applications) B & W to furnish the materials at customary prices, to pay ⅔ of the expense of the Patent & to receive ⅔ of the premiums or profits of the Engines. Mr. R to have the remaining ⅓ & the whole of whatever profits may accrue from the contrivance of propelling boats, B & W receiving the same premiums on these Engines as for others of like power.

5th All Engines ordered by Mr. R for American or foreign use to be paid in London by some proper person who will accept B & Ws draughts at 6 months for the value of the materials & at 12 months for the amount of their ⅔ of the premiums.

6th Mr. R. to engage not to make or cause to be made any of our Engines by any other person than ourselves in America or any other Country.

Mr. Rumsey does not consider himself as at present at liberty to accept these proposals on account of his Engagements with his friends in America. He therefore comes to beg your best advice how to proceed & we join in requesting that favour of you, we refer particulars to Mr. Rumseys narration & remain with much Esteem

Sir

Your most Obed^{t}

hble ser^{ts}

Boulton & Watt

(J. Watt's hand)

Office Letter Book 14, p.19.

[1]James Rumsey took out patent no. 1673 in 1788 for certain boilers and a steam engine. (This was followed by three other engine patents: no. 1738 in 1790, no. 1825 in 1791 and no. 1903 in 1792.)

DOCUMENT 36

To James Rumsey at London. Soho 14 Aug 1788.

Sir,

I returned yesterday from a journey of business, when I found your letter of the 6th Inst with one inclosed to you from Mr. Vaughan, both which I have sent by this days post to Mr. Watt, who is now at Buxton Wells for the recovery of his health.

It appears to me that there is so great a change in your sentiments that I have little hopes of our forming any connection, because you reason as if you expected shadows to counter balance substances.

I sincerely wish you may fully reap the fruits of your own labour & inventions & hope you'l allow us to do the same.

You say its no favour to be supply'd with Engines upon the usual terms — our terms for foreign countries are different from those in this Country for obvious reasons. What ever we have the trouble & risk of manufacturing here, whether it be Buttons or Fire Engines we certainly will not do ourselves the injustice to send to America or any where else without takeing a fair profit, and the Favours between buyer & seller I consider as mutual, particularly as we did not wish to restrain you in your profits upon goods so bought.

You say that in America we cannot make you any concessions, not having the claim, but surely for our knowledge, our Experience, our fair profit & our trouble etc. in manufacturing we have a natural claim to in all Countries; and in <u>this</u> the Law gives us to our natural right & Claim an Exclusive priviledge to make Vend & use Steam Engines upon certain well defined principles.

I own it appears to me that restraining you to use no other Engines for boats but such as are upon our Principles was unnecessary because I know no other will answer the purpose so well.

As to boilers you must allow me to observe that Mr. Watt was the first & only discoverer of the precise quantity of water & the exact quantity of heat necessary to make a given quantity of Steam of a given elasticity as well as the knowledge of the Total quantity of heat produceable by a given measure of such Coals as this country produces & thereby we are enabled to know the utmost Limits of Boilers.

I can't help smileing at Mr. Vaughans reasonings upon that subject, which I own is very ingenious but my opinion is, that when he hath tryed as many boilers as we have Hundreds, he will discover some Mistakes in it. There are many Cases in which your boiler cannot be used & there are many in which it may, & therefore we wish not to discourage you from takeing out a patent for it & will render you any assistance in our power in the introduction of it.

At a former meeting we told you that we can & do, in common practice, evaporate ten pounds of Water with one pound of Coals & inasmuch as your boiler will do more, so ought its nearer approach to perfection be estimated. You may easily determine its ratio by Experiments which are worth a thousand fancys or theorys.

Partnerships ought to be founded on equitable principles & like a pair of Scales be Ballanced, either with Money, Time, knowledge, abilities, or possession of a Market; & upon these principles were our propositions grounded but it now appears to me that our sentiments are not in union & that you have mistaken your road to the Goal in view; however I shall forbear saying more untill I have had a conference with Mr. Watt upon the subject & in the interim remain[1]

Sir

Your Obed^r^ humble serv^t^

Matt^w^ Boulton.

Office Letter Book 14, pp. 23-24.

---

[1]B & W obtained no priviledges in America

DOCUMENT 37

Spanish Patent

To James Watt London 23 May 1789

Sir

After having reflected upon your favor of 17th Ins^{t} I answer, that I as well as my Partners in Cadiz are willing to be united with you in the sollicitation of a Patent for your Engines at the Spanish Court; but as our Government has certain ideas & think differently from other nations & that at present the importance & utility of your Engines as well as the applications, such as grind^{g} wheat &^{c} they have been destined to is entirely unknown, I fear they may perhaps look upon our solicitation with indifference; nevertheless as I know the state of our Gov^{t} I can venture to presume that you being united with my Company at Cadiz & offer^{g} to introduce your Engines with all their applications excepting those against which you have contrary laws in this kingdom; thus obliging ourselves to introduce their diff^{r} uses we may not only procure a priviledge that is desired, & altho it is true priviledges are generally granted to the inventors yet often it is considered whether they are subjects of the nation, so that it appears to me that conforming ourselves to the Spanish stile the solicitation shoud be made in all our names which is the method my company has resolved upon to make the solicitation, beginning with the applications of Engines to grind wheat.

I did not answer you before as Mr Rennie 8 days since delivered me the plans & estimate of all the millwork[1] & as I have got to examine his prices I am very much occupied.

I beg the favor of your letting me have for some days the plans Mr Rennie has of the Albion Mills that my Partners may see in Spain the difference of our Mill with that of London. With Mr Bewicke's complements to you I am

Sir

Your most obed^{t} &
most Hble Servt
Pasqual Mensa

Box 36 March

---

[1]For the Cadiz steam corn mill.

DOCUMENT 38

Translated from the Spanish of Don Garcia Gaslon & Rowe Co.
Cadiz 22 Feby 1791

We have had the pleasure of receiving your letter 4 Jany in answer in reply to ours 6 July & are very sorry this delay proceeded from want of health in addition to your much business & an hope of an end to the differences that subsisted between your Court & ours which God be praised have terminated pacifically so that we can carry on a reciprocral Trade —

The forementioned differences & the occupations of our Ministers have been the cause that our affairs have not had the course we expected & our Partner Pla y Mensa having left some of them depending to be continued in time more convenient returned to this City and has informed us verbaly & by ye Documents he brought of all the Contracts & Agreements made with you[1] & also of the state in which he left at Madrid sundry matters in which we are all interested as well respecting the establishmen(t) of the Mill, as also that the Privilege which his Majesty has granted to us for the Introduction of Engines into his Dominions in Europe should extend to & comprehend also all those in America which in effect he has granted to us with the liberty of Duties for ten years —

In those Privileges which are in the name of the forementioned D^{n} Simon Pla Mensa & Company, tho' neither your Names nor ours, nor those of the other Persons of this place & of Madrid comprehended in the Association, are expressed, we do not the less remain Interested therein, as it will be so expressed when we transmitt the ratification of the Deed which you signed with the said Mr Mensa, who & us two, are appointed by all the concerned, Directors —

Before we cause any Pump to be brought we hope to obtain from the King certain pretentions which we have premeditated and are very usefull to the Mills's establishment, because afterwards it would not be so easy: This Establishment we continualy look upon as more & more usefull & necessary in this Country, as well on account of the bad method now in use for grinding Flour, as for the great opportunities of purchasing wheat & disposing of ye Flour; in consideration whereof & as we look upon you for the whole as Partners in this Enterprise, we hope you will ingenuously specify fully to us the concern you chuse to take in the establishment of said Mill, for the which we inform that each Action is to be 10000

Dollars of 128 quarts, or 7529 Hard Dollars 4 rials & 2 quarts which is the same: & in reply you may specify to us whatever you find convenient on the subject —

We do not know how M^r^ mensa's letters cou'd have miscarried, for positively it is evident to us he wrote to you on various occasions, as well from Biscay as from Madrid —

In the Privilege for the Pumps or Engines, the King chose that he should be at liberty to cause them to be brought for himself & for the use of his Arsenals etc. in this we had no Option, but you had not to have contracted with another; as was agreed between you & Mr Mensa, therefore the sale you made of the Engine to Mr T.[2] is very displeasing & inconvenient to us, for the bad consequences which perhaps may ensue therefrom hereafter to your Interests & to ours, for even tho' you do make good to the Company ye 20 P.C^t^ stipulated on the Price of s^ch^ Engine, we had rather be deprived of this small benefit than by similar Contracts have our good Intentions frustrated, and tho' by ye miscariage of said letters you suffered some suspense, it was less prejudice than that which from the contrary may result to us, for the stay in Madrid of Mr Mensa was not voluntary but necessary to establish & settle with the Ministry every thing conducive to ye Project, & therefore in future if we are to proceed together with concord & good harmony you must deprive yourselves of contracting similar Engines by any hand but ours & with our consent. —

Mr Mensa as well as the Company expected you would have sent the Tarif of the Prices of the Engines for taking out Water,[3] as you promised him when he returned to this Kingdom, but you do not even make any mention in your letters of a subject so interesting to all of us, that we might be enabled to make in this Kingdom, with any Person who may want them, the Agreements we may find proper, & therefore we hope you will not delay giving us this Inteligence any longer —

With regard to what you mention of having compleated the Pump or Engine & that you waited to receive our Orders for shiping it, we think that until all the circumstances and pretentions are finished we cannot nor ought not have it brought to Spain, & that this is convenient for you & for us, because by keeping this Engine without fixing it in the place where it is to be, bad use may be made of it; and it is more convenient that it be in your place than in this, and when the case arrives that it suits us to order it, we shall take good care to do so, to transmit the ratification of the Deeds, and to name a Person to whom you may apply for the Interests which may offer —

We have no doubt that you have taken every precaution in the agreement & sending of ye Engine for T. & it will be very convenient that you let us know all that has occurred in said Contract for our Govern^{t} for this affair since we knew you had concluded it. Beside having given us much disgust we suspect (it) will in time perhaps be prejudicial to you as well as us — We are etc

P.S. We are assured that in Paris in ye Isle of Cisnes[4] there has been erected a Mill with 12 Stones after ye Model of the Albion: we wish to know if you are apprised thereof & what is your Opinion on this particular.

Box 36 Spanish Engines.

---

[1]Negotiatoins for a Spanish patent began in 1788. To expedite matters Boulton sent a Memorial to the King of Spain and in 1790 a 20 years patent was granted.
[2]B & W ignored the exlusive priviledge they had granted to the Spanish company, privately selling an engine to Don Fernando de Torres.
[3]B & W had not at this time even published a price list for the home market. But a draft price list for Spain survives, dated 1789 (Box 36).
[4]The Isles des Cygnes mill was powered by a Perrier engine.

DOCUMENT 39

Proposed Austro/Hungarian Patent

To W. de Kempelin,[1] Vienna ?Jan 1791

Monsr Le Baron Vay & Vaya having done us the honour to intimate that you Sir were the proprietor of a Patent for all the dominions of His Imperial Majesty for the sole erection of steam Engines, & that you were willing to treat with us for the disposal either of the whole or some part of the privileges conveyed to you by that patent, We found ourselves flattered by the offer of which we wish to avail ourselves but having no disposition to leave our country, at our time of life, & with the engagements of business we are under here, we now make you the following proposal which we submit to your consideration & hope will prove agreeable

1st That the patent shall remain your sole right as it now is.

2nd That we shall supply you with the engines of our invention in their most improved state at certain prices to be fixt in a Tarif accepted & signed by you & us. This Tarif to include only the mettallick materials of the Engines & boilers, but no part of the woodwork, pumps, Mill work or other machinery which may be required to be wrought by the Engines. The materials to be delivered at those prices at the port of Liverpool or Hull.

3[d] That we will furnish you with drawings & directions for erecting the Engines in a proper manner & so as to produce the desired effects, without any charge for the same.

4th That we shall provide on your account any pumps or other Machinery you may want at the customary prices you also paying all charges on the same from the places they are made to the place of their destination.

5th A guarantor to our satisfaction to be appointed in London as security for performance of this contract & which shall accept our drafts on him for the value of each parcel of goods payable in 6 months from the date of their delivery at Liverpool or Hull.

6th That we shall not supply any other persons in the dominions of his Imperial Majesty with any steam Engines whatsoever.

7th That you shall engage not to make nor purchase from others, any Steam Engines whatsoever other than those of the ancient or Newcomens Engines or such as have been in use previous to our Invention in the dominions of his Imperial Majesty, & also that by virtue of your patent you shall prevent any other person or persons from importing into those dominions or making any Steam Engines other than what we have described.

8th That we shall communicate any improvements we shall make in our Engines from time to time on being allowed for any additional parts which may be required in consequence of such improvement.

You know the situation we are in as the greatest manufacturers of Steam Engines in Europe, and you have general Ideas of the merits of our Inventions & improvements on these Engines, & you have now Sir an offer of the benefit of our long experience, & consequently of a certainty of success if our proposals are acceptable to you, in which case we shall explain ourselves more fully.

In the meantime permit us to say that unless we entertained high Ideas of your ingenuity as a Mechanician, & of your probity we should not have neen tempted to make you this offer hoping to be honoured with your speedy answer.[2]

(J. Watt)

MB — K/69

[1]Wolfgang de Kempelin took out a patent for his engine in England (pat. 1426) in 1784.

[2]In fact no agreement was made.

DOCUMENT 40

To James Watt London 24 April 1784

Dear Sir,

I have yesterday got an Acct of M[r] Kempelin's contrivance of his fire engine and can assure you you have nothing to fear on Acc[t] of his Patent. This you will readily believe when I tell you that it is an old Scheme of your Friend M[r] Henderson while he liv'd with D[r] Irvin, inshort it is Steam applied to Barkers Mill which is most certainly the worst of all the Applications of Steam hitherto.

Thinking you might have some anxiety to know the Construction of this imaginary formidable Scheme I have taken the liberty to trouble you with this.

And am most Respectfully

Your humble Servant

J. Ramsden.

Box 2 R

Section D

Pirates and competitors at home and abroad

Pirates and competitors at home and abroad

Documents 41-65

The initial cost of James Watt's engine was higher than that of any other engine and this coupled with the scaled annual premium payable for the privilege of using the engine during the term of the 1769 patent led to the continued and increased use of both simple Savery and Newcomen engines. Indeed a number of patented and unpatented improvements were made to the Newcomen engine in an attempt to improve its efficiency without infringing Watt's patent. Between 1769 and 1800 fifty-nine steam engine patents were taken out, only four of them being Watt's. Seven were taken out between 1769 and 1779, twenty-two in the following decade and thirty between 1790 and 1799. Some of these patents such as those by J. Hornblower, jnr., and Symington were deemed by Watt to be infringements, others were not.

The mid to late 18th century atmospheric engineering industry can be divided into two main sectors. There were consulting engineers like Francis Thompson and William Hase on the one hand and ironmasters and founders like John Wilkinson of Bradley and the Walkers of Rotherham on the other. By the late 18th century quite distinct foci of steam engineering innovation had emerged in the Manchester area, Birmingham and the Midlands, The Potteries, Leeds/Bradford, Sheffield/Rotherham, the Newcastle area, Bristol, South Wales, the Shropshire coalfield, Glasgow and London.

A minority of these engine makers also produced pirate Watt engines, some like Alexander Brodie of the Calcutts, Shropshire, solely for their own use, others like John Wilkinson for sale on a large scale[52

& 53]. Whilst some of these British pirates may have gone undetected, Boulton and Watt identified nine pirates who had, between them, produced seventy-four engines for British customers and five for overseas and took proceedings against them.

In addition to home pirates there were industrial spies from overseas seeking to obtain descriptions and drawings as well as knowledgeable British mechanics to order engine parts and construct engines abroad. The case of Baron Stein is one of the best documented[62-65].

DOCUMENT 41

To James Watt Rotherham Aug^{r} 9 1776

Sir,

I make free to inform you that there is at this time two men from Carron[1] in this Country in Order to pick up your New modled Engine. One of them Came to this place being informed that there was one Aplyed in lifting up the watter from the talerise[2] to suply our works; this indeed is intend by next Spring, but being disapointed went further into the country but where I know not. His Name is Wm Beldin master wright at Carron the Other is Wm Wild Loom moulder. He has been for Seven weeks past at a place Caled Ashen[3] Near Birmingham where the Cast work is making for one of your Engines. He is there as a work man & writes one of my Comerads that he is to Continue till he Accomplish his Desire having therby six Shilling weekly from Carron with what he can Otherways make — there is three such Engines propos'd to be within a mile of this place two for Coals and one for Suplying works by Drawing the watter back. Dear Sir I have wrote you the above information with a Sincer Regard to your Intrest & hopes you will not take it otherways and in no ways make me known as it may have atendence to prove to my hurt and you may trust that I shall allways be Ready & Shall Count it my honner to do you a Service as (much) be in my power. Be to kind as to Remember me to Mr John Wilson and James Dickson in Comp^{y} and William Fulton and let not my Name be once mentioned further which will much Oblidge your Hum^{le} Servt

John Meikle

P.S. if you Shall think proper to write to me Direct to the Care of Thomas Chambers at the Sign of the Blackmores head Rotherham.

J.M.

Box 2 M

[1]Carron Ironworks, Scotland.

[2]Tail race.

[3]He probably means Bersham, John Wilkinson's foundry near Wrexham.

DOCUMENT 42

To M. Boulton Ashover March 9th 1781

Sir

I Was Sent for this week to Mr Milnes House Whear I was as much Supprised to hear your letter Concerning the Yatestoop Engine as you was with the information of Us Putting in a Condenser whear their is No Such thing, or anything done their in Such Manner as Concerning your Pattent or aney other thing Concerning your work. The Engine has not been work'd for most part of a year; not Since the new Set of pumps has been Put into the Shaft — nor any thing done in repairs Else Except a 4 Inch Pump which stands in a Small Cistern by the Side of the hotwells which the water runs into and then is to be drain'd to the top of the Cylinder to Supply the Piston with water because Since the Sump was Sunk, 12 yards lower, the housewater pumps Would not Supply the piston and Injection In the Cylinder; which we use a large quantity of Water in the Cylinder, it is Discharged at 5 fountin holes.

But for your more Sattisfaction you are welkam to send a man and View the work through; I know my Duty better than to do anything but what I Can Stand by In those Cases; And am your humble Ser^{t}

Francis Tompson[1]

MB — T1/165.

---

[1]Francis Thompson took out patent no. 1884 in 1792 for a steam engine, but had previously erected a number of pumping engines in the Midlands.

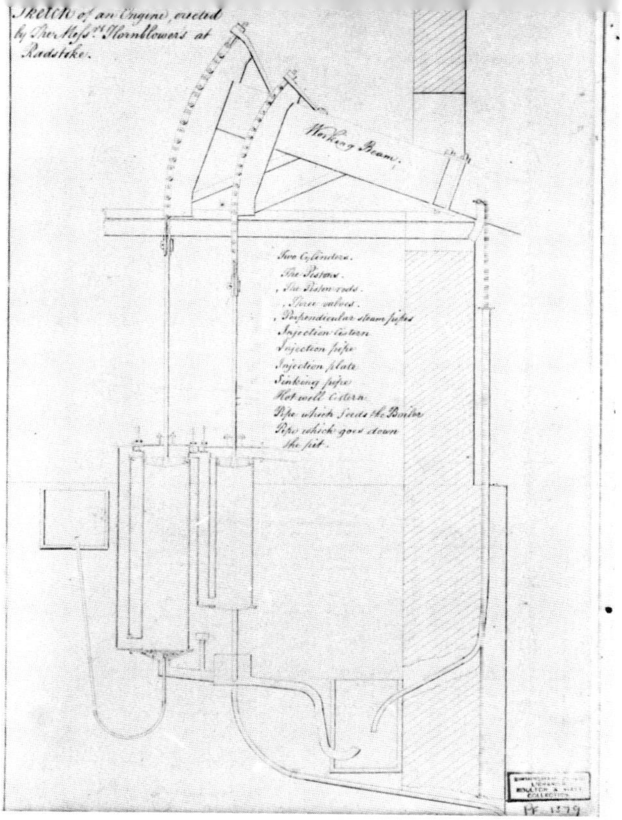

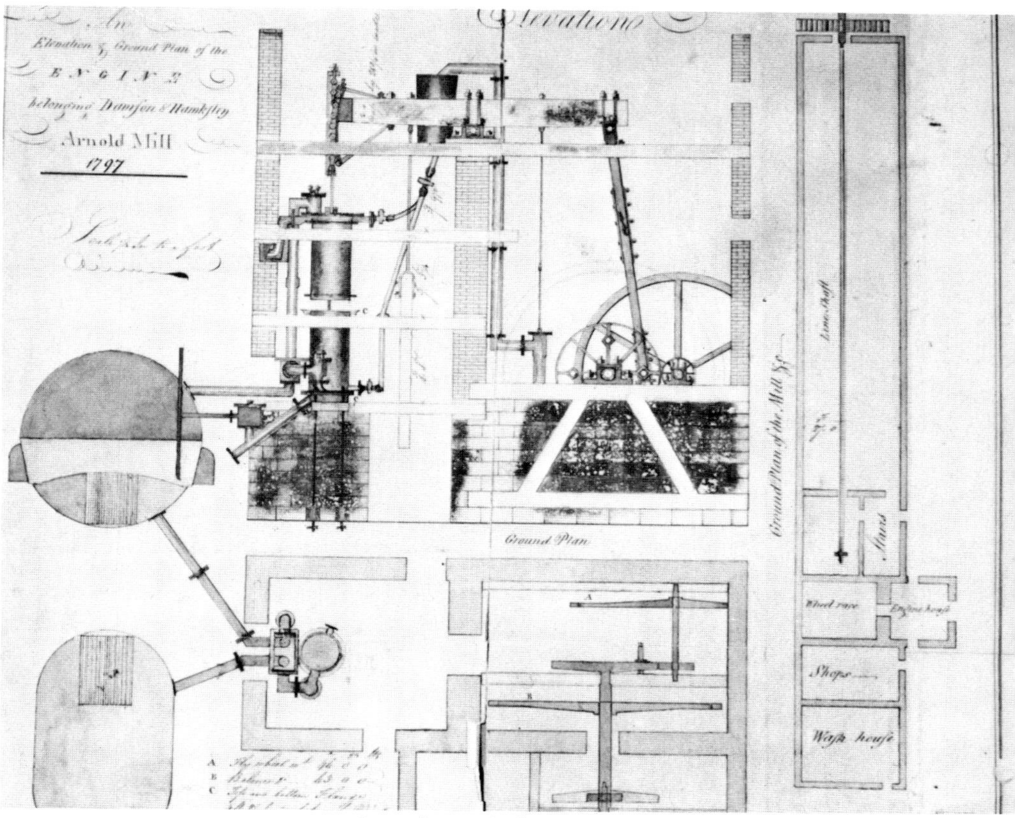

Plates [5 and 6]

… to stand parallel with the Road or River — … be plac'd on upper …

… may either be at the End — or on lower Side — or if better may by digging away …

Depth of Engine pit — from the Surface about 18 feet. —

Water to be raised 20 feet — 33 diam: Pump 8 feet stroke

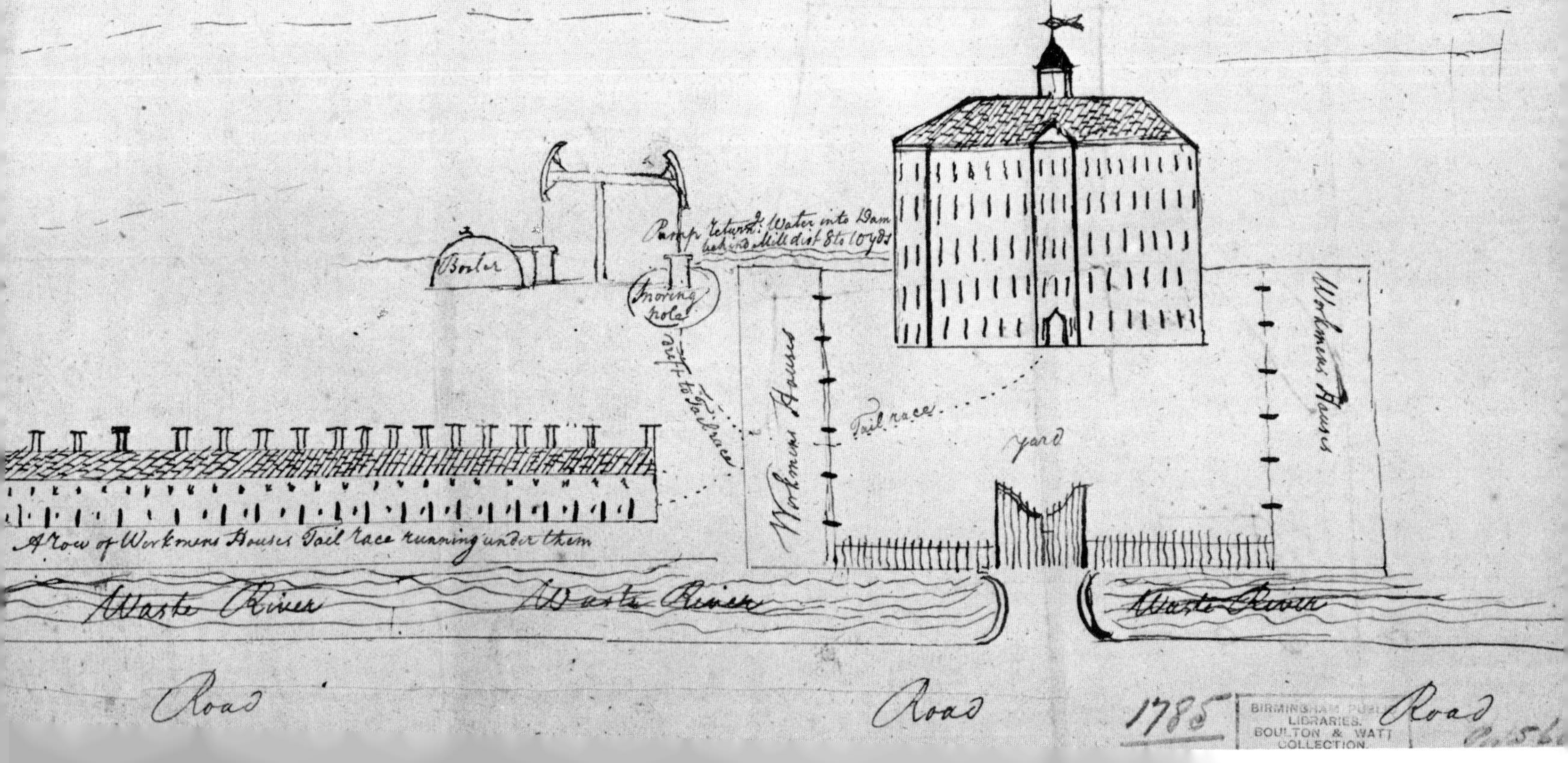

DOCUMENT 43

To James Watt London Monday 23^{d} – 1781.

Dear Sir

I rece^{d} your double letter of Saturdays date this morn^{g} before 8 oClock. The Spirit & the Contents of it exactly accords with my own sentiments, & should I find it necessary I shall make a proper use of it.

At ⅓ past 8 this morn^{g} conformable to appointment I mett at ye London Coffe House Mr Wasborough Senr, Mr Wasborough Junr,[1] Mr Sampson Freeth, & their Engineer Jno Steed. I took Mr Matthews along with me to prevent any after misrepresentations & to witness the conversation. After breakfast Matt^{w} W.S. Freeth, Mr Matthews & Self took to a private room & entered upon the business. I set forth our friendly dispositions towards their Comp^{y} untill I had receivd information that Mr Was^{g} had said & wrote all he could invent in disfavour of our Engine & consequently to our prejudice. He found it was in Vain to deny & therefore pleaded guilty & said it all arose from his mis-information. Upon which S. Freeth said Mr Was^{g} conduct was unpardonable, & that it had been condemned by himself & Mr Pickard, & had given them both great concern — all this softend my resentment & my anger subsided. S.F. then made a friendly preface & wished us to try if we could not devise some plan of mutual accomodation. We then went into ye Subject of the origenality of the invention & they said that it was invented at Bristol by Wasbo^{g}, Pickard & Jno Steed & that Jno Steed had thought of applying a Crank for that purpose 14 y^{rs} ago & had then made the Machine in part. It would tire you to recite all ye Conversation But as they expressed the strongest desire to make peace with us & from other circumstances then upon my Mind, I proposed that we would lay our proofs before two Gent^{n} of reputation respecting your priority of ye invention & that if those Gent^{n} were convinced of your prior right to it, Then they should by a deed declare that right belonging to us & the reasons why, & and in the same deed, for the aforesaid reasons, & to prevent further disputes, they should assign over to us one half of the patent right. We paying one half of ye expence of it, & that we should proceed to erect what ever Engines we might think proper in the same way as if no such patent had ever been taken out. In consideration of wch We 1st would not take any measures for laying the patent open — 2^{d} we would consent to erect the Engine wanted by the

Victualing office to work their Machine,[2] wch is now in forwardness, 3^{d} that we would never take any job out of their hands which they had been first applyd to for, & that we would ourselves erect our Engines for working their Machines in all cases were Coals were not of a less Value than 5^{s} p^{r} Ton. At the same time I ownd that it would be difficult to draw a precise line between us, what we should, or should not do, but it must be left in some degree to our honor unless they could invent some more distinct plan that should prove agreeable to all parties.

If they will consent & confirm the aforesaid plan I then gave as my present opinion that we should erect our Engine for Y^{e} Victualing board to work their Machine. We haveing the right of manageing ye Engine. The following are my reasons for that opinion

1st We shall oblige the board (who are now well informd of our case) in conforming to their request.

2^{d} We shall avoid the insinuations of the Whirling Comp^{y}[3] as to takeing the job out of their hands.

3^{d} As we have our hands full this summer we shall get quit of the most anxious part of the business & do that part wch experience has taught us to do with certainty (ye Engine)

4th I think there is a probability of that Comp^{y} failing so much in their Mechanicks as to bring disgrace to them selves & honor to us in this Corn mill & thereby settle the reputation of both parties.

5 We shall gain some experience in so powerfull a Machine probably at their expence
near 2 I must go to the Board

I am just returnd from ye board where I was desired to accompany Mr Smeaton on Fryday next down to the Kings Mills.

It was also requested by the board that I would relieve them from the Cleft Stick they are got into & that his Majesties business might not be retarded by any disputes we might have wth Wasboroughs & therefore recommended that we might refer all matters of dispute to 2 of our friends to decide & that we might each proceed on with our respective parts To wch I & Was^{g} & Freeth have agreed & I am accordingly to deliver in our proposals on Wednesday next in proper form.

If I write more I shall loose the Coach be assured I will be attentive to our Interest & to ye termination of disputes I hope to your Satisfaction.

I am most aff^{ely} yours
M: Boulton

Box 20

[1]Matthew Wasborough took out patent no. 1213 in 1779 for a method of obtaining rotative motion from a steam engine.
[2]A corn mill. In the event B & W did not obtain this contract, chiefly on account of Smeaton's opposition to a steam engine providing direct rotative power for corn milling rather than pumping to a water wheel.
[3]i.e. Wasborough, Pickard etc.

DOCUMENT 44

To James Watt Etruria 13th Jany 1785

Dear Sir

I am this day informed, that a new kind of Steam engine is intended to be erected soon in this neighbourhood, for the purpose of grinding flint; — which is to raise 1200 gallons of water per minute, to work an overshot wheel of 26 feet diameter. The boiler is to be 6 feet diam^{r}. the lower part of copper, and the upper of cast iron; and the consumption of coals (the smallest coal, slack) 80 pounds per hour. The expence is to be £240, including a patent premium of £30[1] — Its construction will doubtless be known to you, when I tell you that it is Mr Bond's engine — One of them, I am told, has been erected at Macclesfield, and has been at work about 6 weeks. A person from Congleton attended the building of this last, in order that he might be qualified to supply those who may wish to have them in this part of the country. Drawings (were made) of the engine, with explanations, but I have not seen the explanations yet.

Another steam engine has been offered here for the purposes by a Revd Mr Derbyshire[2] of Brierly in Staffordshire, which is to do great matters, but I do not yet know what — only that it works a cylinder of 36 inches diam. with 11 tons of coals per week.

I am apprehensive that these offers are fallacious, and should be glad to prevent my friends from being imposed upon if they are so. However, I thought it an obligation of friendship to acquaint you of these circumstances, and whatever reply you please to favour me with shall go no farther than you wish it.

I beg my respectful compliments to Mr Bolton, if at home and wishing you and yours many happy returns of the present Season, I am

Dear Sir

your faithful & obedient humble Servt

Jos. Wedgwood

Box 36

[1]This engine was not patented. An interesting example of charging in a similar manner to B & W.
[2]Derbyshire left the midlands in c.1784 to go into partnership with the Manchester engine maker Joshua Wrigley.

DOCUMENT 45

To M[r] Chippendale — Decem 4.th 1790

Dear Sir

I am obliged to you for your care & punctuality in the execution of the several commissions I troubled you with & again take the liberty of requesting you to apply to your Brother the Attorney & desire him to apply to the Patent Office for Copies of the drawings & specifications of all the Patents that have passed for improvements in Fire or Steam Engines since the beginning of the present year 1790 particularly for one Invented by M[r]Heslop[1] & another by M[r] Ashworth.[2] These two will be sufficient, with a list of the titles of the others, & if I find I want any of the others I will write again. This I beg he will do without loss of time & request of you to pay him the Expence But I beg & desire that he would not mention our name at all in this business. The more expeditious the more he will oblige

D[r] Sir

Your Obed[t] humb[e] Serv[t]
Matt[w] Boulton

MB — Chippendale/28

[1]Adam Heslop's patent was for an engine consuming less coal (no. 1760).
[2]James Ashworth took out a patent in 1788 (no. 1674) for a 3-cylinder engine.

DOCUMENT 46

To James Watt Newcastle, May 11th 1795.

Sir,

I am informed by Mr. Barnes, the Agent to several Gentlemen in this neighbourhood, that he has agreed with you for two Engines to draw Coals[1], but, in consequence of an advertisement inserted by me some months since in the Newcastle Papers, he had, previous to his doing so, applied to you to indemnify him from any claim I might make to the exclusive right of drawing Coals by the direct application of the Steam Engine, without the intervention of a water wheel; which, he says, you have engaged to do.

It is nine years since Mr. Blair prevailed on me to form a connexion with himself & Cameron,[2] and I believe you are not unacquainted how unfortunate it proved for me. At the dissolution of our Partnership last year, I was under the necessity of taking the Patents and stock in trade in lieu of the money I brought into the business, and I was led to believe that I should be able to recover a considerable part of my loss by the validity of the Patent for drawing Coals, Indeed, I never entertained a doubt of it, till Mr. Barnes gave me the above information.

The very extensive knowledge you have in the line of your profession makes you in my opinion a better Judge, in cases of this nature, than any Lawyer can be. Under this impression it is that I trouble you on the present occasion, and shall feel myself infinitely obliged to you if you will favor me with your reasons for supposing that Cameron's Patent for drawing Coals is not valid.

Fully convinced that I was standing on good ground I have commenced an action against a Gentleman of this neighbourhood for an Infringement of this Patent, but I hope to be indulged with your sentiments before I proceed. An early answer will, therefore, much oblige your most obed^{t} hum^{e} Servt.

Hum^{y} Jeffreys.[3]

Please to direct to me at Wilson's Coffee House Newcastle upon Tyne.

Box 3.

---

[1]Probably the engines for Cowper and Rainton Collieries.

[2]Robert Cameron held patent no. 1414 dated 1784 for a steam engine and patent no. 1525 dated 1786 for a method of raising coals, ores and water.

[3]Humphrey Jeffreys took out patent no. 2294 in 1979 for ore-raising machinery.

DOCUMENT 47

To John Southern Walker Colliery Nov. 16 1795.

Dr Sir,

Your favour of the 4th has been laid before the Cowpen[1] Company & to my extreme mortification I am desired by Mr. John Surtees to say that when he in conjunction with Mr. Row empowered me to renew the Treaty with you for the Rotative Engine he did not know that a part of Jeffries' Engine was actually on the Spot, but now, (every circumstance considered) the Company think themselves bound in honour to go on with it though conscious of its inferiority to yours & to Erect one of your construction the next they want of which you will have timely notice — This resolution of the Company at a General Meeting though may be a proper one for them to take, really leaves me in an awkward situation with you, & Mr. Surtees is so conscious of it that he offered to write you a proper apology & take all the blame on himself, but as the matter could not by that means be at all mended I told him I should not desire it.

Pray be so obliging as to inform me whether the progress you have made in the Castings &c for this Engine will be likely to rest on your hand or prove any inconvenience for I would strain every nerve to bring it into employment here as soon as possible if the Goods cannot be readily applied to any of your other Works — Your favour of 10th is received & will be properly attended to & I am Dr Sir

Y^rs very sincerely Thos Barnes.

Box 6 Barnes (2)

[1] Cowpen Colliery

DOCUMENT 48

To James Watt Bristol 4th June 1796
Enclosing copy of letter from Mr. Featherstonhaugh.

Sir,

Immediately on my return to Bristol I wrote to my Agent Mr. Featherstonhaugh enquiring into the fact of his having erected one of Hornblowers Engines at one of my Collieries on the Wear, to w^h I rec^d the enclosed answ^r. From the kind & polite treatment w^h I met with at Soho, it cannot but be my wish to give you every possible explanation, & in the most candid manner, for w^h I will trouble you to return with any comments you may think proper to make — you will readily perceive that my Agent is a very unintentional, & at all events a most repentent Pirate, for he ascribes a preference to y^r Engines in every respect — For my own part I not only regret from interested motives but also from an anxious wish to encourage at all times as far as lays in my pow'r such transcendent Merit as yours that my Agents have been so uninformed of the superiority of y^r Engines over those of Newcomen's Construction — I sincerely hope ignorance alone can have led them into the error, & not an inclination to derogate from y^r Invention — at least I am not aware of any motives w^h c^d have seduced them to such an act.

I have unfortunately failed hitherto in several Trials with my copying Machine, tho' I have made the ink & follow'd every direction contain'd in the book of Instruction — The Paper appears to me to dry so quickly in removing it from the damping box to the Double pasteboard that the impression is unequal & imperfect altho' I have increased the surface by putting in the Spare interleaving and papers.

I have the honor to be Sir
y^r most ob^t H^{ble} Ser^t
W.H. Lambton.

Copy Lambton May 27th 1796.

Sir

I rec^d the honour of your letter of the 22nd Ins^t. I am sorry to find myself considered by Messrs Boulton & Watt as an Encourager of Pirates upon their inventions. I therefore beg leave to state to you how I come to be placed on that list. We have three Machines known by the name of the Crank Machine, which Mr. Johnson recommended to us to make use of, the Model he said he took from three that had been built at Blythe and Hartley by the owners of those Collieries, at the same time Mr. Wade had two of them erected on this River, not knowing that there was any patent for them, immediately after the erection of the first of ours a Mr. Jeffries brought an action against those Gentlemen for using the Crank only, pretending that he was the patentee — I was applied to and requested to join in the expence of resisting his Action which I agreed to. So far and no farther was I an Encourager of Pirates. Mess[rs] Bolton & Watt were wrote to who very politely said they were the original Patentees, and that if Jeffries persisted they would set him aside, upon which Jeffries thought proper to desist and we heard no more of him. As to Bolton & Watt they never made a complaint about it, therefore I conclude no trouble will arise from them, its probable the favourable reception their Engines met with last Winter when Mr. Watt was in this Country might be the cause of their Silence. The Machine erected at Pensher was done before you left the Country is one of the same we had upon the Hannaton pit, and is known by the name of the double bucket, therefore the charge of our erecting one of Hornblowers upon Pensher must be a mistake, it must be meant against Mr. Nesham who began the erection of one immediately after the erection of Mr. Peareths and which I rather think is hardly compleated. They very justly deserve a preference, their first cost is much less and they are removed at a much less expence from Pit to Pit when its wanted to remove them. When Mr. Watt was down in this Country last Winter I was with him at Newcastle and had the utility of the different Machines fully explained. We were then unfortunately too far gone with the erection of ours at Pensher, otherwise I would have one of his plan. Your list which you were so obliging as to inclose is exactly the same which he shewed me. There is no doubt of their superiority over Newcomens, either for drawing Coals or Water, that is proved by Mr. Brandlings. They have for some time being drawing the same quantity of water with one Engine (of Watt's) that they had two of Newcomens for, which makes a decided preference in of Watt's. All your Engines for drawing coals are of Newcomens construction.

signed W. Featherstonhaugh.

Box 3L

DOCUMENT 49

To James Watt Junr. Manchester 10 Septr 1795.

Dear Watt

Though Rumour has reported the Existence of Air pumps to the common Engines, yet neither I or Lawson could procure any proofs, I shall have my Eyes and Wits about me to discover any Incroachm^t upon your Rights — if Bateman & Sherrat do it, it is certain it is very clandestinely for though I frequently go into their works I never saw an Air pump nor do they generally avow the Right as I have had conversations with both of them upon the Subject. Thomas Livesey of Werneth says Thackeray[1] uses them to a double Cylinder Engine but all his Information though probable is merely Hearsey. I am glad to hear you go on with such Spirit, as I am persuaded you cannot be too alert in your motions, and with Peter[2] in the Office of Head Engineer at home & Lawsons indefategability abroad you must be unfortunate indeed if you do not carry every thing before you.

I find your father is full charg'd with the Air-System[3] for which both on account of Innovation & Simplicity all the Faculty nearly seem to have a complete Abhorrence.

Pray does Mr. Peter design to send the little Boiler soon and Apparatus? Tell him to have it strong enough and in respect to Size he is acquainted in my last with the quantity of Water to be evaporated daily.

Enclosed are Three Bills value £122-2-0 which with sundry Advances made upon your account amount^g to £ 34-13-0 as p note
£156-15-0
annex'd please pass to our Credit, & advise accordingly.

Agreeable to your permission we have return'd some articles which came too late or did not suit the purpose, for part of which Mr. Southern says you will credit us the Cash and the other the Value.

The Copying Machines are not yet arriv'd, but I shall not fail to contribute my little Influence to introduce them to Notoriety. Remember me to all friends and believe me truly,

Y^r friend
G. Lee.

Box 3L

[1]of the cotton mill at Cark in Cartmel.
[2]Peter Ewart.
[3]The Pneumatic apparatus developed by Watt for the treatment of ailments of the lung.

DOCUMENT 50

To Boulton and Watt. Oldham, 11th March 1796.

Gentlemen,

Some time ago I had a little conversation with Mr. Uwart[1] respecting a Patent Steam Engine as it was then, and had always been my intention to apply directly to you, and to make a quantity of Bricks this summer and the next to Build a new piece at the west end of the mill to contain the Engine, and other machinery exactly upon the Plan of my neighbour, Mr. Clegg, but since that time the Truss beam over the Horse Mill[2] from which all the other floors are supported, has failed and almost every day bends a little, so as to oppress the Horse wheel, and has actually stopped the works twice and threatens do it again and that without remidy, and as I could not get Bricks at any price to Build with there being none in the country (but what are intended for the navigations) nor being able to work a Beam in the lobby of our Mill, I called a few Machanicks together who told me that An Engine of Two Cillenders might be worked in the said lobby upon a Plan which is said to be, compared with your Engine, as two to three, but upon further enquiry, I could not learn of a certainty that we have a right to make use of a hot water pump, and I would not even doubt as to the inquiry of another, therefore yesterday I prevailed upon your Mr Lawson to view the Premises along with Mr. Clegg & Thomas Livsey of Werneth, now the fact is this, so far as we have gone it is so fixed that we can work it as a common Engine and the Mill is in the next field to Werneth Colliery and a Coal Pitt was opened last Christmas within less than 40 yards of the Premises,[3] but as the alteration which would be needful at the expiration of your Patent, would cost nearly 50£, and as I could wish to make it compleat at once by your permission upon reasonable terms I have come forward in this open way. — and as the Plan cannot now be altered, we only want the air pump, which if granted and the Price agreed upon, you may either make every part of the Machine which is not already made, or none at all, as you think proper, waiting an answer, when you have seen Mr. Lawson.

I am
Gentlemen
Your Humble Ser[t]
Daniel Lees

NB My Brother and the other Gentlemen of Werneth Colliery have been very kind hitherto, and I believe they will accommodate me with the slack of one Pit, which will inable me to work a common Engine, Cheap as any man in the Kingdom.

Box 3.

---

[1]Peter Ewart.
[2]An example of the common practice of employing animal power in early textile factories.
[3]In such a situation, providing reliable atmospheric engine power could be obtained, Lees was probably wise to reject the idea of a Watt engine.

DOCUMENT 51

To James Watt Wrexham 2nd Dec[r] 1795

Dear Father

No Purchaser having appeared for the Works[1] the Sale did not take place yesterday, but upon the recommendation of the Arbitrators, each party gave in the Sum at which he would purchase from the other. William estimated Johns $^7_8$ at £10,000 & John reckoned *the whole* at £12,000, so that you see whatever variance there has been between them upon other Matters, they were nearly of the same opinion upon this. Each appeared willing to purchase from the other upon these terms, but the Arbitrators rather advised William not to embark in an undertaking of such Magnitude, in which I think they did right, as his Capital would certainly have been insufficient for carrying them on to advantage. John therefore is now sole proprietor of the Works & has paid W. £1800 down for his Share.

John came here on Monday Morning whilst I was at Bersham. Upon my return I sent my Compliments by M[r] Fawcett and informed him that I should be happy to wait upon him if disengaged; He sent a civil Answer by Fawcett who told me he was evidently startled, however as I was determined to keep in good humour, I soon brought him into the same & we had a long recapitulation of grievance, in which he admitted that he had been too precipitate as he had given too much credit to the reports he had heard respecting his brother & me. We dined alone together & the whole of our conversation was upon the same subject interlarded with most plentiful abuse of William and of the Law which he has in utter abhorrence. I told him that was not the case with us for some of us, myself for instance like it very well, which was not astonishing as all our legal proceedings had hitherto been attended with success. We then came to our Accounts, which he said he should settle with Mr Boulton. (I said) to him that might do for the Premiums, but there must be an immediate settlement of the Bersham Account as we did not choose it should appear to the world we were indebted in so large a sum at a time that *he* was actually so much our Debtor. I therefore requested that that account should be given to me & arranged between me & Mr Fawcett to which he assented & also agreed that the ballance should be placed to the Debit of his *own* Account with the Bersham Company.

I have not time to say more; we are all going to dine with him at Brymbo when I shall urge other matters. MRB is arrived & well. J.W. is better than I ever saw him, he wants much to renew his Connexion with us, which I shall not commit us in any farther than prudence may require. With best respects to all in haste

Your dutiful son
J Watt Jnr

Box 33

[1]Bersham Ironworks belonging to John and William Wilkinson.

DOCUMENT 52

Statement of facts in the case against John Wilkinson.[1] 10 Sept. 1795.

Our connection with this Gent^n^ commenced about the years 1775 & 6, at which period his enterprizing Genius and the superior execution of his work, pointed him out as a proper person to do us justice in the casting & boring of our Cylinders & other Engine iron work. He soon discovered the goodness of the principles upon which our Patent was founded & as early as the year 1776, we furnished him with drawings of an engine to be erected at his works at New Willey to blow his furnaces. In the same or following year, at his request, we gave him instructions & drawings for converting an Engine upon the old construction, called Topsyturvy into one of ours. We also furnished him with drawings in the year 1777 for two other Engines, one to be erected at his Works at Wilson House in Lancashire for blowing an Iron Furnace, and the other to draw water from a coalpit upon his works at Snedshill in Shropshire. The nozzles or Regulating boxes of two of these Engines, to wit, New Willey & Snedshill were fitted up by us. We believe that the other two were also, but our books were so badly kept at that period that we cannot speak positively nor w[t] Material, as the Premiums of at least three of these Engines were given up by our letter of the 17th July 1777, of which the following is a copy. —

"We have been very much engaged ever since you were here with Mr. Meeson of Edinburg & some other Engine business, or you would have heard from us before agreable to our promise, for we wish not to keep you in suspence, but to preserve a right & clear understanding between us and to give you every proof we can with prudence of our good intentions towards you. In respect to a general proposition for you to erect such Engines as you may think proper for your own use. We have attempted to make a general table but we soon found ourselves involved in difficulties & that it could not be made satisfactory to all parties. We therefore hope the following proposition will satisfy you. In the first place, we will grant to you & to your successors our Licences to use the two Engines now erected at New Willey, and the one you propose to erect at Wilson house as well, as a devil to be erected where you please during the remaining term of our exclusive priviledge, upon payment of one shilling yearly when demanded provided you do not remove them to situations where coals are of much greater value — In respect to the Engine you propose to erect at Snetshill we will take the Premiums in Coals. Such an Engine going at the rate of 12 strokes per Minute of 7½ feet each, would burn per hour about 230 Pounds of Coals if made with a double cylinder, and a common Engine doing the same work, viz the 44 you intended, making as long a stroke and as many of them, would use about 700 Pounds. One third of the difference, or 106 pounds of such coals as are used at the Engine we claim to be delivered to us upon the spot as premium for each hour the Engine goes, but that only if demanded, which will not be the case if they are of no value, and if they are you certainly can afford to pay them. As that Engine is proposed to be without an outside Cyl[r] it may possibly burn more than we have mentioned in which case one hundred weight is to be deducted from our Premium from every three hundred weight it burnt more, & if you should suppose a

Common Engine uses less, we are willing to submit it to a comparison. This may serve for an Example how much our Premium would be upon any other Engine. We would willingly, if we could, make a more simple proposition, but we do not see how it is to be done."

(Signed) B & W

There is no answer to be found from J.Wn to this letter, it may therefore be presumed that he acquiesced in the proposals therein held out to him and that he told B & W so, at his next call at Soho where he used at that period frequently to stay a week and upwards. That he was perfectly well acquainted at that time & subsequently with B & W's terms of taking one third of the savings is very evident from many of his letters in the years 1777 to 1789 inclusive during which time, several orders for Engines were procured by or negotiated through him.

The following Extract from his letter of October 1778 respecting a Mr. Jarry a frenchman, who wanted an Engine for France may be of some use.

"I am now terribly pestered with Jary. Some time ago I had a letter requesting I would give him directions where to come to me, which he said he was determined on, let me be where I would. I wrote that it would be a fruitless expensive journey & that I could do nothing for him upon any plan this year at any rate. I requested him to be patient until my return & that I would let him know when I should be back at Wrexham, where he might meet me, if he persisted in having an Engine upon your plan but that I would have no concern in any thing that imitated your Invention. He had laboured hard to convince me that as a founder I might execute any order he gave me etc etc."

In the years 1778 & 9 we furnished him with drawings & nozzles for two Engines with 40 inch cylinders, the one to be erected at Bradley and the other at Snetshill & both to be applied to the blowing of Iron Furnaces. I find no minute of any agreement made, at the time, for these Engines but you will see by what follows that the premium of them (& of them only) was settled in 1784 and has been regularly charged by us ever since.

In 1782 we gave him drawings & instructions for a 42 Inch Cylinder to be erected at Bradley & applied to driving a forge, receipt of which drawings is acknowledged by his letter of 22 March 1782. It appears also from the correspondence & from our accounts, that Robt Cameron then in our employment, attended the erection of this Engine on our behalf. For this Engine no Premium has been ever paid, nor any charged by us, as we never came to a settlemtnt about it, but a Memorandum that it was unsettled was made at the bottom of the Account in 1784 & was repeated in every succeding Account until 1788 after which it was omitted. More of this hereafter.

In the beginning of 1782, J. Wilkinson being probably then at Soho, commissioned a 42 Inch Cyl^{r} to blow the Furnaces of some foundry in France (I believe Mont Cenis). Of the order for this Engine, or Agreement for it, no Minutes whatever appear in any of our books, but we have copies of the drawings for it & notice of their being sent off on the 13th June 82, as also a charge against Mr. Wilkinson for the nozzles upon the 15th August of the same year. Upon the 26 July 1784 I find £100 charged to J. W^{n} "for Premium of French Furnace Engine" which I presume means this. Therefore it may be looked upon as settled for, and I should not have said so much about it but that I mean to make these minutes save me trouble in future.

On the 16 July 1783, he wrote to say that he had received an Order from his brother in France for an Engine for raising Water. After stating the effect to be provided, he proceeded to say "If you will be so good as to give your opinion of the size of the cylinder etc with probable expense of the whole & what you expect for the drawings etc to be paid you in consideration thereof — I shall be enabled to give a decided answer upon meeting my Brother at Calais, which I hope to to in the course of the Month." My father in his answer, after stating the size of the Engine, which was a very small one (14 inch diameter & 2 ft stroke) estimated one third of the savings at £30 per annum, proceeds to say "we think therefore that you should give us £100 in full but at the same time submit it to your soverign will & pleasure. We hope you will be the more generous in this case as we had much too little for the last French Engine." Mr. Wilkinson in his reply says "I have no objection to give you £100 — as required by your letter, but it is upon this condition that you send me drawings to enable us fitting up the nozzle at Bersham — Indeed in all my *own* Engines where I am willing to pay your demands — I consider it as a right, and am sure it would conduce more to your Interest to let me do even more than my own where the Engines were convenient to that situation." I can meet with no minute of the drawings of this Engine being sent, nor one word more about it, but they certainly were made & sent, as we still retain the copies, which have very minute & accurate delineations of the nozzles etc (prob dolor!) and have certainly served as directions for him to make other Engines from. No sum is however charged in our Accounts for this Engine and I think it probable it was not carried into effect. Still however we had performed our Engagement & the £100 still remains due to us for the trouble of calculating the Engine, Scheming & making the Drawings.

In the same year 1783 he requested us to prepare drawings for an Engine to be erected upon the lead mine of Maesyfynnon wen at Minera in North Wales, the Cyl^{r} of which was originally intended to be 50 Inches in Diam^{r} but was, I believe, executed only 48 — whatever agreement was made for this Engine appears to have been verbal & most probably was for one third of the savings, as a counter was fixed upon the Beam to tell the number of Strokes the Engine made. Receipt of the Drawings is acknowledged in Mr. Wilkinson's letter of 30th Sepr 1783 and that the nozzles were made at Soho, appears from a charge in our account current for 1784. The drawings for the foregoing Engine were probably not then in his possession, as it does not appear that after the receipt of them, we ever had any thing more to execute for him — We are now informed by Mr. William Wilkinson that this Engine has stood for the last two years & that the counter was removed seven years ago by order of his brother. It seems that J W. had only 7/16 of this mine & that 9/16 of the adv^{rs} passed an Order that we should be paid & have referred the settlement of the business to J.W. as he must best know what sort of an Agreement he had made with us. Whether upon this, or some other account, J.W. has sold out 4/16, so that he now holds only 3/16.

So late as the end of 1783 no Agreement of a regular kind had been entered into for the payment of Premiums, and B & W, as may be supposed, were become very uneasy upon that account. They therefore appear to have pressed Mr. Wilkinson to come to some fixed terms and accordingly in January 1784 a meeting was held at my father's house near Birmingham, the minutes of which are now before me in Mr. Boulton's handwriting & the following is an exact transcript.

"Memorandum at Harpers Hill 1784 Jany 22. Present Mr Jno Wilkinson, J Watt, & M Boulton.

Agreed with Mr. Wilkinson that the coals used at Snetzhill Engine shall be valued the same as at Ketley Engine (viz 2/6 per tons) & that Mr. Wilkinson shall pay B & W: annually in the proportion as the Sq^{r} of Ketley Engine (vis: 58 In Diam^{r}) is to £120 pr year so is the Sq^{r} of Snetzhill (40) to £57 per year — and that Bradley & all his other Engines (except such as B & W have relinquished their claim to) shall be rated according to Ketley, as the squares of their respective diameters and the differences of the Price of the coals & that the coals at Bradley shall be rated at 1/6 p^{r} Ton.

Sq^{r} of 58 Sq^{r} of 40
Hence as 3364 : 120 :: 1600 : £57 p^{r} yr
then say as 2/6 : £57 :: 1/6 : £34 pr yr.

I believe Snetzhill began about the beginning of 1780, but say Ladyday 1780 & that the forge at Bradley Ladyday 1783

	£		£
1780 Snetzhill	57	per year, 4 years to Ladyday 84 —	228
1779 Bradley Furnace	34	——— 5 years to Ladyday 1784	170
1782 Bradley Forge	34	——— 1 year to Do	34
	125		£432
Wilson House Eng	0		
Willey Eng	0		
Topsy Turvey	0		
Snetzhill Coal Eng	0		

Thus for Mr Boulton. Whence it would appear that no Premium was at that time intended to be taken for Snetzhill Coal Engine, as the one estimated at £57 pr annum is the Furnace or Blowing Engine. The Memorandum also was not conclusive with respect to Bradley Furnace or Blowing Engine, for the Premum has only been charged from Ladyday 1783: why it was given up we cannot now discover. The following entry upon our Journal the 30th Sep[r] 1784 will shew what actually was charged.

——— 30th September 1784 ———

Joh Wilkinson D[r] to Trade £262 —.—.

	£
For our 1/3 savings of Snetshill Furnace Engine from Ladyday 1780 to Ladyday 1784 = 4 years at £57	228
And our 1/3 savings of Bradley blowing Engine for One year to Ladyday 1784	34
	262

NB Bradley Forge Engine, say our 1/3 savings of it, to be charged to Mr. Wilkinson afterwards, from ladyday 1783, but the amount to be settled when the other Forges are set to work.

I have got our book keeper Mr Pearson, who is a pattern of exactness to copy this from his Journal & write under it what he recollects about it, this he has done as follows.

"The above is a copy of a Journal Entry in the Books of Mess[rs] Boulton & Watt, under date 30th Sep[r] 1784, the subject matter of which was agreed to between B & W and Mr Wilkinson in my presence, but cannot say the month or day on which the same was agreed to. Sep[r] 30 is B & W's annual date of the settlement of their books, and the books having been open when the agreement was made for the settlement thereof at 30 Sepr 1784, the above was entered under that date, although I believe the same was agreed to in the month of Decr 1784.

(signed) James Pearson."

Our account Current containing this charge for the Snetzhill & Bradley Blowing Engines & morandum respecting the Bradley Forge Engine, was delivered to Mr Wilkinson at Christmas 1784 of which I send you an exact copy, that you may see the way in which it was done. This account was neither objected to nor approved of, therefore if silence implies consent it ought to be supposed that he acquiesced in it.

The £57 Premium for Snetzhill & the £34 for Bradley blowing Engine were again charged in our annual accounts for 1785, 1786, 1787, & 1788 and the same Memorandum respecting the unsettled state of the Premiums for Bradley Forge & his other Engines was still inserted; only the wording in 1787 & 1788 was altered to the following

"NB B & W's ⅓ savings of Bradley Forge Engine from Ladyday 1783 is yet to be charged to Mr. Wilkinson as also ⅓ part of the savings of the other Forges from the time they set to work. The rate per annum of all these Engines to be ascertained afterwards and the whole of B & W's ⅓ savings due therefrom to be by them charged to Mr. Wilkinson's debit."

In the Acct Curr^t for 1788 after the word Forges an "etc" was inserted — thus far the Accounts were annually delivered to Jno Wilkinson & never openly disapproved of, although Mr. Pearson says that he always grumbled about the Memorandum at the bottom. but the 24th of Decr 1788 he made the following remarks in writing and delivered them to Mr Boulton (unsigned) upon the following day after.

"It appears by a half years tryall between Snedshill & Hollinswood Furnaces — Cylinders, Coals etc all the same in Comm^n that there is saved at the Snedshill Engine per annum 262^ton 1^cwt coals } £ s
reckon them at 2/6. Though J.W. has his Engine coals } 32 - 15
free of Royaltys & other than Winding Expenses

Mr Wilkinson has been charged from the beginning of Snetzhill at the rate of £57 p^r Ann for savings and at Bradley 34 — for the blowing Engine. He wishes the matter to be corrected & revised — before he has any Accts from Pearson otherways he is sure it will prove a seed of a mischievous nature —. Mr W is willing to pay a third of the savings and where he gives no kind of trouble & pays to the established rule he concieves he is not unreasonable — With submissions to Mr B Mr W thinks that a consideration for all his <u>own Engines</u> more or less, as say once mentioned by him as the best made leaving Mr. W at his option to build or change at pleasure was then & is now the only line that should have been adopted for the sake (if nothing else) of lasting harmony. The quota day approaches when it is necessary to settle accts with a Curious Genius who begins about savings not adjusted though charged — with a long train of etcs which creates a sincere wish in Mr. W that the next year may begin without any season of unrighteousness or bad policy.

Xmas Eve 1788"

This is written in Mr. Wilkinson's own hand, as is also upon the back of it "to mention the enclosed to Mr. Boulton Dec[r] 24 1788." What notice was taken of it, or what passed in consequence does not appear. There was no account delivered in again untill Jan[y] 1791 & then the Memorandum at the bottom was left out & "Errors & Omissions excepted" inserted in its place, which was again repeated in the next acct. transmitted June 14 1793, which is the last we sent. In both these the Premiums for Snetzhill & Bradley Blowing Engines were charged as usual.

You will please to observe that the Accounts current were always made out by us & that he never sent us any, nor pointed out any Errors in our statement (except by the above note) which among merchants is generally looked upon to be an acknowledgment that the ballance is right. I do not know whether Mr Boulton had been writing to press a settlement of Accts but in a letter to him of the 26th Nov 1792 he there expresses himself "a paper that I delivered at Soho last Xmas 2 years (on a reference thereto) will show my opinion of unsettled accounts. Most commonly they are the forerunner of mischief. I trust ours may be closed without such consequence."

This nearly closes all that we have to say respecting the Engines erected with our knowledge & consent by J.W. I have thought it needless to perplex you with the whole of his acc[ts] Current from the beginning, but I have sent you a general statement of the ballance each year that an account was delivered. Also the Acct Current for the year 1784 & the one we should now have delivered for the two last years, if nothing unpleasant had occurred. You of course will point out, in what way it should be altered & amended.

I also send you a list of Engines made by J.W. without our knowledge & consent. Indeed we were utter strangers to their existence (except those at Bradley) until Mr. W.W. was so obliging as to copy the list of cylinders from the Bersham Books & send it to us. It may not be superfluous to insert here the little correspondence that passed between us upon the subject of Two of these Engines that for the Flint Canal Co & that for Milur Mine —

Flint Canal Co.

Extract of a letter from J.W^{n} 29th Oct 1788

"I shall propose to Mr Williams that we put up one of your Engines 40 Inch diam^{r}. The first question he will ask will be what are we to pay B & W for the use of their method. This question was asked at Soho, but I did not get an answer. Pray give me a letter to Castle head — as I shall transmit him a copy of it. We must begin on some plan directly. You'll observe that we shall not trouble you for plans or men. You will also note that the Expense of Coals are only the Carriage and the Expense in getting by the workmen."

Extract from our Answer

4th Novr 1788

"In regard to the Engine 40 Inch Cyl^{r} on Flint Canal; I really don't know what to say. When Engines of our Construction are not carefully attended to & kept in order they burn more coals than a Common Engine that is well kept & may disgrace us as contrivers & you as a recommender, as has been the case at Crouchley's & some others; and low premiums even in the cases when coals cost nothing hurt us in those cases which can afford to pay. I have consulted with Mr B on the subject and the above was our joint reasoning. If however you should continue to think otherwise and wish to erect the Engine the premium will be £150 and you shall have any assistance you desire in the plans etc."

Extract from J Wn's letter of 9th Nov 1788

"at Chester by the Flint Canal Co it was determined to erect an Engine upon your Principles & to rest the Premiums upon the rule of ⅓ savings — should any other mode be adopted that will now rest between you & Mr W^{ms}. We have a meeting at Flint the 2nd Decr at w^{h} we must attend. I recommend it in the strongest terms that Mr B should come!" etc

Extract from our answer of 14 Nov 1788

"We observe what you say about the Engine which on the footing it is put, I fear will be abortive as the ⅓ of the savings where you sit as fixed upon the coals cannot be a consideration with us & we do not now bargain in that way except when the consumption is very large & coals more than 4/- per ton."

After this we heard no more upon the subject, but it seems that Mr: W^{n} was in this as in every other instance, determined to have his own way & sent the Cyl^{r} (and probably the other parts) from Bersham on the 13th July following. You will consider in how far our proposal, although rejected becomes binding in consequence of their having executed the Engine. —

Milur Mine

Extract from J W^{ns} letter of 25th Feby 1789:

"This morning a person has been with me who wants a Fire Engine for a lead mine belonging to Messrs Panton Pigot & Co near

Hollywell. The depth 60 yards. Bore of the pump to suit 9 Inch plain pipes which they have been by them. Have strongly recommended your Engine rather than an old one & request by return what the cylinder will be in your opinion & what you must have for the priviledge of erection on your plan. It must be settled some way or other whilst I stay here."

No answer was sent to this letter. Whether this was intentional or accidental, I cannot discover; nor does it appear that any meeting with J W^{n} took place about that time. The Engine however was sent off executed by him in August following.

I find no trace of any thing having passed between Jno Wn and B & W, relative to any of the other Engines mentioned in the list which accompanies this nor do I think that any particulars have escaped me in the correspondence that can be made use of, unless the following delineation of his own character by himself, copied from his letter of Octr 9 1784 may be pressed into our service in case of his future malbehaviour.

"I am entirely of your opinion that I ought to steer clear of any contest with Mr D (meaning R A Daniell) & shall thank you most cordially for any plan you can devise that will prevent it. Peace is a most desirable thing & the more so to one of my constitution who cannot be angry by halves. Resentment with me becomes a matter of business & stimulates to action beyond any Profit whatever."

List of Engines made by John Wilkinson at Bersham, for his own & other Persons uses, with & without our consent.

Date	Diam^{r}		
1776	38 inch	New Willey blowing Engine	These Engines were all for J. Wns own use and the Premiums were given up by our of 17th July 1777.
1777	—	Topsy turvy. An old Engine converted into one of ours.	
1777-8	27	Wilson House blowing Engine.	
1777-8	36	Snedshill Coal Engine for raising Water.	J.Wn. See our letter of 17 July 77. No Premium has ever been charged for this Engine. Nor is there any Proof of it having been given up.

1778-9	40	Bradley Old Furance or blowing Engine.	J.Wns. The Premium this Engine has been regularly charged at £34 per ann. from Ladyday 1789. Why it was not charged before does not appear!
1779-80	40	Snedshill Furnace or blowing Engine.	J.Wns The Premium of this Engine has been regularly charged from Ladyday 1780 at £57 per ann. according to the Agreement made at Harpers Hill 22 Jany 1784 and the entry in our Journal of 30 Sep[r] 1784.
1782-3	42	Bradley Forge Engine now a Rolling Engine	J.Wn. The Premium of this Engine never finally settled but a Mem[m] entered upon the Journal in J.Wn's presence 30 Sep[r] 1784, to say that ⅓ savings of it were to be charged afterwards & the amount settled when the other Engines then in contemplation were set to work.
1782	42 Inch	French Furnace Engine.	A memorandum to this purport was also inserted at the bottom of the annual acct from 1784 to 1787 — In lieu of wch, error & ommission excepted we afterwards sub-

			stituted — Intended for Mont Cenis. There is no note of Agreement for this Engine the order for w^h^ was probably only verbal. In the Acc^t^ Curr^t^ for 1784 we charge £100 for Premium of this Engine.
1783	14" 2 ft stroke changed to a 13" 3 ft str.	French Little, or Rochfort Engine	See our correspondence inserted in the Case. J.Wn agreed to give us £100 for the drawings etc. for this Engine. The drawings were sent but the money has never been charged.
1783-4	50" changed to a 52	Maes y fynnon wen commonly called Minera Eng. upon a lead mine in the township of Minera in North Wales.	Jno W^n^ held 7/16 of this mine the time the order was given which appears to have been merely verbal. No agreement nor mem^n^ of Ag^t^ is to be found but we can prove the receipt of the Drawings & our having executed the Nozzles. 9/16 of the Adv^s^ are willing to pay & have referred the settlement to J.W^n^.
	All subsequent Engines have been erected without our knowledge or Licence.		
1787			
Feb. 3	50		Further information
"	38	Mr. Perrez of Cadiz	unknown. No other
"	26		voucher than the Bersham books.

April 8	30	J.W[n] for Willey	
22	28¼	J.W[n] for Bradley	For a lifting hammer near the water wheel.
June 9	60	Mon[sr] Perrier for Mons	For a coal work. Appears upon the Bersham books.
1788			
Aug 31	40	Caemuneth Lead Mine in the township of Minera, North Wales.	Jn[o] Wilkinson is the principal if not the sole proprietor. It has stopped for the last 18 mo but is now to be set to work again.
1789			
July 11	20	British Plate Glass Co.	
"	48	J.W[n] for Bradley	For the Great Rolling & Slitting Mill.
"	42	J.w[n] for Bradley	Now applied to Boring Guns.
"	36	J.W[n] for Bradley	For Blowing Furnaces.
"	42	Suppose another 42 to	be in the same house.
13	40	Flint Canal Co.	Consists of Mr. Tho[s] Williams, Mr. Wilkinson, Mr. Jones of Wepra near Chester. See our correspondence with J.W[n] about this Engine in Oct[r] & Nov[r] 1788.
Aug. 3	24	J.W[n] for Bradley	For boring & turning.
24	41	Milur Lead Mine at Holywell in Wales.	Belongs to Paul Panton Esq & Co. See Jn[o] W[ns] application about this Engine w[h] was never answered.
Nov. 11	45	Bog Mine in Shropshire.	Error our Engine Erect[d] upon that Mine.

1790			
April 27	14	Hollinwood Colliery.	
Aug. 23	14	Spon Green Do.	Wind[g] Engine for W.W[n].
Oct[r] 10	14	J.W[n] for Bradley.	Winding Engine.
Dec[r] 27	14	Nicholas Ashton Esq[r] Colliery near St. Helens Lancashire.	
1791			
April 2	14	J.W[n] for Bradley.	Winding Engine.
July 15	48	Nic[s] Ashton Esq. Colliery St. Helens Lancashire.	Error is a Common one.
Aug. 27	20	Michael Hughes Esq. Colliery at Ravenshead Copper Work between Prescot & Wigan.	
1792			
May 14	30	J.W[n] for Bradley.	
24	40	Count Reden for a Lead Mine in Prussia.	
Sep[r] 17	24	J.W[n] for Bradley.	
	45	J.W[n] for Maes y fynnon Wen.	See this mine above.
1794			
Aug. 4	19⅛	J.W[n] for Bradley.	

1791			
June 13	48	J.W^{n} for Llyny-pandu Lead Mines near Mold in Flintshire.	We do not know whether J.W^{n} had any partner in these Mines, which are said to be very profitable. As coals sell there from 16/- to 18/- per ton-the savings must be considerable.
Aug. 17	38	Do. Do.	
1792			
Jan. 20	18 double	Do. Do.	
1793			
Dec. 3	26 Do.	Do. Do.	
1794			
May 16	52 single	J.W^{n} for Brymbo.	Coals cost 12/- p^{r} ton delivered. Only the first of these Engines was erected on 19 Sepr 1975 & was not yet set to work.
1795			
Feb. 26	24	Do. Do.	
July 9	18	Do. Do.	

Office Letter Book 17, pp 279-285.

[1] Enclosed with a letter of same date to the lawyer, Ambrose Weston, in which J. Watt Junr. asks whether 'you think we have grounds sufficient to recover the Premiums at common law, or whether the better way will not be to stop his Engines by Injunction. Perhaps a distinction should be made between the Engines he has erected for himself & those he has erected for others; at least where we do not know that the parties are disposed to resist our claims we think a civil application should be made.'

DOCUMENT 53

5 Octr 1796

Considerations upon the measure most proper to be adopted in the present state of affairs with Maberley[1]
(In hand of Watt Jnr.)

1. Supposing Maberley will renew the terms proposed by him when at Soho, viz that B & W should relinquish the Premiums upon the Engines already erected by him and that on his part he shall give security that he will not infringe in future. Quere' in how far is it the Interest of B & W to accede to such a proposal?

Reasons for agreeing

1. B & W will avoid a trial for the present with all its attendant anxiety and expences.
2. They will continue to receive Payments of Premium upon the agreements as heretofore without demur
3. They will have more leisure to compleat their new works & to make useful arrangements in their business.
4. They will meet with no obstacles from Maberley in settling the litigated Premiums in Cornwall.
5. A year or two may elapse before any new Infringem[ts] get so far a head as to render it necessary to notice them; perhaps even then they may be suppressed by an Injunction and any further trial avoided during the continuance of the Patent.
6. If the trial should be deferred for a year or two, they will then be better prepared against the consequences of an adverse decision by having their works compleated and such a system established as will bear down opposition, or at best enable them to maintain the lead in the manufacturing of Engines.

Reasons against

1. Granted. But they give up Premiums to the amount of which are now ordered to be paid into their hands & which they would remain possessed of in every case, but that of the verdict of a Jury against them. In case of a favourable Verdict and decision, they would recover taxed Costs, perhaps also other damages. This compromise if necessary, is not honourable nor just in itself. It is sanctioning a dishonest act, for which others have been made to pay the penalty. It will excite discontent and complaints from our Customers and from Pirates with whom we have compromised. It will establish a precedent for future Invaders and will flatter them with the hopes of Impunity. It may be brought in evidence against us upon any future trial to shew our opinion of our own weakness & the encouragement that was held out to Infringers.
2. Some discontents may arise among the different parties under Ag[ts] when they hear of the Concessions we have made and may cause us some disagreements, but still it is not likely they will contest the payment, Cornwall excepted.
3. Granted.
4. True, but they will from

Reasons against

Hornblower who will trumpet abroad the terms of the compromise which will tend to confirm the Cornish People in their obstinacy and in their idea of the badness of the Patent and of B & W's timidity. They will say that if we give up the Premiums in one case from fear, we shall do so in another. If this idea gains ground, it is possible that we may recover little or nothing from Cornwall during the remainder of the term. Besides, will not the Cornish people set up some new Infringer who shall make Engines of a somewhat different construction from Bulls & Hornblowers merely for the purpose of compelling us to notice them & so bring on a trial; and if we suffer the term of the Patent to elapse without noticing them, would it not operate strongly against us upon any action we might bring for the recovery of Premiums, if proof of our having winked at the Infringement, or not dared to subject our rights to a trial, were adduced.

5. Hornblower out of revenge & young Maberley out of pride, will take care to make known the concessions we have made. The prospect of impunity will let loose upon us not one but the whole tribe of London Pirates, who

Reasons against

will not want followers in the country. Bramah has now an Engine nearly ready, which we must either suffer to pass unnoticed or sustain a trial upon. He will be less conscientious in his means than Maberley. It is questionable whether another Chancellor may be equally well disposed with the present to grant Injunctions.

6. They certainly will be in a better state for maintaining their Monopoly independent of the Patent. But they have much to apprehend from a change of Judges & of Ministry, and still more from the death of Witnesses, many of them being men of infirm health. It is doubtful whether upon any future trial they can obtain such clear evidence of the Infringement, at least by such Witnesses as Southern & Barnes.

2. Taking it for granted that it would be neither politic nor just in B & W to compromise matters with Maberley by ending the Premiums of his contraband Engines, in how far would it be proper in them to purchase exemption from trial by giving up the Costs? only?

Reasons for

1. They get rid of the trouble & anxiety attending a trial and they get paid a considerable Sum in hand.
2. The Penalty paid will be sufficient to discourage others, and will tend to strengthen the validity of the Patent right in the public

Reasons against

1, 2, 3 & 4 Granted. But would Maberley be quiet. After having once had the option of better terms, would he now be satisfied with these.

This objection has little weight for if he is tied down by Agreement and under

Reasons for

opinions. The Cession of the costs will be considered rather as an act of lenity than as extorted by timidity.

3. Such a compromise would destroy the last hopes of the Cornish people and make them despair of their cause.
4. It would confound Bramah and all the petty London Invaders, and take from them all hopes of bringing us to trial at another's expence. There is little prospect of their doing it at their own.
5. Even in case of the successful Issue of a trial at Law, with a Verdict & Costs in our favour, it is doubtful whether we should not be more out of pocket than the present Costs amount to.

Reasons against

Penalty, the harm he could do is not apparent.
The Pirates who have been obliged to pay the Costs would remonstrate; but their situation is not similar, as they have not smarted so much by their own folly, as Maberley has.

3. Supposing an honourable compromise not practicable what are the advantages & disadvantages that might result from the favourable or unfavourable Issue of a trial under all the circumstances of B & W's present situation and have they upon the whole more to hope than to fear from a Trial?

Arguments for a Trial

1. The probability of being tried by the C.J.[2] who has expressed a favourable opinion of the cause.
2. Having Counsel who know all the merits & demerits of the case and who are opposed to others ignorant thereof.
3. The Respectability of the Witnesses we have to adduce, the clearness of their testimony and their intelligence of the subject.

Arguments for a Trial

4. The other side have already exposed in their Aff^{ts} the points they mean to go upon. We know their Witnesses who are either ignorant, interested or malicious. We can prove one or other upon all of them. They will get no man of character to assist them.
5. It may perhaps not be impossible to adduce proof of the combination by which Maberley is supported, particularly if he goes to Cornwall.
6. The Verdict of the former Jury so strongly in our favour.

MIV Misc.

[1]The case against Jabez Hornblower and his partner, Maberley, was begun in 1796.
[2]Chief Justice Eyre who had already given judgement in favour of Watt's patent specification in the Court of Common Pleas on 16 May 1795.

DOCUMENT 54

Carharrack
Gwennap
near Redruth, Cornwall
27th April 1797.

To Boulton & Watt

Honoured Sirs,

I have been proposing to write you evry week since I left London and can make no other appology for this delay but that I have been expecting the Adv^{rs} in U^{d} Mines would e'er now come to Some Conclusion about the Engines talk'd of and wanted on S^{d} Mines And as Bull and Trevithick are in wonderfull repute here, I have had Some reason to think Should any thing in the Engine way be done they would be the men — The(y) have often been at U^{d}

Mines & it's certain that Trevithick has made an offer to build a Whim-engine at Ale & Cakes mine at his own expence to draw all the Ore &c of 4 Shafts for one Guinea p^{r} hundred kibles and I believe it's Just as certain it would be of no use for me to take this as an afront.

The first 3 or 4 weeks after I return'd from town, the Patent-breaking party Seemed to be wonderstruck, and in general were as Silent as Church mice: but about the begining of Febry I Saw a letter with an Agent of U^{d} Mines which Said, Sir, I take pleasure in informing you that Mess^{s} B & Watt are intirely cast — 3 Judges are on Maberlys Side — only one on B & W side. The day after this I was told by another that the affare is now coming before Lord Kenyon and he will Soon do their business. Others Said, it is coming before the house of Lords & on the back of this, Mr Penrose of Redruth, told me that Bull inform'd him that La Blon is to have what he requested for Maberly in the court, and which was given up by the Plaintiff, i.e. the "Steam acting upon the Piston". Here I made it my business to tell him that neither the Steam nor anything els was given up, nor ever would in my opinion. I hope Sirs, you will excuse this freedom with your business but a worm touch'd in the middle will rise at least at one end.

Something more Curious Still, was a Speech of Mr Moyles one of maberlys Side: he Said as I was inform'd that the Plaintiffs evidence might know Something about Stars, but little of fire engines, and that I myself swallow'd it down pretty tight. but that is best known to me, I can ashure him I have nothing to condemn me on that account, but heartily wished in the Guildhall that I could be further examin'd and permited to explain myself — I was realy grieved to hear So much Stuff with Maberlys Counsil about the Situation, Particular Size, Construction, &c of Condensors — I am persuaded no Judicious Engineer has yet determin'd the exact Size, form, or Situation of the Condensor, or pump. I perfectly well remember when I hear Mr Watt condensed the Steam in another vessel, that a methood of doing it Occured to me immediatly and in my model, altho the whole of the condensor was undesignedly under water, I never foresaw the effects it produced, till finding my Injecting hole too Large, I Shut it quite and found the engine wro^{t} very well intirely by the application of cold water to the Sides of the Condensor, which Condensor contain'd as much or more Surface than the working Cylinder: and I'm now persuaded, had it been Smaller or Larger, it would have wro^{t} very well. My pump too was much to Large, and as I could not make it less, to remedy this defect Should it be any, I contrived to lengthen or Shorten its Stroke at pleasure, as m(a)y be Seen on the Beam of Said Model.

But to return. The Lords of U^{d} Mines having agreed Some time ago to give £7000 and pay it out of their dues towards erecting the aforesaid engines, but if the Said engines should be Stoped from working by an injunction the(y) would have the money paid back again.

The Adv[rs] have agreed* to accept of the above Sum £7000 towards erecting one Large Engine on Poldory mine, but are not willing to return any mony So paid — the are now waiting an answer from Lords, and if The Lords comply with their request, a Large engine will immediatly be Sett about: but, it is not determin'd whose engine it must be — The adv[rs] have a 63 inch Cylinder for Ale & Cakes engine already cast at Coalbrookdale by Cap[n] Tho[s] Michells order, and I am ordered to make drawings for nozels for it to work Double if wanted — I had much rather they were to be had at Soho, for in that case it would Save me a deal of trouble, but I can make no kind of objections to any things of this sort for Bull and Trevithick are at hand.

The Adv[rs] of Wheal Sparnon mine which is close by Redruth town, Bo[t] a 63 inch Cylinder of Poldice adv[rs] last Good friday — Bull is to have the erecting it, he has already marked out the foundation, and the talk is that he intends to work it without a cover, that is, with the power of the Atmosphere — but why not with the Steam? if Le Blon has it given up to him — I percieve he is Still big with mischief, and <u>will</u> be an engineer in Spite of fate. The principle adventurer in this Mine is a M[r] Franks, Son of a Jew in London.

I was well inform'd last week that Bulls law cost is Some what more than 1800 pounds, that his Journeys to and from London cost £300 — Some mines have paid their part of this law cost in full, others only ½, the U[d] Mines only 200£ as yet.

Young Trevithick came to U[d] Mines last week; and inform[d] me as certain truth, that the engine trial is coming on at the Kings bench in May next — that he hoped, nay believed it will then be finally Settled — Like all of that party he argues that the Patentee has had money enough out of this County — I told him I believed he nor no man in the County is a Judge of the merits or rewards of Such an invention, well he Said, the Judge have done pretty well for us already, by givin us the Steam; we trust we Shall have all our wish very Soon.

If the trial does come on in May as he Said, or any other time I heartily wish it may meet with better Success (if possible) than it has heretofore, and I do wish that the Worthy Adair with you may come off Perfectly Satisfied at last.

I suppose M[r] Wilson inform'd you how through much dificulty and danger we got Safe at Truro — Therefore I omit the history of that Journey — But when I found my wife and family very well, and things tolerably well at the mine, I repented that I did not go to Birmingham with Mr Murdock, my Brother too, have blam'd me on that acc^t^ — Your kindness to me was very great in that and evry other respect, for which I now return my most Sincere and Gratefull acknowledgmints

I hope you excuse whatever you
have Seen amiss in, or with me.
I do ashure you Sirs, that I Saw
myself while in London, Such an ill
behav'd Clumsy creature, as was nowhere
els to be found. I laboured under great
anxiety of mind especialy at dinner time
with respect to the Company I often wished
to be with my equals — or to be by myself.
I add no more, But that I remain Honoured Sirs,
your most unworthy, Obedient humble Servant
Richard Michell.

Box 3 M

*17 April Court day

DOCUMENT 55

To Boulton & Watt. Carharrack in Gwennap.
23 Nov. 1797

Honoured Sirs,

I was asked about 5 month Since at United Mines Acc^t^ meeting if I could erect a double Engine, and not infringe on your Patent? my answer was, that it is beyond my ability, nor did I think any body els Could. The manager was then desired to consult other able Engineers and have their answer at truro at a Canal meeting but the very next day after I was asked the above Question, word was Bro^t^ to U. Mines that Trevithick will erect a Double Engine without infringing on B & Watt and give any security for the performance of the Same; and as well, indemnify the adv^rs^. That he would erect an Engine which Shall do ⅓ part more work than either Engine on the U^d^ Mines.

A few weeks after this young Trevithick proposed to erect a whim Engine at Ale & Cakes mine at his own expence and get it working in 8 weeks. After all these proposals none has been aceeded to; but U^{d} Mines adv^{rs} have since bo^{t} Wheal Carpenter Engine for the purpose of drawing up the Ores &c from the Mine. I am now putting it together, and Trevithick has Quitted these mines.

The 63 inch Cylinder is on the mine for Ale & Cakes Engine and I am ordered to make drawings for another $\frac{\text{in}}{63}$ cylinder to go at Poldory mine to work double. — There is an agreement on the Order book to have an house built to receive it and its propos'd also for me to have the Erection — and according to custom, I expect to have little or nothing for my trouble more than my bare wages, yet for the Sake of my familly, I think it my duty not to refuse it, and as well to obey my imployers, if I refuse it Trevithick or Bull will Snap at it however since Trevithick has left coming to U^{d} Mines he has Complain'd of his not being used well with the agents of Said Mines — He is now trying to take Poldice Engine to work at his own expence, and have propos'd to work it Atmospherically, perhaps on the Pednandray plan — in order to cut off the Savings!

And E^{d} Bull in order that Mine Adventurers Shall not pay your Savings, has erected a Curious Engine on a mine call'd Pednandray close by Redruth town, with one of Poldice old-Engine-Cylinders (63 inch) it is to work with the Power of the atmosphere, and instead of the Air pump and Condensor he has a preponderating Colum of water 33 feet long which forms his Vacuum. Having got it ready to work; Mr Bulls phylosophy fail'd him, for after he had been trying her to work 16 or 17 days Successively to little or no purpose, and the puzeled Engineer had not been in bead for 13 days and nights; about the 18th day, they discovered that a Valve belonging to the long discharging pipe Or Hydraulic-Snout, did better when Cloathed with Duck than with Leather, and hopes are entertain'd that by-and-by it will do better; i.e. burn less coal and work more regular, for it only burns at present 2½ ways in 24 hours drawing 8 or 9 fath^{m} deep with a 14 inch box!

I am pretty well inform'd too, that these Adv^{rs} are gone to about 400£ Extra expence to bring about this Complicated Mischief.

Monday 13th ins^{t}. I was informed the managers got uneasy and Questioned Bull respecting the Consumption of Coals — when he affirmed the engine was doing very well, and that he would Submit it to the Judgement of Mr Moyle at Chasewater; Mr Moyle said it consum'd two way to much — 16th ins^{t}, a friend told me that Mr Carpenter and Mr Penrose had applied to Mr Murdock, desiring he would alter the Pednandray Engine. This information warmed my heart as much as Duncans was — I think this an unquestionable proof that your Savings were not taken and made out from the very worst Engines in this county. I have heard it Said a hundred times that Calculations ought to be made from a New engine on the Old Plan.

Here these learned Gentlemen have a new engine on the old plan, And a r(e)cent proof of the Superiority of your Beam engines as improved on the atmospherick or old way.

Mr Hornblowers very new Engine at Penryn has lately been tryed, I know Gentlemen who has been Solicited to See it, but nothing concerning it has yet transpired.

Trevithick is erecting an Engine at Ding-dong Mine — I Saw the Nossels for it, Casting at Hayle — These Nossels were last to work double, and the parts which Contain the Valves in yours were but one thing in this.

I suppose this is to be an Atmospherick double Engine., unless he is inclinable to turn Bull, having Grazed with him on the Same pasture —

Monday 21 ins^t^, Today I hear Mr Murdock has begun to transmodel the Pednandray Engine — and has taken a great deal of Bulls Jem-cranks to pieces; Amen — Trusting you have, or will have a fuller acc^t^ of this affare from Some abler hand — I add no more, but hope you will excuse freedom, honoured Sirs, in your very

humb^e^ Servant

Richd Michell.

Box 3 M

DOCUMENT 56

To Boulton & Watt London April 20th 1797.

Gentlemen

Please to send to Gardner & Manser's[1] two Exhausting Valves to the drawing full size; & three Gross of Springs to pattern, you will oblige them if they can be sent by this time 3 weeks, as they will have an opportunity at that time to put them in, the old ones are very bad, they complain of the stems getting loose before the valves are wore out. Thrale's new Boiler is nearly compleat, shall be glad if you will inform me by the first opportunity what stays you think will be necessary, the sides are very strong near $^3{}_8$ thick, & curving inwards 3 Inches. I receiv'd your's of the 17th. Sr John's Eng.[2] shall be begun immediately upon the arrival of the materials, & may depend on its being done with all possible despatch. I was yesterday at the Wharf and was inform'd that the Boats had not arrived at

Oxford, being stop'd by the Canal being under some repairs so that it is uncertain when we may receive them; however no time shall be lost whenever they arrive; the framing &c is in a good state of forwardness the whole mill be compleat in the early part of next week.

You will excuse me for troubling you with what I am now going to state; but I think my duty to acquaint you with any thing that particularly occurs in the business; a young man who's name is Dixon, & who has work'd some of Simmingtons Engines, that at the Paper Mill dock head last, has made a small Eng. of the power of about one horse & which he is at present erecting in John Street near Blackfriers Bridge for the purpose of turning some Cards; the Gentleman for whom it is, on being ask'd why he had not one of B & W Eng. answer'd that he has been inform'd he could not have such a one as he wanted in size or he certainly should have applied; for four horses was more than he wanted, or more than he had room for; now I think it would be worth while, certainly, to make Eng^s^ small enough to accommodate every body; not that perhaps any great profits would arise from small engines, but only it would prevent the business going into other hands,[3] for this Same Dixon offer'd proposals to Clowes's to erect one for them upon his plan; and I have been inform'd that it was only from doubts of the performance (of) it, as none had been try'd, that prevented it, Mr. Cooper at the same time recommending them to take that wherein they could not be deceiv'd; in my opinion too much precaution cannot taken against these people who are your avowed enemies. if it is thought that engines of this description may fall into the hands of those that are poor, any inconvenience from that may be prevented, for if we had two or three here constantly for sale, let them be sold for ready Money only, and the risk in putting together afterwards would be but a trifle, however this you will be best able to judge upon.

I shall esteem it as a particular favour if you will please to send me a strong portable Copying Machine, which place to my Acc^t^ & I will settle it with Mr. Woodward, in which you will much oblige your humble Obed^t^ Servant.

Richd Dayus.[4]

Box 6 D

---

[1]Of the King and Queen Foundry, Rotherhithe, London.
[2]Sir John Eamer's 20 h.p. engine for a London sugar refinery. It was ordered in 1797.
[3]Others made the same point although not so forcefully but B & W did not go into the small engine market preferring, with rare exceptions, not to make engines of less than 4 h.p.
[4]A former B & W employee, then an independent millwright but handling much of B & W's London business.

DOCUMENT 57

To M.R. Boulton Leeds 19 June 1802

Dear Sir,

This is my second dispatch of this day & the purport of it is to inform you that Messrs Bairds of Glasgow are here and travelling South with the intention of inspecting the various establishments in the Iron & Engine way to which they can gain access. They applied to Gott[1] to see his dyehouse this morning, upon which he consulted me & I advised him to recommend to them an Inspection of Murray & C^{os}[2] works. They at first seemed to say it would not be possible to get admittance, but upon cross examination, it appeared they had already made the Essay & had got through the boring mill & a considerable portion of the Works before they met with Mr Wood who denied them all further access. They had however already made a sketch of the boring Mill, which Mr Gott says seemed very well done. Poor Wood will think this an unfortunate day. They say they have a letter from some person at Glasgow to my father: the orders for non admission at our Works should be doubly enforced, as they seem to be Gentlemen who are likely to inspect all they can first & ask leave afterwards.

I omitted to mention this morning that Dixon says he has left a com^{n} with Joseph Hughes to ask me, or you, for an Arrear of Wages, which he pretends to be due to him & that Joseph very willingly undertook the Com^{n} saying he did not care whether he affronted us or not, as he was compleatly tired of the place and wanted to make us turn him off.

I understand that Creighton is just arrived here and should not be in the least surprized to see Lawson. If any thing occurs after seeing him I shall mention it at foot.

Give my Duty to my father & remember me kindly to all yours.

J. Watt Junr.

Mr Gott desires you will procure from Lawrence a plan of a Stall, such as he would recommend. The Stable is now building & the information should be sent direct to Mr Gott as soon as possible.

Box 2

[1]Benjamin Gott, Woollen manufacturer of Bean Ing, Leeds.
[2]Murray Fenton & Wood, Ironfounders and engine makers.

DOCUMENT 58

To M. Boulton Newcastle on Tyne 5 Feby 1803

My Dear Sir

Tho' I cannot express the pleasure I felt on receiving your kind letter which I rec^d here two days ago — yet believe me I feel your affectionate Idea of more than a Sincere friend — and in this point of View I have long considered both yourself & Family of which I so long formed a part.

Your acc^t of the Birmingham Eng^ns is too nearly in Character (I am sorry to say) with those here as Since the Loss of Poor Barnes there is no one (I know off) of sience or even common natural abilitys (for what they may naturely have is destroy'd by Wine & Brandy) and they seem going backwards for they are erecting common Engines of large sizes, and many other Cheap (say nominally cheap) ones of various constructions — not one of which that I have seen, are at all equal to the worst ever made at Soho.

The Puff you alude to is merely a paper one[1] — The exp^t was made on the great Canal near Glasgow — it made so little noise, that I did not hear of it till I saw it in the News Paper.

The Engine is made by Symington & turns a Wheel — which made so much Splashing that they were affraid of washing down the Canal Banks — tho' the boat only moved ab^t 2 Miles pr hour — the papers say Lord Dundas is the great Patroniser of the Scheme — which I rather think will Die without further noise.[2]

Symington has been constantly at it for twelve Months past (on the Canal Co.^s account). I am now this far on my road South at least so far as Manchester so that the Petitions I mentioned ab^t the Coinage from Scotland I cannot attend to at present. Here nearly the same scarcity prevails — I shall speak to Mr Chapman who can readily get one sent from this as every one complains.

I made enquiry ab^t Mr Sherriff (when in Scotland) & find he intends returning next Summer. I understand he has failed in the object he went ab^t. a Lady to whom he was to have been united, who declines — I cannot find he had any intentions of again entering into business there.

The conclusion of your letter is distressing — I will hope that your complaints are again aleviated, would to God they could be totally removed. With kind rememberance to Miss Boulton & Miss Amelia —

I remain My Dr Sir

yours most truly

J Lawson

MB — Lawson/82

[1]The allusion was to a report of the sailing of the *Charlotte Dundas,* engined by Symington, from Lock to Port Dundas in March 1802.
[2]He was right. Fear of the canal banks being washed away led to schemes for steam navigation being abandoned for the time being.

DOCUMENT 59

To M.R. Boulton | Leeds (Sat^{y} Ev^{g}) 12 June 1802

Dear Sir

We reached this to breakfast this Morning and upon a consultation held with Gott determined to employ his millwright, a sagacious fellow of the name of Prittchard to discover the abode of Halligan.[1] Pritchard occasionally visits Murrays foundry for the purpose of ordering castings & therefore had no difficulty in introducing himself there & singling out Halligan from the description we gave of his person. Entering into conversation with him he learnt where he lived and returning to us, conducted Murdock to his house, where he found the wife at home and had a conversation of upwards of an hour. Of the particulars I am not fully possessed, but in all respects confirmed Dixons information. She is exceedingly discontented and her health evidently effected by the state of her mind. She cried bitterly, bewailed her own folly in quitting us, had wished herself a thousand times back among her old acquaintance & wanted to know whether Murdock thought we would agree to take her husband, who she said was disappointed in his expectations though not desirous of going back except upon her Account. That his wages were 22/- per Week and he worked all the dry Sand under Dixon. They pay £7 a year for a worse house than they had with us & I believe, rates & taxes. Provisions &c much about the same price. Young Hughes she said lodged with them, but she had never heard him say anything about his reasons for quitting Soho & knew little about his situation with Murray, as he was remarkably close. — Upon being apprized that I was at Leeds, she expressed a wish that I would permit her to call, which Murdock offered to negociate & desired her to tell her husband that he should be glad to see him at the Inn when he had done work.

I have been surveying the environs of this rival Establishment & making enquiries respecting the property & tenure of the neighbouring lands, with a view to seeing whether we could purchase any thing under their very nose that might materially annoy them & eventually benefit ourselves.[2] I find there are about 2 Acres of Land next field to Murrays works, which may be purchased, but the price probably will be £5 to 600 per Acre. I shall learn the exact term. There is a Malt house which projects into their premises, which they have in vain endeavoured to purchase at a moderate rate. It is in the possession of a Widow, who is aware that it would be of some advantage to them & therefore asks a high price. This would enable us to overlook their whole Yard & holding it we might dictate our own terms. I shall employ a negociator to learn the terms asked for it. It is said that Murray & C° have purchased from Marshall the land upon which their works are built, and it is not known that he advances any thing to the new Rotundo[3] for such it appeared to me to be, but I shall get a nearer aspect of it tomorrow. Whether the land upon which the Malt house stands is under lease from Marshall or is Freehold property, I have not yet learnt, but the purchase of the lease would probably answer our purpose. It is generally believed that Murray & C° are making a great deal of Money & that the new works are constructing from their profits, as also a superb house which Murray is building for himself. — Fenton is, as we formerly heard, a man of some property & was Marshalls partner, who provided him with a share in this business to get him out of his own. Mr Gott speaks well of him & thinks he would be ashamed of his partners proceedings although he may not have the power to controul him. Wood is the steady man of business who directs the works: Fenton keeps the Cash & Books & Murray solicits orders superintends the erection of Engines &c. He passes him for a great scoundrel but a very able mechanic & Gott says he has got great credit from his last patent, no one doubting that the Inventions are exclusively his own.

Haligan & Pritchard are soon here supping with Murdock. I have left them to themselves, conceiving that more information may be obtained in my absence than if I remained with them. I had but a short conversation with Halligan, who has been the greatest part of the Evening with Murdock. — He told me that his Wife fretted much & he did not like the country so well as Birmingham, but should have made up his mind to stop here, if she could have been contented. Says he is under no Engagement, tho' most of the other men are, and he has been much pressed to enter into articles. Is employed solely in the dry Sand, which he says they do worse than we do & confirms Dixons report that the nozzles are all made in dry sand. The moulder is dead who made them in green sand & they have no workman now who can do them. Says he does not think much of their moulders although he is sensible of the great perfection of their green sand works which he attributes entirely to the sand. Says it is exactly the same as we had at the foundry, but that we spoiled it, by mixing too much Coal dust with it. The bottom of their green sand foundry (which is a distinct building from the dry sand one) consists for about 4 feet deep of this sand, which they work over & over again & have never renewed since he has been here (13 weeks). When the mould is formed, a little Coal dust is strewed over it, which is all that is used & he supposes this to be consumed by the Iron so that little or none remains mixed with the sand & though its colour is somewhat affected by frequent use, it is not so much changed as might be supposed. Connecting rods, Shafts & Wheels of every description are made solely in green sand, but no pipes or hollow goods. Boxes, where any are used, are of Wood & patched together for the occasion. They melt a great deal from the Cupola, which is about 8 feet high & 20 Inches Diameter & in which they do not use one third of the Coaks which we do to the same quantity of Iron. Believes the Coaks come from near Low Moor. Says the Iron runs much finer from the Cupola than ours does & about 6 Cwt per hour. The worst burnt stuff they can get comes out excellent. Cyl^{r} bottoms, tops & pistons are all made in green sand, until they come to be very large. They had a terrible blow up a short time ago in making a large cylinder. Says they have sent off about 2 Engines since he has been here & have 4 in hand. Employ their pattern makers to erect the Engines. Owen for one. John Hussey the smith who left us some time ago, is not here & he knows nothing of him. John Hughes he says has had two letters from his father since he came, but is very close & communicates nothing of their Contents, except that he said that Dixon had called at our foundry.

I presume that by this time, Murdock has gained a harvest of intelligence. You will see that it will be proper for us to inspect the place whence the sand is got and send you a further dose of this nostrum. We shall also endeavour to learn something more of the secrets of this cupola & obtain an inspection of the Coak and also of the Coal dust.

If any thing has occurred, you may safely direct a letter to me here, as I shall have work for some days. Tomorrow Mr Gott goes with us to inspect some situations in the neighbourhood.

I beg my duty to my father & kind respects to Yours & am truly

Yours &c
J. Watt Junr.

P.S. Fenton is in London & not well. Murray at Newcastle. Halligan has applied to be taken back.
Box 2

[1]a former employee now working for Murray, Fenton and Wood.
[2]B & W had seriously thought of establishing their own Yorkshire foundry. Nuisance value was clearly implied here, moreover Murray's casting techniques would appear to have been more advanced than B & W's. B & W were therefore engaged in their own espionage.
[3]Murray's new round foundry was deliberately fortress-like to discourage spying.

DOCUMENT 60

To M.R. Boulton. Leeds 14 June 1802.

Dear Sir,

I have yours of the 11th — I mentioned in a postcript to my last that Halligan had applied to be engaged. The application was made to Murdock at the close of their drinking bout on Saturday, who advised him to come the next morning and speak to me. He came only yesterday (Sunday) and after stating his wifes uneasiness & ill health and glancing at his own disappointment, proposed to be readmitted into the pale of grace. This led to an explanation of his motives for leaving, to which he appears to have been instigated by Dixon, who wrote more than one letter to him. He said also that others had prompted him, & that some disagreement he had had with one or two of his shopmates had made him indifferent about remaining at our Works. He seemed averse to entering into the details & I did not concieve it necessary to press him at that time. I told him that under the circumstances of his Wifes ill health, his former good conduct and my belief that he had not been instrumental in seducing others, I should be inclined to relax a little from our accustomed rigidity and use my Influence with you to take him back, but as he had shown so little confidence in me before, in his manner of leaving us, I should expect him now to atone for that by leaving the time of his return entirely to me. This he agreed to, & said he could not expect any other terms than we might please to allow him, or usualy gave. I thought it necessary to proceed in this manner, as his wife had previously stated to Murdock that he was resolved to enter into no Articles. I appointed him to call again this Evening & have got an Agreement ready cut and dried for him, to which I make no doubt of getting his signature. I propose to give him a Guinea per Week, as I think it politic that he should come back satisfied. I shall also lend him what money may be necessary to start him & pay Dixon three or four Guineas which he has borrowed, or rather which the other cunningly pressed him to borrow. It may then become a question of policy whether we should take him back or leave him for some time as a spy upon Murray & Dixons' proceedings.

I have been this morning with Halligan's wife who certainly looks miserably. She seems delighted with the prospect of returning and hates Dixon cordially. I hoped by her means to have got his letters, but she assures me in a way which makes me believe it, that they were carried about by her husband in his pocket until they were worn out & then committed to the flames. She says that young Hughes says he is informed by his father that in consequence of Dixon's call at our foundry they would soon see plenty of their old shop mates here. Hughes is very close and she could learn no more from him than this. She had never heard of the five guineas given him by Murray, but believed he had a guinea per Week wages & was to rise at the end of every Year. That he was not yet engaged, but was to be so upon Murrays return. She showed us his bedroom, and as my key fortunately opened his trunk, we had a compleat examination of its Contents, among which a roll of drawings of various parts of our Machinery & Engines deserves most conspicuous Mention. The things are very indifferently done, but the dimensions are written both upon the drawings & upon separate slips of paper. There was about 4½ Guineas, the remains of his fee probably, in his breeches pocket, but no letters. Nancy says he always carries them about with him & neither her nor my ingenuity can suggest any means of coming at them, other than that of seizing his person, which I had therefore resolved to do, but upon enquiry of Nicholson & Upton this morning, I find it can only be done by a Warrant from a Justice in the district in which the offence was committed. I therefore inclose the indenture, that my father may upon the strength of it procure a Warrant from Mr Lane, which will be backed by one of the Justices here upon my swearing to the signature & the Gentleman will be apprehended, when we can at the same time seize upon his drawings & letters. It does not appear that we can make any legal use of the fact of his having made these drawings, as it does not appear that any of them are our property & perhaps he will be entitled to claim them back from us. To avoid this & also to prevent his disposing of them otherwise during the exchange of our letters, I was strongly inclined & indeed advised, to have laid hands upon them, but after having debated the subject, I am of opinion it will be better to leave them where they are as I do not think he will discover or suspect that they have been inspected & if they were removed it might lead to a premature discovery of our operations. They are not indeed of much consequence, as they contain nothing but what Muray already knows, or soon may know through other channels. Hughes has no griefs to alledge against us & has been evidently instigated by his father, who intends coming also as soon as his time is out. He however speaks disrespectfully of us & the place, and I have no hope of bringing him back to his allegiances, at least whilst they continue to flatter & coax him here which they naturally will do until his articles are signed. Under these circumstances there seems little advantage likely to result from putting the law in force against him, but I am not aware of any other mode of getting at his fathers letters, and it will at all events be some punishment to him to be sent handcuffed under the care of the Leeds constable & it will cost him some money to produce the proof of his being of age & in the mean time, he will probably be confined. It may also serve in some degree to criminate his father & enable us to commence those proceedings against him, which I am convinced we cannot postpone & which I hope we shall be able to do effectually from the additional evidence which Dixons recent embassy may furnish. It is certain that he went in search of three or four apprentices who have absconded from Murray, for which purpose he volunteered his Services, having no doubt the double object in view of employing his talents in the seduction of our Men. I do flatter myself that we shall be able to get compleat evidence against this scoundrel & his employers. — I of course shall not see any of them, (indeed Wood is the only one within reach) unless the denouement of this affair should require it. I am in treaty for the Malt house through the medium of a Malster & if it can be had, shall certainly make the purchase without waiting for further communication, as I am pretty confident that the position of it must

enable us to dictate terms. I also expect an answer respecting two other plots of land immediately contiguous, which I think we ought to have, as they seem eligible speculations independent of the ignoble motives which dictate their purchase. Fentons character is not very respectable. His brother is an Attorney well skilled in all the nefarious practices of his profession. Give me the Soho Opinion about purchasing the land, the other I shall take for granted unless I hear to the contrary before I have compleated my intentions, or the price proves too extravagant.

Fame has not outdone the magnitude of Murrays new Edifice. It is a rotundo of about 100 feet in diameter with a magnificent Entrance. The Engine is to stand in the middle and the lower rooms to serve as deposits for Engines & other finished goods. The Upper rooms to be for fitting. It is an excellent building & will not look amiss. It is up to the top of the 2*nd* story. I will make a better sketch of it before I leave Leeds than this, but am now much pressed for time. He employs altogether about 160 Men the greatest part of whom are engaged. Many are Apprentices. He has much trouble to keep them & has been left by several.

I am in some haste

D*r* Sir
Yours &c
J Watt Jun*r*.

Box 2

DOCUMENT 61

Overseas espionage

*S*r Jos. Banks, Bart. Soho Feb 10*th* 1787

Dear Sir

I feel myself very much obliged to you for the fav*r* of your letter & the communication of the Baron's[1] Memorial as well as for the friendly part you have taken.

If I have given the Baron any offence it can only have arisen from the relation of matters of fact, & from throwing difficulties in the way of his accomplishing his design by means discordant to his rank.

You'l please to observe the Baron says that some days after he had obtained admittance into y*r* Brewery[2] Engine by a note he went (a 2*d* time) to pimlico & was there advised to apply to Cartwright whereas you will find it is stated in my account of Facts that he apply'd to Cartw^{t} the first time he went to the Brewery & then took Cartwright with him although ye Baron says in his Memor^{l}. He having come to me. Car^{t} sayd he had told the Baron that Boulton & Watt were in Town & youl observe in his Memorial he says (Car^{t}.) went off to go to Mr Boulton. Hence it is plain that ye Baron knew I was in Town yet nevertheless he went with a Blackguard of the Brewery on the following Sunday when ye Masters were absent & the Engine not going to examine it but was surprised.

If the Baron had gone alone I should not have thought so much of it, but finding him with his 2 assistants making their attacks with Bribery & Blacklead pencils I was convinced he intended to serve us as his two Country men had before done to which I might add some other circumstances of a similar nature.

Example — In the year 1782 the duke d'Arenburg sent a Mr Fastor of Nanswere to me at this place with a letter (wch I will inclose). In consequence of which I took Mr Fastor into my house for a few days & shewd him every civility in my power: and as I expected he was to be employd to look after the dukes works was very communicative to him & not only shewed him some small Engines at Soho but I also hired a Chaise & took him to see some large ones in Shropshire for which Civilities Monsr Le Duke rewarded me by employing ye saide Mr Fastor to erect one of our Engines without our assistance, knowledge or consent & He (like Mr Buckling[3]) advertised in Foreign News papers that he had greatly improved B & W Steam Engine & had nearly erected one upon the Dukes Mine. But finding some difficulties he returnd again last year to England to correct his errors & how he hath succeeded I have not yet learnd.

If I have any thing to reproach myself of in my conduct towards Strangers it is in haveing been too liberal & too unguarded in showing those things which my duty to my Country & to my self required more reservation in.

I have no objection to your shewing my state^{t} of facts or this letter to any or all the Gent^{n} mentioned in ye Barons Memoire or whomsoever you please.

I have not time to enter into disputes & only write this for my own justification.

I think myself happy in the enjoyment of prosperity & will never forfeit it by a deviation from that rectitude which so eminently distinguishes your Character.

I am with most sincere regard

Dear Sir

Your faithfull & obliged Friend

& obed^{t} humb^{e} Servt

Math^{w} Boulton

From an accidental circumstance your letter did not come to my hands so soon as it ought.
please to put ye Dukes Letter in a Cover & return it to me p^{r} post without giving your self any trouble to write.

MB-Banks/13

[1]Freiherr vom Stein, a Prussian mining official.
[2]Barclay and Perkins' brewery, London.
[3]The other German industrial spy referred to by Boulton.

DOCUMENT 62

Facts Relative to Baron Stein

In the year 1779 — Two Persons from Prussia found out Richard Cartwright (one of Boulton & Watt's Workmen, who was then employ'd in erecting the Steam Engine at Chelsea) and endeavoured to bribe him to betray his Masters, but without Success. They then procured a Letter of recommendation to Messrs Boulton & Watt and one of them assumed a Title, which Letter they delivered to Mr Watt (Mr Boulton being from home)

Mr Watt endeavoured to shew them every Civility in his Power. He accompanied them to Soho, shewed them that Manufacture, he then walk'd several Miles to shew them one of the new Steam Engines at work upon the Birmingham Navigation, & such other particulars as were necessary to gratify the Curiosity of Gentlemen travelling for Amusement. He then took them back again to his own House to Dinner, and spent the remainder of the Day with them.

Whilst these Gentlemen were enjoying the information, the protection, the hospitality, and the Time of Mr Watt they were forming Strategems to Rob him of his Inventions, and debauch his Servants, for early on the next Morning they returned to Soho, and bribed a Workman to take asunder some of the principal parts of an Engine, as well as to shew them many other things equally improper. These Gentlemen soon after returned to their own Country (prusia) and one of them undertook to erect one of the said New Steam Engines at Mansfelt in Germany and during the erection he announced in one of the French periodical Publications, as well as in other foreign News Papers, that he had attrap'd all the Secrets of Boulton & Watt's new invented Steam Engines, to which he said he had added many important Improvements of his own Invention and that he had then nearly completed one of these improved Engines at the Mines at Mansfelt, which when finished he said would be equal in Power to 108 Horses, whereas it was not equal to 8 and moreover it did not answer in any Degree the Purpose it was erected for.

This very ingenious Artist (whose real Name we are now told is Master Assessor Buckling, Councelor of Mines of his Prussian Majesty) finding himself mistaken returned last year a second time to England to correct his Errors and by his usual Arts obtained Admittance into many of the Cornish Engines erected by Boulton & Watt, where he endeavoured to seduce some of the Workmen to go out of the Kingdom with him, bribed another to make Models of such parts as he did not comprehend, and did actually take one of the said Workmen (George Richards) with him abroad in the Autumn of last Year who is not yet returned & is of little consequence as he is good for nothing. —

On the 25th of last Month, Richard Cartwright one of Messrs Boulton & Watt's articled Servants, came to his said Masters in London and informed them that three Persons had been that Morning to Messrs Barclay & Perkins Brewery, that they had obtained a permission from one of the Clerks to go into the New Engine House at a time when all the Workmen were gone to their Dinners (except a young Man who did not understand the Engine). One of the Persons, he said, passed under the Name of Count Vidi, and that another began to make a Sketch of the Engine soon after they entered. They desired the Man to stop the Engine which he did. They then requested him to empty the Condenser Cistern, but he reply'd he durst not as he could not afterwards set it to work again, but said that Cartwright could as he had erected the Engine. They then desired him to take them to Cartwright's Lodgings which he did. They requested Cartwright to go with them and he accordingly did go with them to their Lodgings when (he says) he was informed that the Count was Baron Stein. He says the Baron flattered him much and told him he deserved great Wages and gave him two Guineas for his loss of time that Day. He then invited Cartwright to accompany them into Cornwall and Staffordshire to shew them all the new Engines erected in those County's, and if we are to believe Cartwright he endeavoured to persuade him to go abroad with him. He then shewed Cartwright some drawings and asked if he could not correct them, which he readily undertook to do. — As Messrs Boulton & Watt were both in Town at that time Cartwright came to inform them of all that had passed, and 'though they approved of his zeal to serve his Masters, yet they disapproved of his Conduct in taking Money, and to prevent further Mischief ordered him to quit London. Although Baron Stein knew from Cartwright that Messrs Boulton & Watt were in Town yet he apply'd again in the same Clandestine Manner to see Mr Barclay's Engine on the following Sunday, when he must know that the Masters of the Brewery were absent and the Engine was more examinable as it was then not at work. However he was refused.

Mr Boulton had some thoughts of waiting upon the Baron to have represented the impropriety of his Conduct and to have acquainted him that he had transgressed the Laws of this Country in endeavouring to seduce English Artists Abroad, whereby he was liable to be fined £500 and to be imprisoned Six Months. — But the Multiplicity of Mr Boulton's Business and engagements during his short Stay in London put it out of his power. Besides as Mr Boulton had not the honour of knowing any thing of either the Baron or the Count, and as their Conduct was not agreable to Mr Boulton's Notions of Honour, He therefore declined wasting his Time, perhaps it might in the Opinion of the Baron be very proper to steal Messrs Boulton & Watt's Inventions, But surely it was equally as proper that Messrs Boulton & Watt should prevent it if in their power, and which they endeavoured to do.

MB-S3/114

DOCUMENT 63

To M. Boulton 5 Feb. 1787

Baron Stein's Memorial

The simple Exposition of a fact, which has been represented in a very disadvantageous point of view for me, will suffice to prove the falsity of the Conclusions which have been intended to be drawn from it, as if I had been intentioned to apply an Invention of Mr Boulton's in Germany and to deprive him of the Enjoyment of the Advantages attached to it.

Ever Since my arrival in England, in the beginning of January, I have endeavoured to procure myself a drawing of Mr Boulton's fire Engine in Order to have a clear and just Idea of it's Construction, of it's Effects, and of the Utility or Use it might be of, for my Object. I obtained a Copy of a very imperfect design of the said Engine, and a drawing of a Model of an Engine upon the old

Construction. I was assured here that there would be no difficulty to have a drawing made on the Spot itself, I was shewn the Engine of Pimlico, & the person that did introduce me to it, informed the Workmen of my Aim, who made no Contradiction at all.

Meanwhile I had been in the Borough with the Intention to see the fire Engine of a Brewery. I sent to ask for the permission of it to the Accounting House and I entered into the House where the fire Engine was, with a Note from the Accounting House.

An Inspector of the fire Engine at Pimlico, having some days after, refused the Man, whom I employed to make me a plan, I wrote to the Governor of the Society of the proprietors of said Engine, to ask him the permission and I caused the Workman at Pimlico to be informed of it, by my Servant, when the former advised him to address himself to a Workman named Cartwright, who could give him the truest Information on the Subject. He went to him, & that man offered himself, to shew all the parts of the Engine & to give all the Information, that might be wished for, but he asked 8 days time & five Guineas as well for his trouble as for his Maintenance. Having come to me I promised him the one as well as the other. I gave him two Guineas, but instead of fullfilling his Engagements, he, during my Absence & the time during which he was alone with my Servant, laid hold of the two above mentioned drawings, under a pretence of wanting to rectify 'em, and went off to go to Mr Boulton & to represent to him the fact, in a point of view the most odious & for him the most Emeritorious.

This happened on the 25th. of January, and having been informed that same day after diner of Mr Boulton's arrival at Mr Moore's, I desired the latter to procure me his Acquaintance, who desired Mr Parker, the Son, to whom I gave my direction, to apprise him of it. Not having received any Answer, I asked on the 27th Mr Greville for a Letter to Mr Boulton, to whom after having received it on the 28th at Night, I sent it on the 29th. I accompanied it by a Letter of my own, in which I begged of him to fix me a time when I might see him, to Converse with him on the Subject of applying his fire Engine to my Object, and to defend or clear myself of a Suspicion which he had formed against me. I received no Answer from him and it was in vain that I sent to him to ask for it, on the 30th at 5 o'Clock in the Evening. I went to Mr Moore on the 31st, exposed the fact to him, and begged he would engage Mr Boulton to have an Interview with me. This Stop proved equally fruitless, I complained about it to Mr Greville the 1st of february in the Evening in the Royal Society & Mr Boulton being set off, I heard nothing moreabout the Matter, until this Morning, when I was informed of the disadvantageous light in which it had been represented to Mr Jos. Banks, & to other persons of my Acquaintance.

The results of this fact is, that I have endeavoured to obtain a drawing of Mr Boulton's fire Engine, but I have never had the Intention of Constructing a fire Engine by that drawing, because I know

1) that a great deal of it's Effects depends on the perfection of the work, which could never be so well made by other Workmen, than by those which Mr Boulton employs in this Object, since a Number of Years.
2) I have ever since my Arrival, spoke to various persons of my Acquaintance, amongst others to Mr Greville & to Mr Zimmerman, of my Intention to treat with Mr Boulton respecting the Conditions on which he would Construct fire Engines, and the moment I learnt at Mr Moore's that he was arrived, I caused myself to be announced to him, in Order to make his Acquaintance.
3) I have no motive of Interest, nor none of Ambition, my fortune Consists in Lands, the Engine could be applied but to Establishments belonging to the King. No Ambition, because it would be a puerile Means to satisfy it, in Constructing an Engine invented by another person, after plans that might have been taken of it. Mr Boulton complains of my wanting to deprive him clandestinely of the Advantages attached to his Invention, and I on my part, presume that he has spread disadvantageous reports Concerning me.

The reasons which I have now alledged will clear me of that Suspicion, and will serve at least as a Counterpoise to the depositions of the honest Cartwright, who has not known any other way, to make himself useful to his Master, than that of Imposture and theft. I hope that the Assurance I give him of my desiring nothing more arduously, as to treat with him upon the Construction of fire Engines in Germany, will contribute to tranquilize him in regard to my Views. But at the same time I demand, that in Case he should have doubts, he will please to inform me of it, in Order to clear up the fact, or that if there remains none more of 'em in his Breast, that he will please to destroy the Impressions he has given (or raised) against me.

London the 5th febr^y 1787
(signed) Stein

MB—S3/110

DOCUMENT 64

Draft letter to Baron Stein before Boulton had received letter from R.E. Raspe. March 3 1787

S^{r}

I have rec^{d} the honour of your letter of the 1st In^{t} — In which you express a desire to erect one of our Engines on your Salt works under similar conditions with those we made in France.

It is therefore necessary to acquaint you that the first article of our conditions was that an exclusive priviledge should be granted to us for the sole erection of Engines on our principles in that Kingdom which was accordingly done & hath lately been confirmed to us by the Ministers of France.

I know not whether his prusian Majesty adopts that mode of encouraging New & usefull inventions. But as my partner Mr Watt went to London yesterday (& is at N° 6 Green Lettice Lane) I will beg leave to refer you to him for a particular conference upon the subject & what ever you & him agree upon I shall conform to with pleasure.

I wish to avoid a repeatition of some late circumstances which cannot have been more disagreeable to you than to me but I must say that if you were informed how unhandsomly some of your Countrymen have behaved to us & unlawfull in this Country I am sure your anger would rather fall upon them than upon me.

I observe that you propose to make a journey to Cornwall soon. If I can promote your views as a Natural Philosopher, as a Mineralogist or as a Gentleman I shall be happy, But as a Mechanick & as an Engineer you must pardon me if I throw obstructions in your way. — An Engineer from prusia[1] made a visit to Cornwall last year & made drawings of Engines as well as enveighgled one man to leave his Country & go with him to Mansfelt in prusia & attempted to take others for wch he was liable to be committed to pison for 6 Months & to be fined $\overset{£}{500}$ Sterling. I therefore advise that person not to trust himself again in Cornwall. I have no doubt but the Gent^{n} of Cornwall will receive you & treat you as such but they will be cautious in some points.

As I am a man of business & love truth & openess I beg you will consider me as such, & pardon the freedome with which I express my self.

I am very respectfully

Sr Your most Obed^{t} humble Servt

M B

MB-S3/113

[1]K.F. Buckling.

DOCUMENT 65

To M. Boulton London : 1787

Dear Sir,

Guard against a Baron Stein, who is soon coming to Cornwall with the patriotic intention to kidnap away what useful knowledge & Men he may find in his way. — His saying that he means to agree with you about your Fire Engine has no weight with me, nor will it I suppose have any with you. — I have pressed him hard upon the subject but never saw anything like full powers — therefore it is more probable tht he means only to learn (per & nesas) how to correct that Fire Engine upon your construction, which a rascal of the name of Buckling stole away from this Kingdom some time ago, & which by his directions has been built at the expence of the Prusian Government at Rotenburg not far from Hull[1] in Sacony. It has cost them already, not to mention loss of time, a Sum of 26/m Dollars or about £4000 St. in Experiments & yet it worked but poorly & stops regularly once in 24 Hours. He comes with an excellent Engineer & Modeller, a Mr Friederich or Friedriek, of whose great abilities he has shewn me astonishing Proofs. — Amongst several others I was sorry to see two excellent drawings of his of your rotative motion & of the regulator of the Piston. — By the Baron's account they were taken from sight at Calvert's or Thrale's. This Young Man is a native of Clausthal in the Electorate of Hanover, the Son of a very worthy comptrolling Captain with whom I have been very often under ground. — He seems to have been Kidnapt into the Prusian service where even the best prospect they can make him will not be above £150 a Year. His great abilities & practical knowledge would make him a very useful acquisition, if the Baron could be punished his own way & this young Man be prevailed upon to stay with us, together with all his drawings. — I fancy I could manage that business, & should be the more happy in doing it, as I am fully persuaded from what I have seen & heard of him that he would well pay for the trouble & expence, was it only by the Models of the various Stamps in the Harts Forrest, Saxony & Hungary, which we are not sufficiently acquainted with & which he has promised to execute for me if further required. — He has been brought up to the practical parts of mining from his infancy; & has gone tho' all it's different branches from the Stamps up to the function of a Carpenter, Engine Man, Captain & Surveyor, in short he is what few Men in this Kingdom are, & what many ought to be — Sapienti sat. In every other respect Baron Stein is a well bred, sensible excellent young Nobleman. — If he thinks it a liberal exertion of patriotism to enrich his Country with useful knowledge in a sly back stair way, he cannot possibly have an objection to ours if possibly we can turn the tables upon him, & send him home civilly & wiser through punished rashness[2]. — I shall not leave Town untill I have finished my Edition of Born's Process of Amalgamation, which will be published by subscription & of which by the end of this week shall send & recommend to you my Proposals. — In many Respects it will prove of use in Cornwall. — It is even applicable in the great Copper operations. — Baron Born wrote to me that certain Zincous Copper Ores produce by Amalgamation a kind of Prince's Metal. — As I bring with me the whole Apparatus for Amalgamation I shall at my return give full satisfaction on this very interesting subject. Meanwhile I have the honor to be as ever Sir

Your most obedient humble Servant
signed R E Raspe

MB-S3/114

---

[1]Halle.

[2]Stein had, in fact, ordered an 8 h.p. engine but it was cancelled. In the event he was not responsible for the introduction of any steam engines to Germany.

Section E

The engine business – subcontracting

The engine business – subcontracting

Documents 66-73

From the beginning of their partnership until the opening of Soho Foundry in 1796, Boulton and Watt subcontracted many of their engine parts. They operated as consulting engineers and in this practice there are parallels with the organisation of the early Newcomen engine business under the Committee of the Proprietors of the Invention for Raising Water by Fire in the first years of the 18th century, as well as with the consulting engineers in the atmospheric engine market later in the century. Nevertheless, parts were made at Soho Manufactory from the outset and the range produced gradually increased.

The engine partnership was only one of Matthew Boulton's business ventures and he had, of necessity, to delegate the supervision of engine parts production. That this led to problems of throughput is illustrated by Boulton's letter of 1777[66].

Whilst there was some flexibility in the choice of founder for some engine parts, the majority of the cylinders were cast by Wilkinson at his Bersham works. The difficulties of subcontracting whilst each engine was largely custom built are illustrated by Watt's 1780 Memorandum[69]. Modifications were constantly being made and it was Boulton rather than Watt who insisted on standardisation which could lead to more speedy production, fewer errors and greater output. By 1790 the cost of making and altering drawings was fully appreciated by Watt who was, by this time, preparing his son, James, for managing the business[72].

Quality control was a considerable problem during the subcontracting period and nowhere is the position of the consulting engineer more clearly stated than in the letter from the engineer Robert Mylne[73] who observed that the subcontracting engineer was nevertheless finally responsible for the goods.

DOCUMENT 66

To James Watt   Soho Sep 6th 1777.

Dear S^r

Since my last of ye 4th In^t I have rec'd two letters from you of 31st Ultm^o & of ye 1st In^t. I am glad to find ye Tingtang Boxes Casks &c are arived at Truro as I had just wrote a letter to Mr James of Bristol upon ye subject but now shall not send it. I have also the satisfaction to learn yesterday from Mr Johnson that the Cylinder & all ye Bursham goods are already put on shipboard & I hope saild before this time, so that if ye wind is favourable they will arrive timenough for you to see the foundation well laid & all ye rest put into proper train. I sent in my last ye acc^t of ye weights & prices of ye things belonging to each Engine wch we manufactured at Soho, but as to the cost of each it is impossible to send it of some days & indeed I dont think it possible to send it at all, as our acc^{ts} of the cost of them are not sufficiently distinct to render such a state of their cost anyways usefull to you. From what I can see of the matter we shall not be gainers but, le Contrair, our Engine shop went on very bad all the time I was in London, not one nossell was finished & our friend Henderson was fearfull of giveing offence to any of them & therefore did not speak. I am convinced that we should avoid manufacturing as much as possable except the Brains, y^e Heart & the lungs of our Engines.

We have just put on the outside Steam Case & before I seal this letter will see something of its effects[1]. I desired Mr H. & J. Harrison would try the effects of y^e Boyler & the number of Strokes & ye Coal burnt but I never got any satisfactory acc^t. The water gage vibrated so much that there was no judging of ye quantity evaporated & all that I could learn was that one Hundred weight of Coals did not raise half ye quant^t of ∇ that it had used to do in ye old Engine. I then set Webb to try ye Engine for one day & he made it still much worse than Josephs experm^{ts}. But I have this day put a floating Buoy into ye Boyler in a copper Case with small holes of communication with y^e ∇ in ye boy & a brass wire coming through ye lid

	ft	In
the length of our <u>Steam case</u> is	6	10
Diam^r of do.	3	1

	Cwt	Qr	lb
weight of the Case & hoops	4	1	8
Screws for do.			27
4 Bars with holes for ye screws to pass		2	18½
Total weight of all	5	0	25½

Hence you may guess at the weights of all other sized Cases.

Tingtang nossell cost us ab^t eleven guineas besides Fire shop rent & various other general expences Clarks (etc) so that ye profit is even worse than Button making. Bradley castings are not charged to us but the Cornish Engines stand d^r to Wilkinson for them I wish you would again take a critical & minute survey of ye Engine W^1 Prosper near Marazion, either you will find out a deceit or a new principle in ye Old Engine for according to ye acc^t it doth 280,000 C^u feet p^r Bushall but I am confident all is not fair.

Mr Fothergill dined yesterday wth Parrott[2] on acc^t of their daughters, Parrott told him that he expected there would have been a greater difference between y new & old Engine in their effects. for He, Mr Whieldon, Mr Ferchan & Mr Taylour, had tryd both Engines & the new is to ye old as 45 is to 29 hence it is but ⅓ better than y^e old. Fothergill was astonished at Parrots wisdom & so am I indeed I know not wch of the two is y^e greatest engineer. They never thought of y^e circumstance that ours makes 11 Strokes p^r M^t & theirs 8 or 9 & that ours (ours filld 300 feet at 29 Stroks & theirs at 45 Strokes) burns only $\overset{\text{Cwt}}{4}$ & theirs $\overset{\text{Cwt}}{10}$ pr hour. They have not yet p^d us a farthing although we are near $\overset{\pounds}{400}$ in advance for them. Be assured I shall not make any disturbance with Perrins at this season but I dont relish his conduct & yet I think it proceeds more from his Vanity than from any Vilinous intentions but it may be =

dangerous to us. Pray transport all ye improvements you can respecting Boylers & Cataracts. Colvills[3] Nossell is gone to London & Salmon[4] will be finished on Monday & ye others are forward, all have been set to rights respecting Mr Henderson mistaking Chelsea for Richmond. I still think that loading our expansive Engine to near 8 lb upon ye inch is ye best way of equalising ye power & will make more of ye Steam than ye present light load of Belzibub[5]. Its now 8 o Clock p:m: & ye Engine is not yet quite ready for working but ye Steam is now 6 Ins strong & ye Steam Case seems to hold pretty well but no experiments can be made to night worth sending you. However you may be assured that every (experiment will be) accurately made next week that is necessary (for) our decission. I dont see how you can make lump bargains in Cornwall untill all parties are decided as to ye rate of goodness of ours. Example Wheal Union is of such a magnitude that I am confident an old Engine of that power would consume in Cornwall 2000 £ pr annum in Coals more than ours & I dont see how you will convince them but by comparison. Its true you may as in Scotts[6] assume a quany that ye old Engine wd require & then Guarantee that ours shall burn only a given quanty & if more we shall be deducted in that proportion & upon that fix our annual sum. remember there is but one County of Cornwall & dont let us give that away.

Dont forget to well coat your Cylinder & Boyler top. Our Stuffing Box is perfect, not a particle of Steam comes out. Your Children were both well this afternoon. My Wife dont like your 2 pony ordinaries & wishes Mrs Watt safe at home but will write next week. give Mrs. Watt a kiss extra for me who wish you both health & good luck in all things farewell.

M B

Box 20

---

[1]Boulton is describing experiments on an engine at Soho.
[2]One of proprietors of Hawkesbury Colliery engine (Bedworth).
[3]Ironfounders at Torryburn, Kinross. The engine had a 44 in. cyl. and 8 ft. stroke.
[4]of Lawton Salt Works, Cheshire. The engine had an 18 in. cyl. and 3 ft. stroke.
[5]The Soho engine.
[6]J. Scott & Co., owners of lead mine at Malehurst, Salop. The engine had a 27 in. cyl. and 8 ft. stroke.

DOCUMENT 67

To M. Boulton Soho Oc^{r} 29th 1778

Dear Sir,

I have just now received yours of the 23^{d}.

1. I shall observe what you say respecting the Navigation Engine.[1]

2. I acquainted you in my former letters that Bedworth Engine goes on extremely well, but Wm Murdock having been fetched away to make the patterns, no decisive & accurate trials have been made. But Wm Murdock has returned there, and Joseph Harrison is to go there next Monday, when they will make the proper Trials. I shall give Joseph an extract of your letter concerning the trough &c, which he says is sufficiently narrow already, there being no more room than is necessary for the coming at the Screws. He says that Hall has made the Trough at Ketley so small that it must be taken to pieces every time anything is to be done with the eduction pipe. Joseph says that all the Water does come in at the end next the Nossel. The load of the ballance-bob at Bedworth lies above the beam, but that I will desire him to alter. They have taken off about two Tons from the Engine beam & the pump rods fell down with great force at first, enough to tear every thing to pieces, till they made the Engine go more gently out of the house, & it now works exceedingly well, but Joseph doubts whether it will bear to have the ballance taken away — In my last I told you I had seen Parrot, & he was to write to his partners about the Engineer. But at all events, had you not better hire him, as you cannot have too many good Engineers. If you had placed good Engineers of your own at Richmond,[2] Bedworth, &c, as you did at Mr Scott's, (at least till you had been paid or agreements signed) much trouble would have been saved. But you had not them. Whenever you can meet with them, it will be well to engage them. — Mr Meason[3] says he hears but indifferent accounts of the Tory-burn Engine, which he supposes is owing to bad management. — Mr Parrot seems much inclined to settle matters, & I cannot doubt but they will either agree to a trial, or to the proposal made them.

3. A letter came yesterday from Mr Chapman[4] with a remittance in full of his acc^{t}. He said that Mr Henderson had sett off for Hull, having finished the erection of his Engine. As he makes no complaints I presume it goes well.

4. I shall give M. Jery[5] a letter to M. Maquer[6]

5. My letter by last post informed you how backward the Chelsea

people are in preparing. But I shall not detain Joseph Harrison from going to London for the sake of Shadwell & Richmond. If you think these reasons are not sufficient, a letter from you by return of post will reach him before he leaves Bedworth.

6. As to the business with Mr Wiss[7], you know he told you that unless all deficiencies which might arise were made good to him, that he would not purchase. And if you consider, you will find that every other body would expect the same security. For, in fact, you do not sell him your right to such & such Engines, but you sell a certain annuity, for the payment of which you transfer your right to these Engines as a collateral Security. Nobody but the proprietors would purchase your right to an Engine, because your income from thence is not certain, but eventual; that is to say, the proprietors are not obliged by your agreement to work your Engine, but only to pay you at such a rate while they do work it. The 1st question you have to consider is whether it be expedient for you to raise a certain sum of Money, say, £4000 on the Security of Fire Engines; If you determine in the affirmative, it signifies not whether the Engines mentioned as a Security cease to yield their Income or any others of equal value cease to yield, provided always that the Income yet remaining, be sufficiently greater than the Annuities. If you had only one Engine, you might lose a great deal by such a transaction, but as your income from Engines will in all human probability be much greater than the Annuities you will sell, I see no difference to your interest whether the particular Engines that are assigned as a collateral Security fail, or others fail. As to the price of ten years purchase, I believe it is too little, but you know the Scarcity of Money, & the use you have for it. If you can get a better price, I shall be glad of it, but if you cannot, I am sure I can demonstrate it is your Interest to accept of that offer, to that amount £4000 at least. — Mr Watt's offer is sufficient for him & the rest I think ought to lye with B & F.

7. Mr Watt has not sent us any orders about the outer case & Gudgeon for Poldice. The Smiths are at work upon the Nozzles.

No orders for buttons.

Orders enough for plated goods & paintings.

Mr Aubert has received the Telescope & paid for it.

I rejoice to hear of the prosperity of Chase-water.

(James Keir)

MB-Keir/29

---

[1]A 20 in. cyl. 4 ft. stroke engine for the Birmingham Canal Co. to be erected at Spon Lane.

[2]Richmond Water Works engine was an 18 in. 3 ft. one, ordered in 1777.

[3]Gilbert Meason of Wanlockhead, Dumfries.

[4]for Byker Colliery engine, Northumberland. This had a 52 in. cyl. and 8 ft. stroke.
[5]Jary of the Mine du Nord, Brittany.
[6]Pierre-Joseph Macquer, compiler of the Dictionnaire de Chymie which was translated by Keir.
[7]He purchased annuities in B & W engines thereby greatly assisting Boulton's liquidity.

DOCUMENT 68

To Boulton & Watt London 12th Novr 1776

Gent^n^

I have seen Mr Moore[1]. His supposd Improvement on the Steam Engine will keep very well till I see you. The misfortune at Bow afford great room for exaltation among your Enemies — Mr Moore has been told by severale that the Engine will never answer. I hope you will lose no time in putting this affair right as soon as may be — & determine not to leave it untill it has workd some Weeks without Interruption. M^r^ M informs Me that Gainsbro'[2] is dead — His Bro^r^ a painter has been offerd from Cornwall £1000 — for that Invention — that he is at a Loss how to act being no Judge — whether to sell it for that Sum or to take out a Patent.

The great *Sm-t-n*[3] (I hear by several) arraigns your Invention, says he has a better Engine to produce on the old Plan. This I think finishes what I call News.

Among all the Pretenders to Improvement may I be permitted to suggest a new Place for the Regulator. Mr Boltons scheme to annihilate Outer Cylinders Bottom & Nozzle put Me in train of thinking that kept Me awake all Night on Fryday — and the result was a regulator in the Piston woud remove all the foregoing incumbrances — and I think may be applied very readily — However I can not find any material Objection — but do conceive a great advantage therein. A Hint of this Sort is enough for you Gent^n^ to approve of or condemn — and when you have occasion to write let Me know what you think of a Plug regulator in that Situation.

I am very truly

Gent^n^ Y^r^ faithfull Fr^d^
& Servt
John Wilkinson

Box 20.

[1]Francis Moore took out two patents for engines: no. 921 in 1769, no. 933 in 1769, besides several others for ploughs, coaches and a device for shutting doors.
[2]Gainsborough, brother of the painter attempted to take out a patent for a steam engine but was opposed by Boulton.
[3]John Smeaton.

DOCUMENT 69

1780 Memorandums for Bersham (by James Watt)
Decr 4

1st That an inch and a half be left for Joints between the Screw holes of the upper flanch of the Cylinder, and the inside of the Bell mouth.

2nd That 2$\frac{3}{8}$ Inches be left for Joint from the inside of the Cylinder when bor'de to the near side of the Screw holes and thate $\frac{3}{4}$ inch be left solid on the out side of the Screw holes which will make the whole breadth of that Flanch about 4$\frac{1}{2}$ Inches from inside Cylinder —

3rd That when cast, the upper side of the Square pipe at Cyl^{r} top be $\frac{1}{4}$ inch under the level of flanch to allow for Shaving —

4th Let the Same be observed in inner bottom for the same reason.

5th To take Care that the face of the flanch of the Square pipes stands parallel to the axis of the Cylinder.

6th To make a Square Corner at the flanch of the lower ring of Cylinder at b.

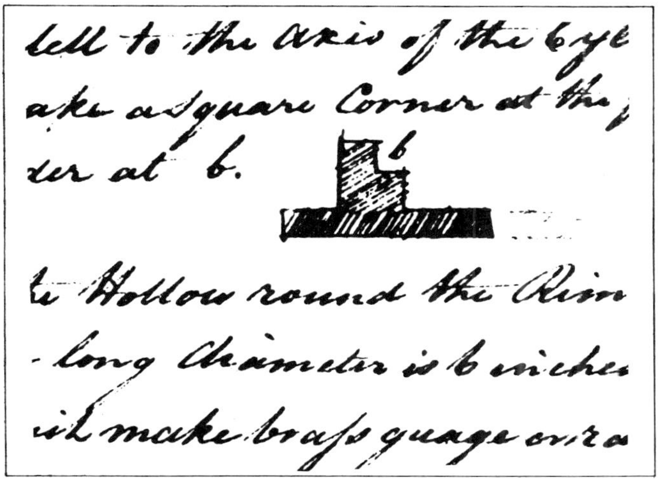
lell to the axis of the Cyl
ake a square Corner at the
der at b.
e Hollow round the Rim
long diameter is 6 inches
il make brass guage or ra

7th That the Hollow round the Rim of Piston be ¼ on an Elipse whose long diameter is 6 inches and the Short one 3 inches to which make brass guage or rather Steel one hardened.
8th That in a 58 Inch Cylinder the Rave rises 1½ inch above the Rim, and in other Cylinders in proportion to their diam^r^.
9th That in a 65 Inch Cylinder the arch of the Bottom and Piston rise 4 Inches above the Straight Line, also in others in proportion to their Diameters.
10 To bore 2 opposite holes through the Cone of the Piston 4 Inches above the outside of the bottom 1½ inches diam^r^ for a 65 or in proportion to the Diam^rs^ for others.
11 To Shave the joint of the Lid of Air Pump and Communicating box, and if not too expensive all the other joints of the Condenser.
12 To take care that the flanches of the Air pump foot and Eduction pipe foot Stand Square to their Joints.
13 That the Air Pump bucket descends so low that one inch of the Cylindrical part remains in the Barrell, not reckoning the Spiggot Part as Barrell, and 4 Snugs be cast in the inside of the Air pump foot to carry an Iron ring of half an Inch thickness, on which the bucket rests when down.
14 To make a Stem upwards as well as downwards for the Valves on Air Pump lid, and a socket for it above about 2 Inches long.
15 To put no holes in the Cylinder lid, the lower flanch of Do., the air pump lid nor in the flanch of hot water Pump where they pass over the square pipes or communicating Boxes.
16 To cast a Gland of double thickness 2½ Inches broad on the ring which screws the Eduction Pipe to its foot, where it passes over the box part and to put a Screw hole in each end of it.
17 To make the Slope above the Wide Parte of Clack door part of hot water Pump, about an inch longer Slope than it has in the drawings.
18 To round or Slope off the lower edge of Air pump bucket ½ an Inch perpendicular to ⅛ horizontal.
19 To take care that the Clack doors of the Eduction pipe foote and of the hot water Pump be rather rounding on their face.

J. Watt Blotting Book
Box 33, p. 65

DOCUMENT 70

To M. Boulton Cosgarne Jany 7th 1782

Dear Sir

I have none of yours today — I wrote to you on Saturday with drawings of the 5th Method of Rotatives, and inclosed I send the complete specification of that method. You should not send the drawings here untill that method be inserted and the parchment stampt if required — but no time should be lost in sending the draught of the specification here as our time of delivery approaches.

I send also inclosed Drawing for Mr Walkers[1] cylinder with Sett of Castings and some directions for the founder which if you think it proper to lett him execute his own should be immediately transmitted to him and I shall send drawing and directions for the Condenser next post, together with directions to Mr Henderson to draw the nozle.

I know not how to advise about M Walkers executing his own Castings. If he does them badly then the Engine will not come up to our agreement and we shall be the losers; granting that he does them well, then he will ask futher orders and we shall have J.W.[2] upon us, besides the more people we instruct in ye part of the business, the more able other people are to compete with us. Lastly he cannot execute them without putting himself to a very considerable expence which he will naturally expect to have paid him by orders from us. — On the other side, if we have them done by J.W.^{n}[2] he will be shewn a bad example as to prices and will grumble at it and possibly may offer to sell cheaper by which means we may be obliged to employ him nolus volens; again if executed by the Dale people[3] there is a risk that they may not be so well done as to be faultless and thence reflections may arise — Suppose we were to consent to his executing them if he pleases, but at the same time to tell him frankly that since you saw him, you, or we, had found it convenient or expedient to enter into a connection in the foundery way, which would put it out of our power in future to throw such orders in his way as otherwise might have been the case, at the same time we shall not restrict ourselves by our interest in that or any other concern as to prevent our employers from giving their orders for the pumps and heavy work where they please, and that in orders in his own neighbourhood It is very probable we might even employ him in the Engine part notwithstanding our connection in another foundery. And that after this declaration you leave it at his own

Option whether he will execute his own Goods or if we shall get them done for him — If we make agreement with J.W.^{n} about Bradley, I think we should take ¼ of it and we should be at liberty to order from the Dale company all the Engines they procure us orders for & if we should like Walkers people we might assign them also a province — I submitt the whole of the above to your discretion only be sure to write immediately and at same time try to get the agreement concluded & executed —

I have been today measuring Tingtang Engine for a new house which I must set about drawing immediately as they are ready to begin to build — Matters have gone on rather Brisker at Wheal Virgin[4] since the last brulée.

I write to Mr Hall to night informing him that I have sent the drawings to you and that you will advise Messrs Walker of our opinion about the casting them themselves.

With best respects to Mrs Boulton & compliments to Mr Henderson and all other friends I ever remain

Dear Sir

Yours sincerely

James Watt

The directions to the founder should be copied into our book and as the copy sent is ill written you had best have a fair one made to send but take care of errors, several of those which were in Mr Hendersons list for Messrs Colclough were errors of copying —

N.B. I have sent drawing for the Gudgeon to Mr Hall.

Perhaps it may be proper for you to write Mr Walker to meet you at Soho, you can talk over matters best face to face — and if he is very desirous of executing it himself I would grant it, for I would not pass my word for any body's work except J.W^{ns}

MB-J. Watt 3/4.

[1] Walkers of Rotherham, Ironfounders, this was a 40 in. 8 ft. blowing engine, ordered in 1781.

[2] John Wilkinson.

[3] The Coalbrookdale Co.

[4] Three engines were ordered for Wheal Virgin in 1780, (East 56 in. 9 ft., Old Wheal Maid 58 in. and Elvan 58 in.) and two in 1782 (West 52 in. 8 ft. and New Wheal Maid 50 in. 9 ft.)

DOCUMENT 71

To Boulton & Watt Truro 18th May 1793

Gentlemen,

Herewith you have an order drawings &c for an Engine of your double sort, and a Cylinder &c of the Single sort for the United Mines, particularized in the Enclosed. —

It is almost immaterial to me who is to have the Casting of these Engines, only this, that as Mr Wilkinsón gave us much trouble about Wheal Maid great Engine by retarding the Work, and Subjecting us to the expence of sending Mr Martyn to Chester in order to prevail on him to load the Materials on a Vessel we sent on purpose & which lay there a considerable time at our expence, and therefore I should not wish to be so Served again. What is given to the Dale Co will I presume reach this place much sooner than from Chester, especially if the War should continue until the Work is completed.

I am Gentlemen
Your most obed Servt
R.A. Daniells

I am sorry to add that in consequence of the high price of Coals and the fall in the Standard of Ores that the Consols are doing miserably, regularly losing at least a thousand pounds per Month, without a prospect of better success — we must therefore hope for some temporary aid from our friends —

I am respectfully Gentlemen
Your most obedient Servt
R.A. Daniells

After a long illness of ten weeks the medical people advise me change of air, in the course of my Tour I know not whether I may not hope for the pleasure of paying my respect to you —

Box 9

DOCUMENT 72

To James Watt Junr. Bir^{m}: Sep^{r}: 16th 1790

Dear James,

I have yours of 13th — When Gentlemen want plans of Engine houses they should first send us plans & sections of what is already built & apprise us of local circumstances, as the making plans 2 or 3 times over as we have to do too frequently is a great loss of time & drawback upon our business. However I do not mean that you should complain to Messrs Simpson & Co.[1] only thus you should advise others what is proper, & they should never be taught to expect plans by return of post, that is generally impossible & always inconvenient.

In respect to the 4 horse Engines they will cost nearly as follows, but it is only the first 2 articles that we have any degree of certainty of, the others vary much according to circumstances, and as we never erect Engines by contract we cannot speak experimentaly of them

4 Horse Engine Mettallic materials	£225
A Copper boiler about	40
The wooden framing & its Iron screws & Straps about	50
Engineers &c putting the Engine together about	40
Freight & Carriage of Metal Materials from hence	
Engine house, Chimney, & boiler seating	
	355

The above Estimate comprehends no part of the Millwork nor any thing beyond the first rotative Shaft. The Engine house & Carriages we cannot estimate, but the latter cannot amount to any great Sum on a few tons of Goods — Our premium on a four horse Engine is £140 in full to be paid in 6 months after the Engine is finished — We do not now make these small Engines for yearly premiums as it does not answer to us — It will consume about 50 pounds of good Coals per hour & require about 10 gallons Cold water per minute.

A ten horse Engine, Metal Materials	£320
Copper boiler about	90
Wooden framing & its Iron Works about	78
Engineers &c erecting Engine about	53
	541

Coals about 1 Cwt p^{r} hour, Injection water about 25 Gallons per minute — But premium £50 per year until Midsummer 1800 — It would be too tedious & unnecessary to give you prices of other sizes, but it will not err much to deduce the prices of the Metal Materials & boilers from these given you, as mean proportionals, but beyond 12 horses the rule will not answer as they alter their Stroke — The wooden framing & Cost of erection must be taken seperately as they do not rise in the same ratio & have more of guesswork in them especially the erection.

The best way in general is to refer gentlemen to us as then we have their letters in case of dispute — I have mentioned before that we never erect engines by the piece as the erection is generaly where we cannot superintend it. But if wanted we should undertake the Metal Materials excepting the boiler, at the prices stated in this letter, but we should be apprized of that in time, that we might deduce more accurately what we were to furnish. Only the working barrel & Clack of the Cold water pump is comprehended, and the Governor is a seperate charge.

I received the books in good order & shall give you credit for the amount. I have heard nothing yet of Mr Hamiltons fossils, but suppose he must have got them — Matt Boulton has been very ill of an ague but is now recovered & I hope on his way home. Streiber is here but I seldom see him — He says he is about returning to Germany.

Jessy[2] is greatly better from her Journey to Cheltenham The rest of the family are well & join in love to you. B & W received your letter today & will reply to Mr Simpson — Mr Reynolds of Ketley was here today he has got a new Patent Engine[3] which is another addition to the encroachments upon ours & Wilkinson has another! If we are not ruined it is not for want of envy & competition, we must try our fate at law with the Hornblowers this winter, I much fear the success — I hope you are recovered from your illness & remain

Your Affect^te^ father
James Watt

We move into the new house next week

[1]Cotton spinners of Manchester. They ordered a 6 h.p. engine.
[2]Jessy Watt.
[3]That of his protégé Adam Heslop.

DOCUMENT 73

To Boulton & Watt. London, April 7th 1795.

Gentlemen,

I have lately received yours of the 4th Instant — I dont know how it is but there seems to be a strange kettle of fish with this same, John Wilkinson Esquire.

You know I never gave him any orders or commission whatever for works of any sort for the New River Company[1] — We have received things, coming I believe from his establishments; but they were always received as from you in fulfillment of those Orders given to you and you only.

On this principle I have demanded & still we are in want of two piston Rods, for which you have received payment in a bill or account at large, of the whole Engine. I know nothing, and am not to know any thing of J. Wilkinson or any other person; and therefore shall have nothing to do with him.

Besides that matter, — There was another under the hammer. We stood in need of a new Bucket & Clack, for the old Engine, I explained it to Mr Watt here. The present one is almost worne out and ready to give way. Much time is elapsed, and I am referred to J. Wilkinson — Surely, this is not furnishing those Engines with the means of working, whose going affords you Revenue. — That is an object to be paid for. The former, (above mentioned) is paid for — I expect both immediately, and without loss of time — and that you will do your own business yourselves, and not leaving us to the fractiousness & correspondence of persons with whom we have not, nor wish to have any concern. —

I am your very humble
Servant Robert Mylne.

Box 3M

[1]Robert Mylne was engineer to the New River Co. The first engine, a 32 in. 8 ft. was ordered in 1786. This was followed by another of the same dimensions in 1794. Further engines were ordered from 1811 onwards.

Section F

The engine business – customer complaints

The engine business – customer complaints

Documents 74-82

For customers unable to diagnose the cause of a technical fault in their engine the difficulties were legion. It was less easy to persuade Boulton and Watt that they had a responsibility to send a mechanic for the suppliers were able to allude to the possible cause being mismanagement. Moreover, there was a shortage of skilled mechanics and they were generally fully occupied. A mechanic could only be transferred from another job at the risk of neglecting the other customer[79].

For customers in areas remote from good iron founders there were additional difficulties in obtaining replacement parts. Sun and planet wheels were the most liable to break[79] but sometimes an engine seemed to have more numerous faults[78]. The quality of the iron could be poor, there could be errors in the pattern or weaknesses in the design.

It was supposed that once Boulton and Watt produced the majority of their own parts quality control would be easier. But customer complaints continued, if somewhat abated,[80], although in the case of Josiah Spode, the potter, Watt, jnr., had doubts about the honesty of some of the points, suspecting Trevithick of attempting to persuade Spode to replace the Boulton and Watt engine by one of his own high pressure engines[82].

DOCUMENT 74

To James Watt St. Clements Wells, Lothian, 10th Decr 1787.

Sir,

Hearing that you was in this Country, we take this opportunity of informing you that the Engine[1] erected here does not work with that sweetness that others of the same Construction does, that wee have seen — Viz the rotative wheells or that one of them that is fixed to the Connecting rods in the revolution rubs the Beam that the fly wheell moves on, But particularly when useing the old phrase when the beam returns into the house that is when the piston descends to the bottom of the Cylinder, She always makes a great Clank, which puts us in fears that some parts of the machinerie is not properly put together, or that this strain may be the means of hurting the other parts of the Engine in time.

And you being in this Country we Judge it necessary that you will Call here to Examine her, when if wrong you will then give the directions for putting her to rights,

We will expect your reply in Course and if you Can name the day when you Could be here so as we may have the Engine working for your inspection. We are

Sir
Yr very Obed^{t} Serv^{ts}
For Aitchesons & Brown[2]
Will Leitch.

Box 4A

[1]a 14 h.p. one.
[2]distillers.

DOCUMENT 75

To James Watt St. Clements Wells, Lothian, 21st Feby 1789.

Sir,

Our Mr Leitch wrote your Co^{y} a few days ago Informing you that both pairs of the Revolving wheels[1] was broke. The first put on lasted on the 10th Inst^{t} the Second pair was Broke to pieces in 10 days after being at Worke. Since writting your Co^{y} has seen Robt Muir who has put in an Iron Teeth into the Wheel that was broke & Sett the Engine to worke again but She is going so very ill has been Oblidgd to Stop her again to see & finde out the Cause. We are sorry to be under the necessity of troubling you but none of us here understanding the Engine so well as we could wish & finding her go so often wrong of late makes me conclude that Some part of her must be in a worse Condition than we Can find out as the Wheel that is mendd may not last long and at this Season we cant Admit of the Engine Standing. Would wish you forward another pair of these Wheels the very first opertunity. Robt Muir tells me you are now Constructing these Wheels in another way beg you to Send us what you think is best. As they may be long by the Way in Comming from your place to this if you think they could be sooner got and as good in this Country as from your place will thank you to Mention who is the properest person to Cast them here my reason for being so very pressing is that should our Engine stop for any Considerable time our whole works will be at a Stand for want of Grist and we have the Duty to pay on our Stills work or not. Your Answer as soon as possible will much Oblidge.

Sir your Most Obed^{t}
Humble Ser^{ts}
Wm Aitcheson

Box 4 A

[1]Sun and planet wheels.

DOCUMENT 76

To James Watt

Wh^{e} Chapel Road, London.
30 Jan. 1787

S^{r}

I am exceedinly sorry to send you p^{r} the Birmingham coach so bad a sample of your own work, and likewise to inform you how much we have been distressed by the breaking of the main Gudgeon of our Engine[1], which has stopt our manufactory for several days, by which we sustain great loss, besides damaging the Engine in many places; I would not have troubled you with these Iron barrs, If I thought a simple narration of the fact could have been believd; but to the eternal disgrace of the manufactorer such is the heart of our gudgeon, and as such I send it to you that you may proceed against him as you think proper, but in my opinion speaking at large pro bono publico such a rascal should not go unpunished. I assure you S^{r} your Engine has greatly fallen in our estimation since our suffering by this notorious neglect, which may happen again at an early period; that we are oblidg'd to have recourse to another expensive engine to regain that confidence we have lost in this — I hope we shall not trouble you with another account of accidents, but I assure you S^{r} the consequence of stopping a Corn Distillery with 2000 Hogs^{ds} in the yard is so great, that we are not without our daily fears — We receiv'd yours of the 27th Ins^{t} and you may safely rely on the friendship, and good wishes of S^{r}

Your Hble Ser^{t}
S D Liptrap & Son[2]

Boz 4 L

[1] a 10 h.p. one.
[2] distillers.

DOCUMENT 77

To Boulton & Watt

Bog Mine[1] Shropshire.
27 Octr 92.

Gent^n

It is very disagreable only to have occasion to Complain, But so it is after so long confinement, that had I not posted home the day I did, the Mine would have been drove to a total Standstill from the Engine[2] brakeing so often in my absence then and Since more than 20 times, which has stop'd me from one day to a week at a time so that I have not work'd the Bottoms this 6 Months, and haveing them Clear for 3 days at a time for near 12 Mo^s. I do not now intend to attempt it till the Rattle Traps are removed from the Surface and the pumps drop'd in the New Shafts — within two months I shall be ready. I mean without disturbing the present to put the Capstin Shears and all I can upon the Surface in readiness being most preparing or is prepar'd.
The Cisterns are ready large and Comodious and the Ground is fine to but I mean to put them back and bring no more forwards than is quite necessary into the Shafts. There is a Ladder Shaft between the Engine and drawing Shaft so that at every Stave the party may step from one to the other.

The upper lift is to be a forceing lift of 20 fathoms. The middle lift 20 fathom a lifting Fire. This is deaper than our present Bottoms below which I mean to Compleat 16 fathom before I cut the Vein and when I do; as fast as the Engine can draw the Vein will be dry.

I don't intend a foot of Timber or a pound of Iron in the present shaft shall be used in the New. But as everything can be fixed dry till I cut the Vein I mean immediately to fix the upper lift in its Cirstan & the next lift in its Cirstan. For these I have nearly enough pumps on the Surface. Then take up the present upper Fire and make it the lowest lift in the New shaft. Then Cut through the wall of the Vein and let in the Water. This will be an immence expence to me as they are heavy things but I will be clear this time.

The Cylinder is 45 Inch From the Centre of the Axces. To the Center of the Cylinder is 14 foot from the Centre of the Axes, To ye extent of the Lever Arch 12 foot 6 Inch and the Working Barrels ⅛ under 12 Inch. I now and then may want the house lift of 6 Inch 48 Yards but that Seldom and I think I can provide against it tho would not be without it. Pray how deep will she go at 10 or

12 Strokes a Minute — chiefly 8 or 9 — by your Calculation.

I have put the Chief of my hands on Dead Work such as Sinking Shafts, driving Drifts to a New Capital Vein (in the whole to Grass); in five or 6 places directing of Water under ground this being the time to find out and cure defects. And have Erected 3 New Whyms in Short stand. Prepared to make a figure when I can begin tho I know not what to think, as the almost Constant Brakeing (even when G. Taylor was here it was too often but he was Ready at Curing it) makes me think there is many Internal defects which should at this time be compleatly examined could I get a proper hand to assist me. The Working Gear is so Complexed that it is too much for common Capacities. In my opinion if Simplified it would be much better.

I have now with me a man once Servt to you by the name of Price, Recommended by G. Taylor and others but I think him unequal to the Task of managing these or undertaking what will be wanted as I question his abilities as a Smith tho he Receives from me 25^{S}/- p^{r} week.

I waited upon Mr Wilkinson at Bersham whom I expected would say on parting all that has passed, I will take back the Iron now comming useless, and give you what you want for your alteration of *good* Iron but on my hinting it to him he said it was foreign Iron Bought on purpose for me tho I can prove it English and such as they have given over the use of. Thus robbed of two years Rest by being brought to the Brink of Ruin and Ten to one that I avoided it, and suffering at a loss, will be better spoke to some Months hence when the Mine will loudly proclame what she would have done but for the Various imperfections of the Machinery. (For I defy the world to blame me); I am now in a delemma with all I want before me — The tempers that has been shewn makes me doubt But I appeal as an honest Man and Gent^{n} and ask you as such what I should do? Your answer fixes my proceedings I want a proper hand to put all things right-a first Rate not a 3^{d} or 4th. The time wanted will be Short. I have got a hand from Cornwall at the Timber works seems Clever. And that nothing shall be wanted have a Shop in the Timber yard. Spanish Swedish & other Iron, and engaged good Smiths at Shrewsbury as the Bog is not so comfortable place tho I now live there. Thus cost what it will I will endevour to be at Home somehow this alteration and send the things as near finished as possible.

I was in hopes of a Conference on my Journey but was oblig'd to drive home and meant to settle.

Now haveing laid before you Candidly my Situation I ask for your Advice and Call for your Assistance. Say openly what I may expect and the sooner the more you will oblige me as some way I must have the business done not to Ruin but Render easy the fortune of

Y^{r} Ob^{t} Servt
J. Weston.

Box 4.

---

[1]a lead mine.
[2]This presumably refers to their engine no. 2, a 45 in. 8 ft. pumping engine ordered in 1789. Engine no. 1, with a 26 in. cylinder was ordered in 1777.

DOCUMENT 78

To John Southern Cockshead Coll^{y} Staffordshire, Jany 11th 1795

Sir

I received your letter of the 13th Decr and when you write to me again, please to direct to the Care of M^{r} Ra Hales Cobridge and it is most probable I shall get the letter a week Sooner —

The Great Engine[1] hath frequently been out of order and Stopped us Sadly. The first thing we had broke was the lower Great cast pin of the parallel motion and the Great perpendicular link also the two little Pins at the top & Bottom of the little perpendicular links and bended the piston rod very badly all at same timbe. If remember right the next thing we had broken was the Great Cast pin that Carries the inner pump rods from no other Cause but being not able to Carry the weight. A little time after this the air pump top broke across the middle, after that the Cast gudgeon that goeth through the V bob Shaft broke, Also in the Course of last summer we have had four pump rod Joints failed though they were approved of by George Taylor and made as Strong as your Selves advised, also some small screw pins that hold the plumber blocks on at the inner end of the Beam. The Last of failures was the Great Cast pin at the top of the Great perpendicular links which broke and Stoped us near 90 Hours at a time when we had 9 Strokes p^{r} minuet for the Engine to draw without the least intermission. It lasted the Engine a longtime to get the water out again you may think. This happened at a time when Coals were most wanted and when we had entered into various engagements has been a great loss and damage and ruined us for this winter, on the whole the loss by the frequent failiers has not in my opinion been less than 1000£.

We are Got rid of most of the Cast pins that were put up at first and with your leave will be rid of them all as they are not to be trusted to. We can make enough Iron ones very easy and they will hold. Besides to appearance the wrought Iron ones will save the Cost of them Selves in Oil and Brass's in a year. The Cast Iron pins Seemed to be made of bad metal that is Soft and weak and were the Brass's wore it seemed freckled & rough and all full of pin holes. The Engine house Shakes Sadly it rocks Backwards & forwards. The Cause of that is when the inner end Bucket sets to, it gives the house a pluck up hill or outwards. There seems a fault in the plan of the house. That is the window and Steam pipe holes in the Second floor and the window in the upper floor being perpendicular over each other it makes a weak place in the house it cracks Sadly though we have Screwed the Cylender Beams and fastened the Spring beams as well as they Can be fastened; to prevent it rocking as much as posisble. We have put a Cateract to the inner lift to keep it from returning out so quick & hastily which we Can perceive to ease it a great deal, but it hath been lately that the water would not admit of the Engine going with a Cataract.

I have Got the Deeds for the little Engine in my possession but have not got all the parties to Execute them yet. I will Get them Executed as soon as I conveniently Can but what the parties will Say to the frequent failiers of the Engine I Cannot Say. I must leave that to them and Mess[rs] Boulton & Watts to Settle they Cannot expect to be paid unless they perform their Contract.

I am sir

your most Obed[t] & very Hum[e] Serv[t]

John Martin

PS I had like to have forgot to have informed you about the peston Corbe being broke and hath been twelve months or more but we have a part made-a wrought Iron one which will nearly be finished today. I should have answered your letter Sooner but have been so Constantly engaged about the Engine that I have not Sooner had time to do it.

Box 6C

---

[1]Two engines were ordered in 1793. One was a 49 in. 8 ft. pumping engine, the other a 6 h.p. winding engine. This complaint refers to the former.

DOCUMENT 79

To Boulton & Watt. Stockport, 17 June 1793.

Gent^n^

We have just had a bad misfortune with our Engine[1] all on a sudden the Connecting link between the Sun & Planet wheels, broke, and the Planett wheel flew off and has done terrible execution. Shall be obliged to you to forward with all possible dispatch by the first fly waggon the necessarys to set us going again. The planet wheel is broke and quite useless the shaft of the fly wheel is broke in two pieces. The link was not sound, half the substance we discovered when broke was hollow and disunited before. Pray let the next be sound. Would you recommend us to have two links one on each side the planet wheel if so please to send the necessarys for such an alteration. What adds to our misfortune is Mr. Lawson being left the Countery and we suppose it is not possible to be supplyed with the necessarys here. I hope you will not delay us one minute as this Engine has been a little unfortunate from the first, and all our hands are standing for the time till we can replace what is wanting, say a new planet wheel, fly wheel shaft & connection link or if you recommend it, two, and necessarys for fixing the same. I remain

Gent^n^ Your Most Obed^t^ Servt

J. Bury.[2]

Box 6B (12)

[1] an 8 h.p. one.

[2] a cotton spinner.

DOCUMENT 80

To Josiah Spode[1] esq^r
Stoke.

Soho 6th October 1803.

Sir,

We have received your favour of the 3^d Inst^t and are sorry to learn that you have derived so little benefit from your excursion to Parkgate — In reply to the remarks of your Engineman, we can do little more than repeat what we have already fully expressed both verbally and by letter, that the source of your difficulties lies in the mode of connecting the Engine[2] with the Millwork; and until this is remedied, so as to enable the Engine to form its Vacuum under a lighter load it will never start pleasantly. Either of the methods pointed out in our letters of 19th July & 7th Ult^o would easily & compleatly attain this object, and as they are such as any common Millwright might execute, we do not see the reason why they have not been carried into effect. If independant of this cause, there exist other defects in the Engine, we should be inclined to ascribe them entirely to the mismanagement of the person to whose care it is confided, as we have numbers at work in different parts of the Kingdom upon exactly the same construction, which give entire satisfaction and are subject to none of the defects complained of in yours, although they have not all, like yours, had the benefit of every successive improvement that has been made subsequently to their first Erection. The facts stated in your letter of the great friction, heat & wear of the brasses, can only be accounted for, by supposing the cutters and pins to be driven and screwed too hard, a fault against which he has been repeatedly & particularly cautioned. If the piston is packed once per week <u>properly,</u> that will be found sufficient, & the consumption of grease will principally depend upon the more or less perfect manner in which that is done. As to Oil, if not wasted, a very small quantity should suffice. — Under similar circumstances we have frequently found a change of Engineman to effect a compleat cure for all complaints.

Our Mr Murdock who is now in Manchester will probably call upon you on his return, and if you wish it, may give your Millwright instructions upon the spot for effecting the proposed alteration. Should this not be done, we should recommend you, as the next alternative, to dispose of the Engine, and if Authorized by you shall use our endeavours to find you a customer. In the mean time we shall be much obliged by your remittance of the amount of this & of the Colliery Engine[3], the first being nearly half a year, and the latter a quarter past due and the delay attended with peculiar inconvenience at the present moment.

We remain very respectfully Sir
Your obed^{t} Ser^{ts}
(signed) Boulton Watt & Co.

Box 2 W

---

[1]pottery manufacturer.

[2]a 30 h.p. rotative engine ordered in 1801. A 6 h.p. engine was ordered in 1802 but the problems concerned engine no. 1.

[3]Fenton Park Colliery in which Spode was a partner. Two B & W engines were erected: no. 1, a 30 in. 8 ft. pumping engine was ordered in 1791; no. 2, a 48 in. 8 ft. pumping engine was ordered in 1802.

DOCUMENT 81

To Boulton Watt & Co Wolseley Bridge 9th Oct. 1803.

Gent^{n}

Your letter date the 6th Ins^{t} is sent me to this place, where I am to remain a few days under the care of an eminent Surgeon, having brought on a Stricture by the Cavalry exercise — I am thank God much better and expect to be home on Thursday if not sooner, at which time I will remit you a bill for the amount of my Engine at Stoke, and I will thank you to adress a Letter to William Heath Agent to the Fenton Park Colliery desiring him to inform the proprietors you expect immediate payment, (<u>not as if I requested you to write</u>) and on my return I will call a meeting and put it in a train of payment. I observe the contents of your letter respecting my Engine at Stoke, I don't know immediately from what cause, but the Engine hitherto has been very unfortunate, it has scarcely worked Six hours without something breaking or being out of order, and has given a great deal of work to our Smith & Carpenters. Your people seem to blame the man who attends it — if he dont do his duty its want of experience, & not carelessness or intentional — as I believe him to be a very honest man and willing to do his best. I understand Mr Murdock has called and examined it, perhaps he can inform you more particulars till I am acquainted with having been a good deal from home. I am very desirous either of the methods you point out should be adopted, & leave it to your better judgment which — and will thank you to let it be done as soon as you possibly can, and at the same time if you could send me an experienced Engineman to look after it for a Month, that will give a proof whether the fault has been in my Man, I will pay any expence attendant, as the Engine in its present state is of no use to me. If the Engine should be found not to answer the purposes its intended for, what would be the expence of one to work on your other principle with a beam — say a eight or ten horse power, and could it be placed in the same building and to attach to the same Millwork, and what do you suppose the probable loss would be on the parting with the present one, and how soon could you make me one of the above description. As you must be well aware — one of some sort I must have in its place — if this should not be made to work regular, your answer will oblige

Gent^{n} your Obed^{t} H^{ble} Servt
Josiah Spode

Box 2W

DOCUMENT 82

To Gregory Watt — 10 Oct 1803

To this letter[1] we have sent no answer, as we are waiting for the promised remittance, without which, we are not inclined to give ourselves further trouble. Mr Heath was written to yesterday and requested to lay our letter before the Proprietors.

You may probably recollect that the last named person has been a great advocate for Trevithicks Engines and has most probably been the cause of the misfortunes that have attended ours, which it appears to be the object to get replaced by a Dense Steam one. He was sent over by Spode to examine Trevithicks Engine & at that time received a bribe of 50 Guineas from the said Engineer, which it is now incumbent upon him to get repaid, and this he proposes to do by making his employer (Mr Spode) use these Engines and pay Trevithick a premium of £4 or 6 for each Horse power.

One of these Engines he has already erected. I saw it in its disjointed state & Murdock saw it at work. It is exactly the same as the one made by my father 38 years ago & as the small carriage Engine made by Murdock in Cornwall in 1785 or 86, which was there seen by Andrew Vivian and probably by Trevithick also, from which it has no doubt been stolen. The latter's specification is a most unintelligible performance and if Mr Spode chuses to make these Engines, you may inform him that he may save himself the expence of paying any Premium, as they are certainly *publici juria* ever since the expiration of my father's act of Parliament.

The extreme danger attending the use of such an Agent as dense steam has prevented us from attempting any thing with it, and the recent explosion at Greenwich has justified our precaution. I shall send you some particulars of it tomorrow which may be of service to Mr Spode and to his neighbour.

His last letter is more to the point than any we have received, and we do not feel averse to complying with the conditions. We have held another consultation upon the two methods and that of the friction wheel is preferred, which renders it essentially necessary either that we should have the pattern of the Bevil wheel here, or an exact drawing of it and of the Shaft, but we should much prefer the former.

We can see no objection to Mr Spodes writing to the person who has the pattern to desire it to be sent to us, and we think if he had been very earnest in the business, he would have done this upon receipt of our letter of 7 Sepr without further prompting.

Thos Pearson is at present erecting the Colliery Engine and we believe will soon have finished it.

With kind regards to M^r & Mrs Hamilton I am Dr Gregory

Yours afftely

Watt Junr.

Box 2 W.

[1]The letter from Spode [81]

Section G

The engine business – profits and premiums

The engine business – profits and premiums

Documents 83-90

During the subcontracting phase, the main element of Boulton and Watt's profit was the annual premium charged for licence to operate the engine. The premium was payable in half-yearly instalments and, for reciprocating engines, was computed as a proportion of the saving in coal consumption of Watt's engine compared with an improved atmospheric engine. Obviously the potential profits accruing to Boulton and Watt were higher in areas of high coal prices, thus we find that estimates of coal savings figure prominently in Watt's private calculations[83]. For this reason Boulton and Watt were not keen to sell large pumping engines to coalfields where, by comparison with the Cornish copper mines where coal prices were high, the profits would be low. Equally it was not generally to the coal owner's advantage to purchase a Watt engine since the higher coal consumption of an atmospheric engine was of little account. Only on the north eastern coalfield where mines were deeper than in other coalfields did the Watt pumping engine have any great advantage over Newcomen's.

So remunerative were the large pumping engine premiums[85] that Boulton, in a period of minor liquidity crisis, sold annuities based on the income to be derived from certain engines in order to raise additional capital for his other business ventures[84].

For rotative engines the system of charging was different, being computed on a standard per horsepower rate per annum. Boulton and Watt assumed constant working, but it became clear that a number of purchasers either did not use the engine regularly, requiring it for supplementary purposes only perhaps, or were not working the engine to full capacity but were having to pay a premium at the full rate[87].

From about 1790 onwards customers were offered the alternative of paying an annual sum or paying a compounded premium for the term of the patent at a discount[89]. Smaller manufacturers generally opted for the former.

DOCUMENT 83

Messrs. Waterhouse Wainhouse & Comp^{y}[1] Hallifax — 1784

Want an Engine = to the power of 2 pair of corn mill stones to drive a cold calendar and a mangle, with a logwood machine for both rasping & chipping — Mr. Jonathan Crowther Millwright at Leeds is their millwright — Suppose each pair corn mill stones = to 6 horses then both pairs = 12 horses = 396000^{lb} to 1 foot high — say 400000 say double engine 8 feet stroke both ways & 20 strokes p^{r} minute 7^{lb} on y^{e} inch.
400000 ÷ 8 = 50000 ÷ 20 Str = 2500 ÷ 7^{lb} = 328 sq´ inches. = 21½ dia^{r} cyl^{r} = 140^{lb} coals pr hour at 20 str p^{r} minute. com^{n} engine = 420^{lbs} p^{r} hour. total savings = 280^{lb} B & W part = 93^{lb} p^{r} hour × 12 hours = 1116^{lb} p^{r} day = ½ ton which at 8/- p^{r} ton = in 300 days = £60 p^{r} annum. cost of such Engine cannot be ascertained but suppose from £400 to £450 exclusive of the house & mill work.

Messrs. Staples and Caufield Lessees of Drumglass Colliery[2] want an Engine with Cyl^{r} 36 inches dia^{r} — Their Coals are sold at the pits at 46 Winchester bushels measure at 6/8 supposing each bushel to weight 4½ stones, the s^{d} quantity weighs 25⅞ Cwts of merchantable coal = to 5/2 p^{r} 20 cwt — and their small coals are therefore worth about 4/- sterling p^{r} ton. at 12 strokes p^{r} minute the Engine would burn 235 pounds each p^{r} hour & the Common Engine 705 pounds p^{r} hour, total savings would be 470^{lb} of which B & W part 156^{lb} × by 8 hours = 1248^{lb} = 2⁴⁄₁₀ p^{r} day × 365 = £40.3 p^{r} year — taking it by Messrs. Colcloughs Engine, wch is a 32 for which he pays us £80 p^{r} annum this would be £100 but as Messrs. Colcloughs Coals are 8/- p^{r} ton & Messrs. Staples only 4/- it would then come to £50 p^{r} annum — The Cyl^{r} was afterwards altered to 32 inch. ordered Jany 28th 1785.

J. Watt blotting Book, p. 91 — Box 53

---

[1]They did not purchase an engine.
[2]Ireland. Engine ordered through the engineer William Chapman.

DOCUMENT 84

To C. Vere[1] April 1779

D^{r} S^{r}

I rec^{d} ye fav^{r} of your letter dated 27 March. I am glad to find that you intend to Visit Astley in May where I hope to have the pleasure of attending you & the satisfaction of seeing all your affairs at that place finaly settled as Nothing new hath occurd since my last at Astley & as I hope to see you in a few days I will wave that subject at present & go to one far less agreeable.

I have just looked into our acc^{t} with Birch Lane House & find this day that your acceptances & advance is ab^{t} 10700 over & above our remittances.

By which it appears to have lately increased instead of diminishing as it ought to have done the reason of wch is partly owing to our haveing been too sure of receiving from one Gent^{n} 6000 wch was promised in Jany last & to our M^{r} F[2] haveing been too free in paying off our Xmas Ballances, & likewise to the great scarcity of Money in ye Country & to the Exchange being unfavourable foreigners, but neither these or any other excesses I know are admittable at a Banker's shop. I sent Chase water deeds into Yorkshire to be executed but they did not arive there untill 2 days after their Quarterly Meeting & now Mr Fenton writes me that he cant execute them untill their next meeting wch will be tow^{ds} ye end of April & then he will either bring them or send them. So soon as I receive them I shall have it in my power wth your consent to receive the afores^{d} £6000 & I am not without hopes of receiving so much more in the course of the summer as will discharge ye Greatest part of the sum which your house hath been pleased to Cred^{t} us. The Value of your Security is very much increased since it was made for besides all those Engines wch were specifyd in the deed being now actually at work & most of ye agreements executed, We are erecting as many others (wch will be at work this summer) as will increase our annuitys to nearly double what they now are & consequently the Value of your Security. I therefore hope that neither you nor Mr Low nor Mr Will^{ms} will be so severe with us as to bring on any (interruption)[3] in our affairs which otherwise will in the course of this summer be in a comfortable train. I think I can find purchasers[4] for the following Engines at the prises annexed. I hope Mr Matthews hath rece^{d} some money of ours which we expected he w^{d} be able to do about this time & if he hath he will pay to y^{r} house

700 on our acc^{t} & I hope in the space of 7 or 8 days you will not be a farthing in advance as we shall be able to remit you a further sum in Cash this Week.

We had used to be supplyd with a little Cash from Webster in a regular way but from the failure of Heath of derby I know all the Bankers in that neighbourhood have been inconvenienced & thereby he hath been obliged to send us Bills instead of cash but that will be discontinued untill his inconveniences have subsided. I find M^{r} Z Walker was under the necessity from Mr Fothergills absence of signing a letter to your house for us. I beg to inform you that he is our principal & Confidential Clark & request you will in all such matters consider his signature as our own.

(M. Boulton)

Chase ∇	7000	
Chelsea & Shad	2500	
J Scott	1220	
Reynold of Ketley	1000	
Hawksbury	1500	wch they offer but we contend for more
Birm^{g} Navig^{n}	210	
Ld Gower	150	
Chapman Newcastle	770	
Hull	300	
Lady dumfries in Scotld	1200	
Wrens nest Forge	500	
	1- 6350	

MB — V/82

[1]of Lowe, Vere, Williams & Jennings, M. Boulton's bankers.
[2]Fothergill, Boulton's partner in the toy business.
[3]illegible.
[4]of annuities.

DOCUMENT 85

To Proprietors of North Downs[1] 1784

Gent^{n}

Whereas an Article of agreement haveing been made between you on the one part & my self & partner James Watt on the other for the Erecting of 4 Fire Engines upon N D & other Mines therewith controled. It is therein agreed that in consideration & lieu of our saving of fuel you have covenanted to pay to us the several sums after the following manner

£		Fath
1660	p^{r} year untill the average depth of the Mine	48
1870	p^{r} year untill ye average depth is	56
2080	p^{r}	64
2290	p^{r} y^{r}	72
2500	p^{r} y^{r} untill the extent of ye power	84

But in consideration of the difficulties ariseing from the present ruinous State of the said Mines & the great expence that will attend the putting them into proper course of working & of the Candor & fairness which You have shewn in Settling of the above terms — I do hereby promise for my self & s^{d} partner that untill the said Mines are in course of working to profit that we will abate in the first station 60 p^{r} y^{r} In the 2^{d} — 80 — 4th 90 & ye fifth 100£ p^{r} year. Wishing you success in your undertaking.

(M. Boulton)

Box 9

[1]Copper mine, Cornwall. A 48 in. 8 ft. engine was ordered in 1784 and later countermanded.

DOCUMENT 86

Scroll of Letter to proprietors of N. Downs — 1784.

If the Advent[rs] shall cease or discontinue to work the Engines or any of them then the paym[ts] so convenanted & agreed to be made to B & W cease & be discontinued.

The paym[ts] which are covenanted to be made to B W are 1600-1800 &c & therefore according to these Words the whole paym[t] will cease when any one of the Engines cease to work which is certainly not y[e] intention of y[e] parties — but if any one of the Engines cease to work & then it is the intention of the parties that such a portion of the aforesaid annual paym[ts] shall cease as bears the same proportion to the whole sum as the power of such Engine or Engines so stoped bears to the whole power of the four afores[d] Engines — (or you may specify in the Article the proportion[1] value of each Engine.

(M. Boulton)

Box 9

DOCUMENT 87

To Boulton & Watt West Bromwich 2 Aug[t] 1791

Gent[n]

It has not been in our power to work your Engine[1] to Proffit for more than 2 Years, & therefore have not Work'd it at all; we find that were you to Permit us to work it to draw Water, without any molument to yourselfs, it wou'd not Answer, we are Inclin'd to Work a Hammer with it, & if you'l Put a Moderate Charge on it by the ¼ or by the year will work your Way, otherwise we must be under the Necessity of Working it the Ordinary Way; we beg your Answer, & hope your terms will be Moderate, otherwise its not in our Power to Work your Way. I remain for self & M[r] Wright

Gent[n]

Your very hble Serv[t]
Rich[d] Jesson[2]

Box 4 W

[1]a 36 in. cyl. 8 ft. stroke engine at Wren's Nest, Salop, ordered in 1778.
[2]of Wright & Jesson, Ironfounders. This did not however prevent the partners from ordering two more B & W engines for their works at Barnet Leasowe, Salop.

DOCUMENT 88

To Boulton & Watt West Bromwich 23 July 1792

Gent[n]

The terms you gave us of our Intended Engine much Surprised us; we did not expect you wou'd have requir'd more than for that in Shropsh[e] as call ourselves Customers, & most likely you will get more by our recommendation than by us.

The calculation you gave us, was when your Engine ran fine, & for an Engine made in former times; since which there is a great Improvement in the Common Engine. We are Led to believe by those that never deceiv'd us, that a 48 Inch Cylinder in the Com[n] way will not take more coals in 24 Hours than 8 Tons. We also know experimentally that your Engine will take ⅓ or a ¼ more than you calculate, owing to our not being able to keep it in that compleat order that you do; our Shropshire Engine cost us £1200, we are led to think that one in the Comon[n] way will not cost more than about half that Sum, & this can be converted to Work your Way, at the expiration of your Patant. For all these Circumstances & Planted as we shall be on the Coal Pit bank; it is not our Intrest to accept of your Proposals. We are certainly Inclin'd to give you the preforance Notwithstanding we Sink £500 in your Engine; but did not expect you wou'd ask more for this than that in Shropsh[e] waiting your Speedy reply we remain
for M[r] Wright[1]
& Self Gent[n]

Your very hble Serv[t]
Rich[d] Jesson

Box 4 W

[1]They did not order a B & W engine for their West Bromwich works.

DOCUMENT 89

To Boulton & Watt — Leeds 4th Octob.r 1792

Gentlemen,

We are favour'd with yours of the 29th ult: by wch we understand that the parts for our intended Engine[1] may be ready to be sent from you by the latter end of May: it would oblige & serve us provided they could be ready at an earlier period.

We would rather wish to pay annually £80 than the Sum of £400 in lieu of that Annuity.

No person has any concern with us in Trade; the parties to any agreement with you will be Ths Coupland of Potternewton in the Parish of Leeds, & Thos Wilkinson of Leeds, Distillers. We are respectfully

Gentlemen

Your hble Servts

Coupland & Wilkinson

Box 6 C

[1]a 16 h.p. engine.

DOCUMENT 90

Apedale[1] Near Newcastle, Staffs.

To John Southern — May 26th 1794

Sir

I was from home when your letter came to Newcastle for a Week after, but have sent the Contents to Sir Nigel[2], as also of the Other letters sent by you. Sir Nigel thinks it is not right to pay the sum of forty pounds a Year as a premium for the Engine[3], as it draws only water, and was so very ill executed, it would not do the business in was intended for, the Brass working Barrel was exceeding bad, Was obliged to have one Cast soon after it was new, and the Gears are continually breaking, we have been obliged to take 21 yards of trus up to shorten the lift, as it would not do the work, we only draw the Water about 25 yards deep at this time; we have a very good Engineer Joseph Vaughan, I suppose Sir Nigel will be willing to pay Twenty Pounds a Year as a Premium but will stop it before he will pay more he says —

A Small Engine hath been put up by Mr Onions[4] close adjoining, that draws water 36 yards deep, and could do our business very well the whole expense not exceeding £150 besides the pump Trus — but desire you will write to Sir Nigel Gresley and what he Orders shall be punctually performed by me.

I am Sir

Your Obed^{t} Hble Servt
Edwd Eardley

Box 36

---

[1]a colliery.
[2]Sir Nigel Gresley.
[3]an 8 h.p. engine ordered in 1790.
[4]of Banks & Onions, Ironfounders.

Section H

The engine business – engineers and mechanics

The engine business – engineers and mechanics

Documents 91-102

The erection of an engine purchased from Boulton and Watt was undertaken by mechanics who were paid in cash direct by the customer. In this way the partners managed to reduce their outgoings considerably as the customers paid for engine parts by a bill at three months. In regions where a number of orders could be expected Boulton and Watt established a resident engineer who acted as supervisor in engine erecting and also as agent for potential customers. William Murdock was the agent in Cornwall for some years[91], William Creighton resided in Manchester for a while[92], whilst James Lawson [92] was based at Soho, but travelled widely in Yorkshire and Lancashire.

It was clearly in Boulton and Watt's interests for their engines to be well maintained since an efficient engine was the best advertisement. Consequently they occasionally advised on the recruitment of enginemen[93] or allowed a Soho mechanic to be retained long enough to train one[94].

General mechanics were sometimes consulted about possible engine orders but in these cases they would write initially to the senior clerk, John Southern, rather than one of the partners[95].

The shortage of mechanics and the consequence that one or two members of a team might be taken off one erection job for a few days and sent to another one, could be exasperating for the manufacturer whose works was at a standstill whilst the engine was being installed[98 & 99].

Mechanics usually accompanied the first engines to be exported but when they failed to return, being offered lucrative posts overseas, Boulton and Watt declined to send any more. Sometimes good relations were maintained with former mechanics, however, and in the case of the one who was tempted to remain in Italy, Boulton and Watt were able to persuade him to supervise the erection of their Spanish engines[101 & 102].

DOCUMENT 91

To James Watt Truro Sep 2 — 1786

Dear Sir

I have the pleasure to tell you that we arived here this afternoon in good health, except a little head ache from lying in a bad uncomfortable bed last night at Bodmin.

I arived at Exeter on Wednesday night & set out from there on Thursday afternoon to sleep at Oakhampton — about 1 Mile on this side of Exeter I met 3 of the Foxes[1] going to a general Quakers meeting at Gloscester & in the same Coach was Wm Murdock. He got out & we had a parley for some time. He said He was going to London to get Men but I soon found he was going there with his Steam Car^{g} to shew it & to take out a patent, He haveing been told by Mr W. Wilk^{n} what Sadler[2] had said & he had likewise read in the news paper Simmingtons[3] puff which had rekindled all W^{ms} fire & impations to make Steam Carriages. However I prevaild upon him readily to return to Cornwall by the next days dilligence & he accordingly arivd here this day at noon, since which he hath unpacked his Car^{g} & made Travil a Mile or two in Rivers's great room in a Circle makeing it carry the fire Shovel, poker & tongs. I think it fortunate that I met him as I am persuaded I can either cure him of the disorder or turn the evil to good — at least I shall prevent a Mischief that would have been the consequence of his journey to London.

If Mr Wilkinson doth not send all the Cornish goods directly I find there will be some diagreeable Clamours & therefore wish you would send a Spur to Bersham by the post — Mr Wilkinson may send the Bill of Lading to Wilson with his possitive ordors not to deliver till p^{d} for, for it will be a real loss now the Houses are ready to be any longer delayd.

I am to dine with Wilson tomorrow & wth Murdock on Monday.

Almost al the Cornish Quakers are gone to Gloscester, with fr^{d} Rich^{d} & Catherine & Vivian is gone to Wales.

The General meeting will not be untill the first Thursday in Oct^{r} & so soon as that is on hope to return.

I hear that W^{1} Virgin Adven^{trs} mean to order a new Engine for W^{1} Maid as that part of the Mine is richer than ever but cannot be much longer worked without I hear they are determind to have a double 63 In.

My best Comp^{trs} to Mrs Watt & Miss de Luc hopeing to hear that you are safe returnd & recov^{d} the effects of your Journey. Adieu
I remain

Dear Sir

Yours most sincerely

Matt^{w} Boulton

Box 20

---

[1]Members of the Fox family, prominent mine adventurers.
[2]James Sadler who later (1791) patented an engine which was made at Coalbrookdale.
[3]William Symington who patented his engine in the following year.

DOCUMENT 92

To J. Watt Junr. Manchester Nov. 25 1800.

Dear Watt

In consequence of your Suggestion that a few days earlier than Christmas would secure to me the pleasure of a general Meeting at Soho, Ewart & I have fix'd upon visiting you in the Mail on Wednesday 17th December — but if we risque losing the Company of any of the parties we will be with you a week earlier. I fix'd upon the 17th because I expect I can conduct my Sisters[1] into Lancashire from Birmingham and that is the earliest period they can set out.

Lawson has been gone a week; I did not perceive any symptons of a Relapse, and I am in hopes Exercise, vegetable Diet and abstinence from fermented Liquors will exempt him from any further attack. Perhaps a Line from Dr. Cn[2] to his Brother or some confidential person might be a judicious Measure?

Creighton has not yet made his Appearance — it is of less importance as you are so full of orders because it answers the purpose of declining them without the Declaration of the cause. Am I asking an improper Indulgence in requesting you will permit him at leisure hours to make a Drawing of our Building[3] with the Beams & pillars in perspective before they are conceal'd by the Arches & Gearing. I have no person who is an Adept in that Art and I think the Effect would be worth representation upon paper. If he makes any Stay here perhaps a Lodging would for every Reason be preferable to an Inn and he would always find a welcome, unceremonious Fireside & Table with me, but I think a separate

Residence may be better for your interests. A Line when convenient will be acceptable to

Yrs truly
G.A. Lee.

M IV L

---

[1]Sophia and Harriet.
[2]Creighton.
[3]The new iron framed Salford Twist Mill. Creighton made the drawing which is reproduced in Jennifer Tann, *The Development of the Factory*, p. 138.

DOCUMENT 93

To O. Henderson, Cornwall. Birmingham July 26 1780.

Dear Sir,

I have yours of the 22d. Am very sorry that you have such Blockheads of Engine men at the United Mines. I will take upon me to affirm it; and I am sure that you can back it, that no mine can thrive where they have not good Engine men.

As the Adventurers have given us a power to dismiss such Engine men as do not do their duty to our satisfaction, I think that upon the present occasion we should fail in our duty both to them and to ourselves if we did not make use of that power, and there never could have happened a time when we could have exercised it so impartially, as we neither know the person nor name of the offender — It is become highly necessary that we should keep to our promises and maintain good order among the Engine men.

Therefore we desire you to enquire into the merits of this cause and if you think the offender highly blameable, to consult with Mr. Wilson and if you are both agreed in the nature of his offence, to complain to the Captains and lay before them what you have learnt. <u>And also at the same time to make a formal requisition in our name that the offender be dismissed from attending the Engine or pitwork at the United Mines.</u> We wish the enquiry to be gone into with all due impartiality, and with due attention to mercy, but without respect of persons. We wish to govern with justice and levity, and hope no person will take offence at our acting a part which must be for the interest of the adventurers. Fiat Justitia ruat Celum.

If you are desired please to recommend another Engine man, but not otherwise lest it give offence. We do not mean by this to find fault with your bringing Harper to them in which you acted very properly, we only mean to say that we have no power to force men into their service though we shall be always willing to give them assistance when desired.

Please make our compliments to Mr. & Mrs. Wilson and also to the Captains in the United Mines wishing them prosperity.

For Mr. Boulton & Self

I am Dear Sir

Your Obedient Servt

James Watt.

Office Letter Book 5. pp. 11-12.

DOCUMENT 94

To James Watt — Cuckney, Notts. Octb^{r} 15, 1789

Sir

this comes with my servise to you and to let you know that me and Mr Gorton[1] do not agree therefore I am determined not to stay any longer with him; If I leave Mr Gorton he will be in no distress for a man to look after his Engine[2] for there is a man that hath been with me all the time I have been here which can look after it Equally as well as myself, I have been with him two years and have never Neglected his Business an Hour since I came but have always had my Engine in good order, Neither hath he any thing against me Sir. I should take it as a great favour if you would draft me to any other place to Retford or where you think proper and I will serve you to the utmost of my power. If not, I will take to Myself.

From your Hu^{ble} Ser^{t}

En^{d} Giles

Box 4 G

[1] R. Gorton, cotton spinner.

[2] a 14 h.p. engine ordered in 1787. It was eventually repurchased by B & W.

DOCUMENT 95

To J. Southern Gregory[1] new Engine, Derbys Augst 12, 1785.

Sir,

Your last letter came Safe to hand, and I take the oppertunity to Acquaient you that Mr Milens[2] and Mr Kirby as binn inquiring of me to Know the Expenc of one of Mr Boulton and Watts Engine in the room of the old one. The Expect Mr Hodgkison Down from London on Sunday 14 instent, which the have a Great Disire to have alteration made in the old Engine and to have things Settld while Esqr Hosgkison Down —

Diameter of the Cylinder 42 Inch

Diameter of the Working Barrel 12 Inch

The hight that the Engine raise the water 62 yards with to lifts, the Engine makes a 6 Foot Strock, and works at the rate of 9 Strocks a minute — the old Engine works about 16 hours each Day this Summer, Joseph Alan the Engine man Declares that the Consume 20 tun of Coales per week take one time with a another the year throu, Stephen Thompson account is 17 tun a week but the Coale Book an him Cant Balance at that rate. The agents of the mine as made Sevrel atemps to have the old Engine triad but Stephen Thompson as binn to to Kowen for them Since the have had the new Engine — The have had one veary whet winter Since i have binn at Gregory mine which the old Engine had 23 hours work per Day and i was informed that the Consumed 30 Tun of Coales a week — I have the pleasur to acquaint you that the new Engines[3] Goes on Exceeding whell at this time. in april 25 the miners begun a Sinking the Shaft Deeper under a 6 inch Bore and raised the water up to the Great pumps that Stands in the veain 11 yards from the Engine pit as the miners Kept Sinking the water Kept increasing on the 6 inch Bore which, the Engine as worked at the rate of 5.6.7 Strocks a minute this 3 months past. in June 8 the lower Balance Beam box break and lost its Balance and Break the Great rod below the 2 Balance beam, I get the Balance box mended, and a new rod in and the Engine to work again on the fourth Day — the miners as Don Sinking at preasent as Set by the 6 inch bore, i have, Get the pumps out of the vain and Set them in the Engine Shaft. The Engine is working at the rat of 5 Strocks a minute at preasent but the Water Keeps Decreasin which i Expects the Engine to have but 3 Strocks a minute — the Engine raises the Water 39 yards 1 foot — I have Compleated all my troblesom Jobs without Frances

Thompsons Assistance and the Engine working well at this time, — With Respect to Frances Thompson machine[4] i belive that it goes on veary Sloley i hear that one Mr Barber[5] as Get a patent for a machine near of the Same Construction. Mr Southern i Shall take Care to give you all the information relating Frances Thompson machine an is Consarns, I have not had the oppertunity of giving you the Diameter of the Cylinder an air pumps of Thompson machine at preasent, — I can give but a veary indifferent Character of the men at the old Engine, I hope to have the pleasur of Seeing you in Darbyshire in a Short time which i Shall be glad to no, when you will make it agreable to Com — I hope you will take the first oppertunity in Sending me answer Relating the old Engine for mr Kirby as a Great Desear to Know what the Expenc will be to have the Engine alterd —

I am with my Compts to you who
I hope is well
I am your Most Obedint Humble
Searvent John Stratford

The price of Coale 9 Shillings a tun in the Summer, 10 Shillings a tun in winter.

Box 4 S

[1]Gregory lead mine, Ashover.

[2]Milne.

[3]a 46 in. engine ordered in 1779.

[4]Francis Thompson was engineer at Gregory. He was also an engine designer and in 1792 took out a patent (no. 1884).

[5]John Barber who in 1766 obtained a patent (no. 865) for a method of raising water out of mines. This was followed by another in 1776 (no. 1118).

DOCUMENT 96

To Boulton and Watt, Gregory Mine, Augt 24 1791.

Sir,

I make Bold to acquaint you that I rote to Mr Jno Southern about a month ago but not Receiveing my Answer at preasent makes me veary oneasy that i have not received my answer, I make bold to rite again hoping that i shall soon receive a answer from you. Since my Return to the Engine I have had no conversation with Mr Milens conserning me or the Engine tho he as sent me word that the will rise my wage if i like to continue with them, but i have not given them answer at preasent Weating for answer from you, The Engine continue working veary well but veary slow only at the rate of 4 Strokes a minute for whant of Day water for injection, I speak to Mr Kirk about the Cylinder being coverd as the Engine is working with all the pipes and Cylinder bear if sixpence will do it he will not let it be don, tho at the same time he fines him self grived as the Engine makes them no more Saving, as thear is so much Grumbling about the Engine, I hope you will be pleased to let me have a fresh place for what Content can a man have to work for so Disagreeable people. in my last letter to Mr Southern i sent you a short account of the old Engine, but not Receiveing answer from my letter i have sent another account.

Diameter of the Cylinder 42 Inch
Pumps 12 Inch, Raises the water 62 yards
with two lifts the Engine making a 6 foot Stroak
working at the rate of 9 stroakes per minute
From the Surface to the level is 40 fathom

I am

Sir your Obedient
Humble Servant
Jno Stratford.

Box 4 S.

DOCUMENT 97

To Boulton & Watt, Manchester, Dec. 17 1797

Honored Gentelmen

These feu lines coms to inform you that I have started Connel and Kennedys[1] Engen[2], And she Works verrye well and she starts work for Good on the 18 of this Month. Therefore I will be Done the Back end of this weak, So if you please to sene me an Anser ane let me no whether I must come Home or to Go to any other Plase I will be mutch obbliged to you, I have not seen Mr. Lawson this three weaks and upwards So no More at Present.

From your Moast obedient Humble
Servent
Michael Wheatley

Box 3 W

[1]McConnel & Kennedy, Cotton spinners.
[2]This was a 16 h.p. engine.

DOCUMENT 98

To Boulton & Watt London 5 June 1792.

Gentlemen,

I have received the engrossed Agreements, that on your part, executed, & I see no Objection to return mine but till we have finished, & settled other Matters, there can be no Hurry for that which relates to an after period — I have Summoned for a long Time, all the patience possible, but must now acknowledge my Resources begin to fail — So much fiting, chipping & boring, altering & grinding, that was it not for the Recollection of whose Hands I am in, & that in them I cannot suspect Deceit, I could not bear the Continuance of this Semblance of work, without the appearance of Conclusion, — Mr Watsons people[1], a few days 2, — once or twice 4 & at other Times 3 Men, began 23 March, a space, in my Ideas, long enough for them to have more than made every Article[2], instead of only Screwing them together — I beg Pardon if my Notions are unreasonable, of which however, I am afraid, I cannot easily be convinced.

It strikes me, too late perhaps, that where Engines are generally fixed, the Proprietors being Men of Fortune, & neither present, or at all used to working people, the progress which is made, & the Expence, are not Objects — but my Situation being very different, & myself in the working Line, the Comparison may affect me in other Respects than the Mind: — I wish therefore to have the work done & before we begin, to know the whole Charge, which will indeavour to have Settled as soon as possible, & then the Agreements for the remaining Term of your Patents may be exchanged, I am

respectfully
Gentlemen
Your most hb^{e} Servt
S Spence[3]

Had my old friend Mr Matthews lived,
I should have had some Consolation in Communicating
with him, whether my Observations be real or imaginery.

Box S.

[1]B & W engine erectors.
[2]for the 8 h.p. engine.
[3]Sawyer Spence, lead roller.

DOCUMENT 99

To Boulton & Watt — Sirhowy[1] near Newport, Monmouth.
13 Nov 1805

Gent[n]

We are now about Building the Engine House to receive the Forge Engine had from you — the Building is up for the Spring beams & We Expect the Roof to be Cover'd in by the end of this Month.

We have to request you will send us a proper person to put up the Engine[2] as Usual and we will give the Needful attention in having things ready.

I never write to you but I have to Lament your not leaving your Man to finish our Blast Engine for want of which chiefly I Conceive is the Cause why it has & still remains liable to so many Stopages much more so than the Sirhowy Engine which Varley finishd.

I am for Sam[l] Homfray & C[o]

Gent. Y[r] Ob[t] Serv[t]

Rich[d] Fothergill

Box 36

[1]Ironworks.
[2]Engine no. 1 was a 24 in. 5 ft. blowing engine ordered in 1799.
Engine No. 2, the object of this letter was the same size.

DOCUMENT 100

To A. Weston. Birm^{m} May 23 1792.

Dear Sir

We have yours of ye 21st and observe the contents. You know that we proposed making some present to Robert Cameron[1] out of mere motives of compassion in his situation, though we are not without grounds of complaint against him[2]. The agitation we were kept in prevented us from thinking further on it while in town; but as we understand that he is still in the same miserable situation we shall be obliged to you to call upon him & to present him 10 guineas from us, and enquire into the real state of the detainer now existing or likely to take place against him, and as if from yourself as a means of evading his difficulties whether he would like, if liberated, to erect Engines for us on the continent of Europe & on what terms which you might say you would mention to us. You must be very cautious as he is very able & a wrangler which has contributed to put him in his present situation.

It will also be proper to learn whether he lies under any hands or engagements which might operate against his going abroad, or bring us into trouble by employing him. I believe he is honest but at times frantic with passion & sometimes I believe insane.

Mr. B. is gone to look out Mr. De Luc's letters, which shall be sent you by packet to Mrs. Matthews to morrow but must be returned as I shall want them.

I remain with esteem

Dr Sir

Your Obed^{t} Hble Servt

James Watt.

Office Letter Book 15 pp. 159-160.

[1]Cameron was in goal.

[2]Cameron patented a steam engine in 1784 (no. 1414) and two years later a method of raising coal and ores from mines (no. 1525). He was in partnership with Jeffreys for a few years.

DOCUMENT 101

To Malcolm Logan
Cadiz

Soho 30 July 1794.

Dear Sir,

We yesterday received your letter of the 19th June from which we learn with much pleasure your safe arrival at Cadiz; your compliance with our wishes[1] in this respect will ever be gratefully remembered by us and we hope will eventually be attended with more satisfaction to yourself than you seem to apprehend.

We are much mortified to learn the inaction and wasting reflections you mention in the Gentlemen at Cadiz; we cannot but lament their want of attention to their own interests and their untoward conduct towards you, but we hope and trust that you will not suffer your energy to be damped and that you will cooly and steadily persevere in your exertions until you have brought the engine into motion. We consider our reputation as being in great measure at stake in this business and we are aware that whatever may be the difficulties that are thrown in our way the world at large will blame us for its want of success. Under these circumstances our chief dependence is upon your perseverance and experienced skill; we know you have already overcome similar obstacles and we doubt not that you will again succeed.

We have already written to our employer here Mr. Fermin de Tastet[2] to recommend more activity from the directors at Cadiz and above all for them to furnish you and Murdock with the drawings. We doubt not that that letter has before this produced its effect but shall again write today to enforce it.

We are happy to hear that you are satisfied with what Murdock[3] has done; we have not been altogether pleased with the information we have received respecting him from various quarters, but we are willing to believe they have been exaggerated, and to place reliance upon his future prudence and his compliance in every respect with your orders. We wish it more upon his own account than ours as our long friendship for his brother naturally interests us for one who so nearly belongs to him. We can and do make allowances for the difficulties he has had to encounter.

When you are in possession of the drawings, you will get the mistake, if any, in the foundations rectified; and as the Engine is merely waiting for the house, whilst that is getting ready, you may be useful to them in the preparations of the millwork for we have mentioned to Mr. de Tastet your ability in that line, in which we doubt not that you will be very useful to them.

We sincerely rejoice in your good fortune at Naples and hope you will long continue to enjoy the King's bountiful patronage; we know that his Majesty was not averse to your journey to Spain as he wished to see a Steam Engine established in his Brother's dominions, we are therefore assured that he will not grudge you the necessary time for compleating the job. It cannot be our wish that you should in any degree endanger your pension by prolonging your stay unnecessarily but we hope you will see the engine fairly started before you leave Cadiz which indeed cannot be long if the Proprieters are in earnest.

We beg you will write often to let us know how you go on, and you will not let any ship sail without sending some intelligence.

With best wishes for your success

We remain

Dear Malcolm

Your Obedient humble serv^{ts}

Boulton & Watt.

P.S. The writer J.W. Jnr desires his best remembrance to you; his father is not at home, being upon a journey into the Country to dispell the melancholy which the loss of his youngest daughter has occasioned.

Office Letterbook 17 p. 67.

[1]Logan, formerly a B & W mechanic, had settled in Italy in the employment of the King of Naples, but had been prevailed upon to travel to Spain to superintend the erection of a 50 h.p. corn mill engine.

[2]The London merchants through whom the order was placed.

[3]William Murdock's nephew. He had a reputation for drunkeness and unruly behaviour.

DOCUMENT 102

To James Watt Junr. Cadiz January 28th 1795

Sir

Not hearing from you in such a length of time as from the 30th July last — I am impasinte to hear from you as things remain in the sam state as befor, no sines of the hows begineng. I hav not seen Mr Torris[1] since James Mordock[2] left this place and ther is no settelment of wadges for they will not pay the wadges you promised me in your letter and I canot live on what they give me. I therfor beg you will settil it as sune as possible and let me kno if James Mordock is come to you. In the coures of 10 days ther will not be one articel wanting for the engine on mead — I was caled on by Mr Witheril[3] at sevil for to see his Mill for a engine of 8 horses. He have a water mill at presant to mill the lether; 4 feet head and fall of water the whill is 20 feet high 1 foot 8 In wide and too pair of stocks on it. The shaft for Stocks should make from 10 to 12 revelutions pr minot and that will make from 20 to 24 strocks per minot, the corn mill must not go veriy quick for they do all in this Cuntrey go slo and they canot be put owt ther old way and the bark mill I think will go the same as the stoks the nutt of the corn mill must lift up when the stoks work but I leve it to you for to calculate. The suner you get it redy the better. I thinck ther will be 3 or 4 more hear ordered as sune as they see it and if sune I thinck to have leve to put it up. Send the drawings as sun as possible. I have ben caled on by the nepolitan cownsual by ordor of the minester at naples to kno for What cas I staied so long. The following is the coppiy of a letter I sent to hem as I was ordored to writ to him.

Hon*ble* Sir

Having been informd that you did me the Honor of making Enquiries concerning me of his Nepolitan Magestys Consul at this Place I take the Liberty of informing you that I remain here still in perfect Health. I have been unluckily detaind a great deal longer than I expected owing to a variety of unforeseen Accidents the principal of which is Caus'd by a Mistake, comitted long befor I came, in the Foundation of a Building destind for the reception of an Engine I was to put up in the Carraccas and which renders it absolutely Necessary to begin the work aneu before I can proceed any further in the Business. And my chief Motive for wishing to stay and compleat the same is the regard I have for my former Employers and Manifactores of the Engine who woud think their Reputation injurd were it not to be properly fenishd. But as my greatist Desire is to return to Naples as soon as I possibly Can with propriety I shall not (be delayed) a Moment in doing so — immidiately I have ordors to that effect and permission from the Court of Spain.

I am with profound Respect

Hon*ble Sir*

Your most ob. humb.

Servant

Malcolm Logan

the peopel ar all saing the engine is to be builded in the island. So the foundation will be right the second time I will send a drawing of Mr Sulherals[3] mill ... by the first Ship from Cadiz

Box 3 L

---

[1]Torres. Malcolm Logan had travelled to Cadiz to erect a corn mill engine.
[2]William Murdock's nephew. He left Spain in disgrace and settled in the West Indies as a mechanic.
[3]Wetheral.

Section I

The engine business – Soho foundry and Soho engine manufactory

The engine business – Soho foundry and Soho manufactory

Documents 103-123

Soho Foundry was opened in 1796 and from that date the firm had the capacity to make all their engine parts — at least in principle. Boulton and Watt had been joined in the business by their sons who both owned and managed Soho Foundry from the outset, whilst Soho Engine Manufactory, that part of the business run from Soho Manufactory, was owned and managed by both fathers and sons. M.R. Boulton gave a synopsis of the preceding twenty year period in a letter to his cousin, alluding to his initial dislike for business[103].

Cost accounting was recognised as an aid to business by only a small minority of entrepreneurs in the late 18th century. Watt Jnr. was among them for he sought advice on the matter from the cotton spinner, G.A. Lee. Lee's reply[106] highlights a problem of particular significance in the industrial revolution, namely that of how to account for machinery made on the premises.

Soho Foundry was modified and extended in the early 19th century[107] and, in the same period, an elaborate costing exercise was begun at both the Engine Manufactory and the Foundry[108-121]. Careful observation of the time taken to manufacture different engine parts[109] together with comparative estimates of castings from other foundries[111] assisted the foundry manager in making agreements for piece work with master craftsmen[115]. These agreements illustrate the operation of a system of partial subcontracting, a common practice in foundries.

DOCUMENT 103

To Walker Z. Junr

April 14th 1797

Dear Citoyen

From former specimens of my irregularity as a correspondent I think you will not be inclined to ascribe my long silence to any diminution of the friendship & attachment which has & I hope will long subsist between us. Impute it rather to its true cause, to that procrastination which unfortunately is inherent in my disposition. You see my friend I rather ingeniously expose my failing than risk the loss of an iota of your esteem, confident that the acknowledgement of it is with you the most probable means of procuring a reconcilement if any be needed. You will probably have learnt from other sources better than I can now describe it the state of your circle of acquaintances in this part of the world. In a word all are well & busily driving our separate wheelbarrows. You will (not be a) little surprised to find that I am a very regular attendant in the counting house & immersed in business. Like a person hesitating on the brink of a cold bath I found that the only means of conquering my aversion was to plunge in; my experiment has so far succeeded. Mr. J.W. jun^{r}. & myself with the occasional advice of the old gentlemen have the entire Management of the Engine business & for the last 12 mos. I have not had respite from it as you will judge from an epitome of our labours.

The attacks of nearly 20 invaders of our patent rights in different parts of the kingdom have been successively repelled & brought to submission. The principal one however chose not to fall so easy forcing us to a trial at law which obliged us to entertain in town for three weeks nearly 20 witnesses comprising all the celebrated Engineers of the kingdom, one of the Scotch Professors even came to the assistance of his countryman upon this occasion...[1] in the most flattering (manner) upon the testimony of the respectable Men of Science of this Country. It has as you may suppose much strengthened the validity of the Patent. Your ungrateful Cornish acquaintances are now the only recusants & we are preparing for hostilities ag^{t} them. In the midst of our troubles we have completed about 1½ Mile from home a very extensive foundry & such other establishments as will enable us to defy competition after the expiration of our term. Our Connection with the Iron King at Bradley[2] is dissolved & by our new works we are independent of all such Iron tyrants. This hasty sketch will put you in possession of the leading events & occupations of your two friends who, au reste, you will find little changed since we last parted allowing for the gravity of 4 addit^{l} years & some add^{l} cares. We expect a visit from you upon your return to America: it will be equally conducive to your satisfaction & interest. An Interview with my father will very much facilitate the arrangement (of these plans)[1] & prevent any disagreement because of an expression regarding[1] your views. With the very best intentions I know my fathers aversion to epistolary commun^{n} often deprives his friends of their benefit by caus^{g} him to delay their execut^{n} till the favourable moment is past. Many other lessons which I cannot suppose have escaped you urge the measure which I know will be agreeable to all parties here. In whatever manner I can render you any assistance, I beg you to command me with^{t} reserve. Your friends J.W. jun^{r}. & sen^{r}., Foreman Southern &c unite in the sincere wishes for your health & welfare & we hope you will avail yourself of the first opportunity to give us tidings of both.

Yours sincerely & affect^{y}
M. Rob. Boulton

I leave it to your discretion to make the purchase of such part the library in question as you shall think eligable & to procure any esteemed publications relating to Belle Lettres

Chemié
Physique
Mechanique
Politique

A list of the french publications I now possess will accompany this list from whence may better judge what will suit me.

Nov 4th 1797
Deliv^{d} Z.W.

MB Z. Walker/20

[1]Illegible (press copy)
[2]John Wilkinson.

Delivered to Z. Walker junr Novr 6. 1797.

Information respecting Steam Engines for your private government. They are made of two denominations Rotative & Reciprocating Engines. The former are applied directly to the turning of Machinery & the latter to the raising of water: they are also of two constructions, the double & single.
Rotative are always of the double construction,
Reciprocating are made single or double according to circumstances: in general if the power comes within of a 50 or 60 in. single, they are employed, unless local peculiarities (such as require great weight of dry Rods &c.) render the other more particularly applicable. The powers of Rotative Engines are estimated in horses: say number which acting *together* would be equivalent to do the same work.

We in general suppose a Horsepower to be equal to the raising of 33,000 lbs one foot high p^r minute. The fact is no Horse is capable of doing so much for a continuance of 1 or 2 Hours & to do the work of a 10 horse Engine working 24 Hours, 40 Horses must be kept as 4 setts of 10 each would be necessary. An Engine of 8 horses is equal to grind & dress p^r Hours 8 Winche^r Bushels or 1 quarter of Wheat 16 Horses do double.

The Engines used in Breweries in London are from 8, 10 to 20 Horses power In Cotton Mills from 10 to 30 Horses accord^g to the number & construct^n of the Spingles employed. There is a difference of ab^t 1 to 10 between Water & Mule Spinning, 10 of the latter Spindles requiring only the same power to turn them as one of the former.

MB Z. Walker/20

DOCUMENT 104

To M. Boulton Muirkirk Iron Works April th 3rd 1796

Sir I yesterday Received a letter from my Brother Wm Wilkes of the Eagle Foundry of Birmingham who inquird after your place at your foundry for me as a foreman of the Shop, but waiting some time he received an answer that you had a manager that was (to) Conduct the whole of the Foundry; one Mr Story who I am informd has great Abilities — but my Brother informs me that you are still in want of Loam & Dryd sand Moulders Boath of which I have been well accustomed too & worked in for twenty years, or green sand as may Occur. I have been use to assist in casting your Engine work & your stock Roll with my Brother at the Eagle foundry — the last Place I Served at in England was Mr Wm Seagers of Brierly hill near Stourbridge where I executed at Considerable quantity of your Engine work at his foundry[1] — I served Mr Seager two years & a half as his forman in his work; but Leaveing him for Considerable more wages I Come to Scotland — but finding the place I am now in a Deal Colder than in England which causes (me) to have the Rheumatick Complaints I would rather return to my own Native Country again if I Can get an opertunity worth my attention — I should have no Objection to serve you as a Loam or Dryd sand or green sand Moulder under the Direction of Mr Story — upon Honerable Terms — I am now serveing Muirkirk Company & have a good Master and a good place. I have a Setled Salary of eighty one pounds fifteen shillings a year & a free house, fireing & Candles found me free without any Deduction whatever. But the Coldness of this part Dont agree with me & I hae had the offer of two guineas a week with the same Liberties as I have mentioned from one Mr Burrow at his foundry at Crawfords Dyke Greenock. But Dont think to leave Muirkirk except I come to England and I should prefer Comeing in or Near to Birmingham as my friends are Near there. Mr Foreman as some Personal knowledge of me as while I was with Mr Seager. I was at the Soho several times — I believe I can say I Can excute my work as well as my Brother Willm Wilkes of which I am older. Sir I will kindly thank you for answer to this specifying the Term you will give a man — if you are not all ready supplyed with hands — as I had a letter about three weeks ago from a Gentleman in Liverpool that I am persenal acquainted with which offerd me the same wages I have here, and I have not given him a final answer nor cannot at Preasant till I hear your Terms as I would wish to be nearer Birmingham if I Could —

Sir I am waiting your answer and Remain
Your Most Ob^{t} & Humb^{t}
Servant Thos Wilkes

PS Please to Direct to Thomas Wilkes at
Muirkirk Ironworks by Douglass mill
Airshire North Brittain

PS Sir if I should meet with your Aprobation and you think me worthy of your place I hope you will Consider Traveling expences as I have a wife & Family —

Box 3 W

[1]Seager & Thompson had been considerable sub-contractors for B & W engine parts.

DOCUMENT 105

To J. Watt Junr. Soho Octr 21st 1797

Dear James

I should approve very much of sending Greg to Saxony were the climate such as is recommended for assuring his health[1] & safe from a visit of the french. The road to Naples by land seems at present very insecure, nor is Naples itself quite safe; to go by sea at this season through the Bay of Biscay is what I should not feel bold enough to advise I think therefore we must content ourselves with Cornwall and the point is to find out some means of employing him there so as not to wholy lose his time. What I have thought of is for him to make frequent visits at all such Engines as he can have access to & to send us frequent accounts of what is going on. Also to be as much with Murdock when out upon business as possible. To reside at Penzance for the benefit of the Wedgwoods company & to study German & Italian as far as Books will serve him — I have not mentioned the scheme to him yet as I do not wish to alarm him — I think him at present in still better health than when you left him & he continues his morning exercise on horse back. I have wrote to Dr Beddoes by this post as I hear he does not intend coming this way at present.

The Messrs Hopes of Amsterdam were here yesterday & there was no preventing Mr B. from taking them to the foundery, but I have said so much to him on that head that I hope he will not transgress again soon. The Mint is now a common show.

We have also had Mr Gould from Neath[2] who as we understand has agreed to settle the premium on the terms offered on giving up interest on both sides — He is an agreable young man but not very knowing in the foundery business. We found ourselves obliged to show him the foundery, he was led rapidly through the fitting shops, but left longer in the foundery — He found great fault with the Stoves as not lined with Iron plates to *contain the heat* & not having the fires in the inside, thought the boring machine a bad one with other similar observations.

We had a remittance from Thackeray yesterday — I do not recollect any other news —

I am sorry to hear that your headaches still continue & would advise you to use the warm bath constantly while you have the opportunity you should ask D^r B?[3] also whether drinking the Bath waters would be of any use, or other tonics. I have received considerable benefit from Oxygene for the asthma & my cold leg & were it not for its bad effects on my Stomach would be disposed to continue it for some time longer. All friends at Soho & Heathfield are well those at the latter place send their love to you & I remain

Dear James
Yours affectionately
James Watt

Isaac Perrins is here & says trade is very brisk at Manchester — Mr Boulton told me lately that a Mr Champion of Bristol had been here & told him several Engines were likely to be erected at Bristol for manufacturers. I think you should call upon M^r C and learn more about them & if necessary call upon some of the proprietors, especialy as orders have been very slack here lately — yesterday there were orders for 30 Cop.[4] machines for Richardson & Harrison & 12 for others.

MIV Watt 8

[1]Gregory Watt was suffering from consumption.
[2]Neath Abbey Ironworks had a 40″ cyl. B & W Engine (ordered in 1792).
[3]Dr. Thomas Beddoes, chemist.
[4]Copying (Watt's patent letter copying machine).

DOCUMENT 106

To J. Watt Junr. Manchester March 11 1797.

Dear Watt

I am glad to learn you design to visit us in Summer — consider such plans as irrevocable and they will always be executed — otherwise the persuit of Business like the Maladie du pays never ceases to haunt the patient till he has been some time out of it. I wish I could communicate to you any Information respecting the mode of Keeping Manufacturing Books — in the construction of Machinery we never yet could reduce it to regular piece work or divide the Labour of Making & Repairing it in such a Manner as to determine the distinct cost of each — In the Manufactory I have attain'd rather more accuracy but yet far short of my hopes & wishes — but the Circumstances are so peculiar to it that I conceive it would be impossible to communicate the Mode in such a Manner as to be applicable to any other Business. The System such as it is has been form'd upon observation upon the essential and inferior points class'd in their respective Order of Importance — originally executed with a Minuteness incompatible with Convenience — such as taking Stock every week — but the first Object appears to me to be close personal Inspection for which there is no Substitute & then a moderate Understanding will analyse & arrange with as much Accuracy as is usually necessary — you must therefore descend to the Drudgery & you will infallibly succeed. ...

Y^{r} friend

G. Lee.

Inclosed is a letter for Southern — Upon Examination the packet is too bulky for the post & you will receive it by the Coach wch comes to the Swan in Birmingham. I believe I agree with you perfectly about the probable Event of paper Money — we are upon the Brink of that Gulph wch has hitherto swallowed up every Individual & Nation that ventured to approach it.

M IV L

DOCUMENT 107

To Gregory Watt   Soho 8 June 1803.

Dear Gregory,

I am much obliged to you for your two letters with your mineralogical observations in Ayrshire, but shall not have leisure to offer any comments at present, being a good deal occupied in consequence of the absence of Mr R. Boulton & Mr Southern both of whom are in London. We are also going on with new buildings & Machinery both here & at the foundry, and the scire facias having issued against Matthew Murray[1] we are making preparations for the trial, which we expect to take place early in July.

Our own Engine house here is undergoing a compleat Metamorphosis, and we are adding a new Engine and shop to the Foundry Establishment; the latter 134 f^{t} long is now roofing in. It is intended for the fitting of the little Engines, and I hope will enable us to get on with at least one per week, which will scarcely enable us to meet the present demand. Some other alterations of minor importance are going on there which I hope will all be finished in the course of this summer, and if it should afterwards be found expedient to add a Shed for making boilers & a Counting House, that establishment will have reached, what I should vote to be, its ne plus ultra.

I am sorry to observe by your letters that you still retain remains of your cough & languor; if I had been aware of it sooner, I should not have sent you the commission respecting M^c^Georges[2] affair, which however I hope will rather prove an amusement to you than otherwise. He seems a wicked chicaner and to have formed his opinion of our docility from that of our Agent. It is much to be lamented that Creighton will not summon up a little more energy in his dealings with such like gentry: he might save himself and us a great deal of trouble. The same things rarely occur under Lawson's administration. The Soho family continue much as usual. Upon the whole I think Mr Boulton mends. The D^rs^ house has made little progress yet, but I believe the orders for the alterations were finally issued about a week ago. Lawrence takes possession of his house in about a fortnight; when the Lady comes, seems not so fully determined! Meecham has already entered upon the holy State; I do not learn that it has improved his Jaundice.

Lee, I apprehend, has full occupation between his new wife & his new Engine, which he has been erecting the last week, I have not heard with what success — Lawson has got my blackey upon trial in the Gig at Manchester. I accompanied him as far as Stafford; the performance was admirable & he says, continues so. His own horse is at the foundry, with your Colt, which improves greatly: Mr R. Boulton's is also there with his Mare, whose new birth is daily expected. Lawrence has at length fulfilled his promise to purchase two new horses for Heathfield & John Frost is gone off to London to bring them down. Present my best respects to all the Hamilton family. Ann is at Heathfield & very well. I am Dr Gregory

Yours aff^tely^

J. Watt Junr^r^.

Box 2.

---

[1]Matthew Murray of Leeds whose 1802 patent was contested by B & W.

[2]McGeorge of Edinburgh ordered a 4 h.p. engine in 1802.

DOCUMENT 108

Classification of Castings for Apportioning Moulders Premiums — June 1800

1st Loam & Cylinders made in the Pits

1st Cylinders, Air pumps, Condensers made in sand or Loam in the Pits, Working Barrels. Clack seat Pipes, Plain pipes above 24 Inches Diam^r Cylinder Bottoms, Tops, Pistons above 40 Inches diameter or occasionally made in Loam from other Reasons at 20/-

2^d Dry Sand

Articles

1st Class — Cylinders, Air pumps, Condensers, Working Barrels & Bored goods generally including Hot & Cold wat^r pumps at 20/-

2nd Class-Cylind^r Bottoms & Tops, Pistons, Air pump Buckets & Covers, Rotative wheels and others made in Dry sand.

3^d Class-Boiler Steam pipes, Socket pipes, Stop pipes & side pipes or perpendicular Do. Eduction pipes Nozles and side pipes when in one Piece.

4th Class Nozles in separate pieces.

5th Class Rotative and other shafts centre Gudgeons out end Do. when in Dry Sand.

6th Class Connecting Rods.

3^d Green Sand

Class 1st Steam Cases, Dampers and frames. Covers for Cylind^r lids.

Class 2^d Manhole pipes, Feeding mouths, Dead plates and frames, Bonnets false faces. Stop Valves, Throttle pipes, Stuffing Boxes and weights.

Class 3^d — Plummer Blocks and Glands, Saddle plates & Glands, Dogs and weights, Governor Balls, Valve doors, Bottom plates for Air pumps and Condensers & Condenser Covers, Cranks & pins for Do.

Class 4th — Air pump & Cylind^r Glands, Covers for pistons & Air pumps, Centre & Outer end Gudgeons, Air pump Buckets, Pistons, Rotative & other wheels, Blow pies & Sockets.

Class 5th — Fly Wheels.

Box 45

DOCUMENT 109

Moulder's Estimate of time taken in the manufacture of a 20 horse engine

20H	One Moulder & Assistant Days
Cylinder	4
Inner Bottom	1½
Outer Bottom	1½
Lid & Gland	1½
Fals cover for do.	½
Piston & cover	1½
Steam case	4
Nozzles uper & under	3½
Bonnets	3½
Perpendicular steam pipe	2
Eduction pipe	2
Socket pipe	1
Condenser Socket	1
Brackets for W Gear	¾
Condenser	3
Cover for do.	½
Bottom plate for do.	¼
Cover for foot valve	¼
fals face	¼
Air pump	3
Cover & gland for do.	¾
Bucket	¾
Bottom plate	¼
Dogs for Cylr &Air pump	¼
Weights for W Gear	½
Blow pipe	1
Main gudgeon	1
Plummerblocks & glands	1
Dead & lintale plates	¾
Fire doors & frame	¾
Damper & Frame	1

20H	One Moulder & Assistant Days
Gland for Beam Straps	¼
Soule plates for motion &c	½
Outer end soule plate & glands	½
Top glands for do.	¼
Gudgeon for do.	½
Connecting Rod	2½
Sun & plannet wheels	2
Rotative Shaft	2½
Plumer blocks & glands	1
Plug rod	1½
Fly wheel	5
Balls for governor	½
Throttle pipe	¼
Boiler steam pipe	2½
Bonnet for do.	¼
Stop pipe Bonnet & Valve	2½
Short pipe to joine do.	1
Plain boiler steam pipe	2½
Socket pipe for do.	1½
Hot water pump & door	1½
Could water pump & door	1½
Stove plate for Cylr	½
feed cistern	½
do. for top of hot water pump	½
Safty pipe cover & gland	1½
Marshall pipe & cover	¾
feed pipe & weight	¼
Bearer & grate bars	¾

If 2 Boilers

Safety pipe Cover & gland	1½
Manhole pipe & cover	¾
Feed pipe & weight	¼
Bearer & grate bars	¾
Dead & lintall plates	¾
Fire doors & frame	¾
Damper & frame	1

If crank be required with seckondry shaft and connecting wheels

Shaft	2½ days
plumer blocks & glands	1
Connecting wheels	3

If cast iron Main Beam and Gudgeon
For the Motion moor time will be Required.

If the Boiler Steem pipe from the Nozzle to the stop pipe and from thence to the socket that joins the 2 Boilar exceeds 9 feet 4 days should be added to the time noted for Boilor Steem pipes

This statement is suposed to be a 20H which may serve as a Meidum from 2H to 40H. The above time is from 7 to 8 at night-that is without over time 12 Moulders at Soho Foundry.

The above is suposing the castings all perfect without any allownce for blanks.

Box 45

DOCUMENT 110

Copy of Agreement with W. Harrison for fitting the small engines 1801.

William Harrison Proposals

I will undertake with the present hands working with him & others that are to be got to fit 2, 3 & 4 horse Engines complete in a Workmanlike manner to be judged of by Mess[rs] Boulton & Watt, put them up, try them, take them down, black & put them in the Magazine ready for Packing up & keep all his Tools in good repair for £20 each, providing he has a constant supply Castings & forged Work. The great & small Drills to be used by W[m] Harrison at over hours, & when J. Allports Lathe is used by W.H., J.Allport shall shall be employed on W.H's Acc[t] for a whole Day together, The tumbling bar Lathe in the Mill to be at his Disposal at all convenient Opportunities. If after any of the Castings are begun to be turned or bored should prove Wasters, the Wages of the Men for the time employed in turning or Boring such Wasters, to be divided between both parties.

If the Engines are not put up & tried, three pounds to be deducted from the aforementioned Price.

This Agreement to be good for five & twenty Engines. Every article is fitted by W H & put into the Magazine Weighing, Packing & Loading not included in his Contract. An Allowance made in the fitting for hemp, tallow oil, Coals & Interest of Capital expended in the Shed & Machinery &c for trying the Engines.

Box 6H

DOCUMENT 111

Piece work Prices at Butterley Furnace 1802

Dry sand 20 to 21/- per ton, proposed to raise it to 22/-
Green sand 7/- to 20/- according to work.

Low Moor Prices

Loam Castings	under 10 Cwt	3/- per Cwt
Do.	10 to 20	2/6 "
Do.	above 20 Cwt	2/- "
Green sand	Millwork Wheels &c	7d. per Cwt
Bath & Stove Work & casting Boxes		1/6 per Do.
Bars & bearers flashed		4d. per Do.

The Moulders stand to all misfortunes in the running, the Master stands to saleable goods only.

Prices at Gorsecote

Loam work	50/- per ton
Dry Sand	21/- Do.
Green sand	10/- Do.

The Moulders finding labourers & all assistance.

Prices at Halifax

Loam	2/- per Cwt
Dry sandwork	1/- per Cwt
Pipe Work	1/- per Inch
Wheelwork	7d. per Cwt

Box 45

DOCUMENT 112

The Humble Petition of the men belong[g] to the Foundry to those Employers for there Later neglect of not fulfilling there time 1802.

Gentl[n]

We beg the favour of our Over-time to be granted us again which we hope you will not Denie us and our future Conduct shall be to give Every Satisfaction that in our power lyes — and Gentlemen we beg leave to Accept our thanks with Gratitude for all favours we have Receiv[d] and by us as Under Signed — May 25 1802.

Edward Collier
James Phillips
Daniel Pearsell
James Do.
Jos Marsh
Bath Young
John Scott
Edward Jones
John Greatise
Thos Bates
Samuel Griffiths
Jos Turton
Will Turton
Samuel Williams
John Leas
Thos Withers
W. Williams
F. Walker
Thos. Ray
Jos Harris
Jos Burns
Thomas Mousley
Joseph Stephens
Jos[h] Croft Sen[r]
James Stokes
Will[m] Dagg
Thos. Edge
Thomas Willis
Thomas Phillips

Will Burns
Will Fiddler
James Middleton
Samill Pearseall
Will[m] Talbutt
Owen Owens
John Burn
Benj[mn] Will[ms]
Will[m] Medows
James Ewer
John Hewitt
Will Day
Richard Webster
Samuell Wallford
Jos Anger
John Nelson
D. Wakefield
B. Croft
A. Croft

Box 45

DOCUMENT 113

Particulars of the sum paid the Piece Men and Day men in the Foundry Department for the Week ending 30th Oct. 1802.

Persons			Piecemen	Day Men	Total
			£ s	£ s d	
8	Joseph Croft for himself, 6 men & a boy		6. 6.		
2	Samuel Williams Do.	and one man	1. 15.		
3	Halligan & Owen Do.	and one man	2. 10.		
2	Isaac Croft Do.	and one man	2. 2.		
3	Thomas Mousley Do.	1 App[ce] & a boy	2. 0.		
4	James Middleton	2 App[ces] & a boy	2. 5.		
3	Daniel Pearsol	1 man & 1 App[ce]	2. 10.		
4	G. Croft	2 App[ces] & Chipper	2. 10.		
5	Collier & Young	2 men & 1 App[ce]	4.		
					25.18
	John Bunn	Air furnace man		16.-	
	James Phillips	Do.		14. 8	
	Jos. Hemmings	Do.		14. -	
7	Benj[n] Williams	Do.		14. -	
	Jos. Turton	Do.		14. -	
	Sam[l] Grigg	Do.		18. -	
	Jn[o] Scott	Do.		13. 6.	
	Wm. Ravenal	Smith		16. 3	
	Wm. Ravenal junr.	Do. App[ce]		7. -	
5	Francis Walker	Striker		12. 6.	
	Jos. Eanzer	Smith		15. 2	
	Jn[o] Greatrex	Striker		14. 1.	
	Thos. Ray	Pattern tender & furnace builder		1. 2.4	
2					
	Thos. Horton	Assistant to Ray		12. 5½	
1	Thos. Edge	Stove keeper		14. -	
	Wm. Day	Loam Mixer		15. 3	
2					
	Jn[o] Price	Do.		18. -	
1	W[m] Burne	Green sand Core Maker		12. -	
1	R[d] Jones	Scrap picker		10. 3½	
1	Jn[o] Hewitt	(to be discharged)		18. -	
	Wm. Mousley	Secretary to A. Storer		4.	
20	Day men				
34	Piecemen				
	Pd. for unloading coals 64 tons at 5d.			1. 6. 8	
	Do. Iron 78 tons 11 cwt 3 qr 2 lb at 2½d			16. 4	
	Wheeling sand into foundry			10 10	16.14.4
54	Total of Men & Boys				£42.12 4

Box 45

DOCUMENT 114

Arrangement of the Foundry Men[1]
1802-1803

Dry Sand

	Name	Status	Wage
1.	Joseph Croft Junr —	articled	21/- + 3/-Prem.[2] per week
	Joseph Marsh	App^{ce}	at 11/-
	Joseph Croft Senr	articled	14/-
	Joseph Stevens	Do.	14/-
	John Nelson		13/-
	Isaac Burne		13/-
	Thos. Betts		12/-
	Richd Davies		13/-
	Alex^{r} Croft	under age	4/-
2.	G. Halligan	articled	21/-
	Owen Owens	Do.	14/-
	Ed^{d} Niblett	Do.	13/-
3.	Isaac Croft	articled	18/- & Dble Prem.
	Wm. Talbot	Do.	13/-
4.	James Middleton	articled	14/-
	Thos. Wells	app^{ce}	10/-
	Richd Webster	Do.	7/-
	Middleton	under age	
5.	Samuel Williams	articled	17/-
	William Meadows	app^{ce}	11/-
	John Lee		10/-
6.	Thomas Mousley	articled	19/-
	Charles Betts	app^{ce}	8/-
	Wm Mousley	under age	3/-

2. Open Sand

	Name	Status	Wage
	Daniel Pearsol	articled	16/-
	Samuel Pearsol	articled	13/-
	Thos^{s} Withers	App^{ce}	9/-

3. Green Sand

	Name	Status	Wage
1.	Edward Collier	articled	18/-
	James Pearsol	app^{ce}	11/-
	Saml Walford	Do.	8/-

2.	George Croft	articled	21/-
	Barnabas Croft	app[ce]	7/-
	Tho[s] Phillips	Do.	8/-
3.	Bartholomew Young	articled	19/-
	Francis Walker		12/-

Labourers in the Foundry to be found by B & W

Loam Mixers

Wm. Day		13/-
John Hewit (To be discharged & Price to take his place)		13/-

Core Maker Green Sand

Wm. Burne		12/-

Stove tender

Thomas Edge	articled	14/-

Scrap picker

Richard Jones		13/-

Pattern tender & Furnace Builder

Thos Wray	articled	16/-
Thos Horton		13/-

Smiths

W[m] Ravenall		15/-
Do. Junr.	app[ce]	6/-
Jos. Eanzer		14/-
John Greatrix	striker	13/-

Air Furnace & Cupola Men

John Burn	articled	16/-
James Phillips	articled	16/-
Benj[n] Williams	Do.	14/-
Jos. Turton	Do.	14/-
Jos. Hemmings	Do.	14/-

Chippers not yet provided for

John Scott	to chip for the Green Sand men	13/6
Saml Grigg	since articled as Air furnace man	13/-
Jn[o] Price	Now Assistant to the Loam mixer	13/-

Joseph Croft	9 men & boys
S. Williams	3 Do.
Halligan & Owens	3 Do.
Isaac Croft	2 Do.
T. Mousley	3 Do.
Ja^{s} Middleton	4 Do.
D. Pearsol	3 Do.
G. Croft	3 Do.
Collier & Young	5 Do.
Piece men & Assistants paid by them	35
Loam Mixers	2
Core Maker	1
Stove tender	1
Scrap picker	1
Pattern tender & stove builder	2
Air furnace and Cupola Man	5
Smiths & Assistant Chippers unprovided for	3
Men working by the day & paid by B W & Co	19
Total of Men & Boys in Foundry Department at the commencement of Piecework August 1803.	54

Box 45

[1]It will be noticed that the totals for each department do not agree exactly with those in document 113.

[2]For Premium see Introduction, p. 13.

DOCUMENT 115

Conditions upon which Piece Work is let to James Middleton, 1802[1]

Boulton & Watt to provide a Man to mix up the Loam necessary to directions given by James Middleton, they are also to provide a Man to heat the Stoves & to work the Air furnaces. Middleton & his men assisting to charge.

James Middleton to find labour of every other kind necessary for the Moulding, drying, casting, taking out of the sand and Foundry, chipping & weighing the goods. Viz to bring the sand into the Foundry, to sift & mix it up, to bring the Boxes into the Foundry and remove them to the bank when done with. To ram and finish his own moulds, prepare his own Cores & Cakes (Loam being found him ready mixed up in the Loam house) & to mix up & apply his own Blacking. To give directions to the Man who heats the stoves when & how to heat them. To work the Crane for his own purposes & to keep it in order as far as he is concerned. To do all the work necessary at the time of Casting (except what relates to the Air furnace). To take the Castings out of the Boxes and the Cores out of the Castings. To remove the Castings to the place allotted for Chipping, to pick out the large Scrap Iron and remove the rubbish to such part of the bank as he may be directed. To keep the part of the Foundry & Stoves which he occupies, clean & orderly & to use his boxes, core barrels, and all other tools and Implements in a workmanlike manner. To chip and weigh his Castings & deliver them upon such part of the bank as he may at the time be directed.

To stand to all Wasters whether apparent at the time of Casting or afterwards in the boring or finishing, in case the defect arises from the Casting. To point out to A. Storey all such defects as he may himself notice in the Castings & to attempt to conceal none by lead, cement or any other contrivance.

All Castings, the surface of which is uneven from any defect of the Mould or improper application of the Blacking, to be considered as Wasters. To keep exact to such dimensions as are prescribed to him, and to finish his work in at least as good a manner as has hitherto been done endeavouring to improve it so that no Foundry may do better. The chipping to be done to the satisfaction of A. Storey and of the Superintendents of Soho and Soho Foundry. Middleton to break up his own Wasters.

Patterns to be carefully used and any damage arising to them from neglect or wilful bad usage to be made good at the expence of

James Middleton. The patterns to be delivered to him in the Foundry by the Pattern keeper, and to give him notice to remove them when done with. Alterations of Patterns or Castings not amounting to more or less than one Cwt not to affect the present proposals which are to hold good for one twelvemonth.

Soho Foundry
24th July 1802.

Box 45

[1]A dry sand moulder.

DOCUMENT 116

Conditions upon which Piece Work is let to Edward Collier[1] & George Croft, 1802

Boulton & Watt to provide one man to make the cores and in case he is not able to do the whole the moulders to assist him at their own expence; Boulton & Watt also to find men to heat the stoves and work the Air furnace & Cupola the moulders assisting to charge the former.

Collier & Croft to find all labour of every other kind necessary for the moulding, casting, taking out of the Sand & Foundry, chipping, weighing & removing the goods to the Bank. Viz, they are to bring the sand into the foundry & to prepare it for moulding, to bring the boxes into the foundry and remove them to the Bank when done with; to ram & finish their own moulds, bring the Coal dust across from the Mill & apply it to their Moulds; to work the Crane for their own purposes & keep it in order as far as they are concerned. To do all the work necessary at the time of Casting: to remove the Castings when cool to the place allotted for shipping; to pick out the large Scrap Iron and remove the rubbish to such part of the Bank as they may at the time be directed; to keep the part of the foundry & stoves which they occupy clean and orderly and to use all boxes and other Implements in a workmanlike manner.

To chip & weigh their own Castings and deliver them upon such part of the bank as they may at the time be directed.

To stand to all wasters, whether apparent at the time of casting, or afterwards in the fitting, in case the defect arises from the Casting. To point out to Mr. Storey the defects they may themselves notice in the Castings & to attempt to conceal none by lead, cement or any other contrivance. All castings the surface of which is uneven & the edge or corners not sharp to be considered as Wasters. The Wasters to be broken up by themselves. To keep exact to the dimensions of the patterns and in case any excess of weight is given to a casting either through design or neglect to forfeit five shillings in every instance. The work to be finished in at least as good a manner as hitherto, endeavouring to improve it, so that no foundry may do better. The chipping to be done to the satisfaction of Mr. Storey and of the superintendants of Soho & Soho Foundry.

Patterns to be carefully used and any damage arising to them from neglect or wilful bad usage to be made good at the expence of Collier & Croft. The patterns to be delivered to them in the foundry by the pattern keeper and to give him notice to remove them when done with.

Alterations of Paterns not materially increasing or diminishing the trouble of moulding not to affect the present proposals which are to hold good for one twelvemonth.

Bartholomew Young engaging to exert himself to finish his work in at least as good a manner as his fellow workmen, is to be allowed to participate in the conditions of this proposal.

Soho Foundry
6th August 1802.

Box 45

[1]A green sand moulder.

DOCUMENT 117

Conditions upon which Piece Work is let to
George Halligan & Owen Owens[1] 16 July 1802

Boulton & Watt to provide a Man to mix up the loam according to direction given by Halligan & Owen. Boulton & Watt to provide a Man to work the Air furnace, the moulders assisting to charge. Boulton & Watt to find a man also to heat the stoves.

Halligan & Owen to find all labour of every other kind necessary for the Moulding, Drying, casting, taking out of the sand & Foundry, chipping & weighing the Goods. Viz To bring their sand into the Foundry, to sift & mix it up, to bring their Boxes into the foundry and remove them to the Bank when done with. To ram & finish their own moulds, prepare their own cores & cakes (loam being found them ready mixed up in the loam house) and to mix up and apply their own blacking. To give directions to the man who fires the Stoves when & how to heat them. To work the crane themselves & keep it in order as far as they are concerned. To do all the work necessary at the time of Casting (except what relates to the air furnace). To take the Castings out of the Boxes & the Cores out of the Castings. To remove the Castings to the Chipping Shed, to pick out the large scrap Iron & remove the rubbish to such part of the Bank, as they may be directed. To keep the part of the foundry & stoves which they occupy clean & orderly, and to use their Boxes, Core barrels and all other tools & Implements in a workmanlike manner. To chip & weigh their Castings, deliver such as go down by the cassons as near to the top of it as they can, and all others upon such part of the bank, as they may at the time be directed.

To stand to all Wasters whether apparent at the time of Casting, or afterwards in the boring or finishing, in case the defect arises from the Casting. To point out to A. Storey all such defects as they may notice themselves in the Castings and to attempt to conceal none by Lead, Cement, or any other Contrivance. All Castings, the surface of which is uneven from any defects of the mould or improper application of the Blacking, to be considered as Wasters. To keep exact to such dimensions as are prescribed to them, and to finish their work in at least as good a manner as has been hitherto done, endeavouring to improve it so that no other foundry may do better. The Chipping to be done to the satisfaction of A. Storey and of the superintendants of the fitting Departments at Soho & Soho Foundry.

Alteration of Patterns or Castings not amounting to more or less than one Cwt not to affect the present terms which are to hold good for one twelve month. Patterns to be carefully used and any damages arising from neglect or wilfull bad usage, to be made good at the expence of Halligan & Owen. The Patterns to be delivered to them in the foundry by the Pattern keeper & to give him notice to remove them when they are done with.

Soho Foundry, 16 July 1802.

Box 45

---

[1]Dry sand moulders.

DOCUMENT 118

Calculation of the Expence of Moulding Cylinders, Air Pumps & Steam Cases 2 to 6 Horse for Sliding Valve Engines July 1802 Jany 1808

Joseph Marsh offers to do the Cyl^{rs} 2 to 6 H at 2/- per Cwt; the Air pumps at 2/6 & the Steam Cases at the same.

These terms being offered to Halligan & Owen they agree to accept them with an allowance of 3d per Cwt on the Cyl^{rs} for Chipping, getting out of foundry & weighing & 4d upon the air pumps & steam cases, leaving the Piece price as follows — Cylinders 2/3 per Cwt, air pumps & steam cases 2/10 per Cwt.

Revised Calculation of Cylinders 4 to 6H Sliding Valve Engines / 3rd Jany. 1808

Each of the above sizes require as stated in the Calculation of July 1802, 1½ day of the Moulder & his man.

George Halligan 1½ day at 3/6		5. 3
His man at 2/6		3. 9
		9. -
But as 1¾ day seems an ample allowance of time, let the surplus go against the Piece work allowance & add only for Risk of Boring 1/12	=	9
		9. 9

	Cwt	qr		
4H =	3	1	at 9/9	= 3/-
6H =	4	1	at 9/9	= 2/3½ per Cwt

Calculation of Cylinders 8 to 10h for Sliding Valve Engines / 3 Jany. 1808

2¼ days of G. Halligan at 3/6	=	7	10½
2¼ days of his man at 2/6	=	5	7½
		13	6

Add for Piecework 1/6	=	2	3
		15	9
Add for Risks of Boring 1/12		1	3
		17/-	

		Cwt	qr			
8H	=	6	1	at 17/-	= 2/8½ per Cwt	When any 12 & 14H are made of the same length of stroke, the price will be the same
10H	=	7	-	at 17/-	= 2/5 per Cwt	or say 17/- each.

Box 45

DOCUMENT 119

Calculation of the time taken by Samuel Williams in moulding Cylr tops, bottoms & pistons from Patterns in Dry Sand. 1802.

During the Quarter beginning 1 Apr 1802 & ending 30 June Do. he appears by the time Book to have worked 80¾ days, of which A. Storey thinks about 8 days have been occupied in moulding a Crank & three main gudgeons, so that he has been occupied 72¾ days upon the Cylr tops, bottoms & pistons. The whole weight of Castings of those descriptions done by him in the above 72¾ days appears to be Ton 8 Cwt 11 Qr 1 lb 20.

His wages are 2/10 per day	=	£10. 6. 1½ in 72¾ days
His assistant Lee's wages are 1/8 per day & he has been 86¾ of which 78 may be reckoned to Cylinder tops, bottom &c	=	6. 10
		£16. 16. 1½

	Ton	Cwt	
Premium in	8	11	
	20		
		171 Cwt	1. 5. 8½
			£18. 1. 10.

171 Cwt : £18. 2 :: 1 Cwt =
2/1^{d} per Cwt.
171 Cwt ÷ 72¾ days = 2 Cwt
per day.

Samuel Williams proposes to take the Cylr tops, bottoms & pistons of all sizes at 1/3 per Cwt which appears attendantly low.

His wages are	2. 10 per day
His assistants are	1. 8
Amount of both	4. 6

1/3 : 1 Cwt :: 4/6 : 3 Cwt 2^{qr}11 lb

So that to earn his present wages he must do one Cwt per day more than has yet been done and 1½ Cwt more than he did in the quarter from Apl to June 1802.

Box 45/1

DOCUMENT 120

Statement of the Quantity of Castings done by the Moulders in different Quarters. 1800-1802.

	1. Joseph Croft	Goods Cwt	Wasters Cwt	Total Cwt
1800	1 July to 30 Sept.	383	—	383
	1 Oct to 30 Decr.	230	7	237
1801	1 Jany. to 30 March	468	—	468
	1 Apl to 30 June	473	6	479
	1 July to 30 Sept.	338	—	338
	1 Oct. to 30 Decr.	253	—	253
1802	1 Jany. to 30 March (Beams)	588	—	588
	1 Apr to 30 June	359	—	359
	divide by 8)	3092	8)	3105
Average Quantity made per Quarter during the Premium		386		388

1 July to 17 July	Premium	71		
17 July to 30 Sept.	Piecework	866	106	1043
		937		

Observations
Whilst upon Premium Work, Joseph Croft had only three regular Assistants, viz J. Croft Senr., Jos. Stevens & Isaac Burne, with extra Assistance of seven or 8 labourers in getting the Cylrs. out of the Pit and sometimes of 3 in ramming.
Since he has been upon Piecework, he has had 8 regular Assistants during the greatest part of the time, viz the above three, Rich[d.] Davies, Jos. Marsh, Jn[o.] Nelson, Thos. Betts & Alex[r] Croft, who have done all his chipping.
Joseph Croft has drawn £7.7 weekly for himself & Man, the whole of which it is supposed he may have paid exclusive of his own Wages — The Ballance of £19. 5. 7 due to him at settling may be considered as clear gain during the 10 weeks of Piecework = nearly £2 per week.

2. Samuel Williams
NB Joseph Marsh was employed upon Cylinder tops & bottoms previous to S. Williams and the following is the Amount of goods done by him.

		Goods	Wasters	Total Cwts
1800	1 Octr to 31 Decr.	219	3	222
1801	1 Apl. to 30 June	206	6	212
	S. Williams as under			
	1 Octr to 30 Decr.	159	4	163
1802	1 Jany. to 30 March	179	21	200
	1 Apl to 30 June	191	3	194
	divide by 5)	934	5)	991
Average weight of Cylr. tops bottoms & pistons made per Quarter by Marsh or Williams		187		198
	1 July to 30 July Premium	61	—	61
	1 Aug. to 30 Sepr. Piecework	335	12	347
		396		408

Observations
In the Quarters in which the Work was done by Premium, Williams had only one Assistant viz John Lee; but in the Piecework Qr. he

had during part of the time, three Assistants viz John Lee, W. Meadows & S. Pritchard, who also with him did the chipping. Samuel Williams has drawn various sums weekly according to the number of Men employed under him and it is supposed the whole has gone to pay them & his own Wages.
So that the Ballance coming to him of £2.3.2 is the whole he can have cleared & it is doubtful whether he may not owe a part for Chipping. — supposing it clear again, he would have made it in 8 Weeks = 5/4 pr. Week.

3. Halligan & Owen
NB William Williams was formerly employed upon this Work, therefore the following Quarters of his have been extracted.

		Goods	Wasters	Total Cwts
1800	1 July to 30 Sepr.	288	—	288
	1 Octr. to 30 Decr.	257	—	257
1801	1 Jany. to 30 March	277	31	308
	1 Octr. to 30 Decr.	246	—	246
1802	1 Jany. to 30 March	214	14	228
	I Apl to 30 June	197	48	245
	divide by 6)	1479		6) 1572
Average Weight of Cylindors, Air pumps & Condensers made per Quarter by W. Williams leaving out the two in which so many shafts were done.		246		262
	Owen Owens			
	1 July to 17th July Premium	42		42
	Halligan & Owen			
	17 July to 30 Sept. Piecework	289	46	335

Observations
In the Premium Quarters; William Williams had 2 Assistants, viz Owen Owens & Ed[d] Niblet.
Halligan & Owen, have had one Assistant viz Edward Niblet — have done all their own Chipping.
Halligan & Owen have drawn barely Wages for themselves & Men & have done all their own chipping.
The Ballance of £9. 5. 10 due to them is clear gain = £4. 12. 10 each. In 10 Weeks = 9/3 each per Week.

4. Isaac Croft
NB No Accurate comparison between the Amount of Nozzles made by him when at Premium & those at Piecework can be drawn, as when at Prem^{m} he was chiefly employed upon Side pipes, to check etc.

		Goods	Wasters	Total Cwts
1800	1 July to 30 Sep^{r}	222	2	224
	1 Octr to 31 Decr	258	2	260
1801	1 Jany. to 30 March	198	19	217
	1 Apl to 30 June	240	3	243
	1 July to 30 Sepr	243	31	274
	1 Octr. to 31 Decr.	221	14	235
1802	1 Jany to 30 March	239	28	267
	30 March to 30 June	191	15	206
	divide by 8)	1812		8) 1926
Average Weight of Nozzles & Pipes & Shafts made by Isaac Croft per Qu^{r}		226		241
1 July to 17 July	Premium	40	5	45
17 July to 30 Sepr. viz 155 Cwt Nozzles, 121 Cwt Roll Frames plates & Shafts 30 Cwt.	Piecework	306	4	310
		346		355

Observations
Isaac Croft has only had one Assistant during the Premium time viz William Talbot. He has had the same at Piecework, but has paid for all his Chipping & for occasional Assistance in moving the heavy boxes. It is supposed that the Sum drawn weekly by Isaac Croft during the above quarter has been sufficient to pay his Man, himself & all his Chipping.
There is a ballance of £10.-.10½ due to him upon 10 weeks work = £1 per week clear gain in addition to his Wages.

Statement of the Quantity of Castings done by the Moulders in different Qrs. Continued

5. Thomas Mousley

		Goods	Wasters	Total Cwts
1800	1 July to 30 Sep^r	205	28	233
	1 Octr to 31 Decr	165	20	185
1801	1 Jany to 30 March	183	16	199
	1 Apl to 30 June	164	—	164
	1 July to 30 Sepr	160	22	182
1802	1 Jany to 30 March	158	27	185
	1 Apl to 30 June	150	7	157
	divide by 7)	1185		7) 1305
Average Weight of Pipes of various descriptions made by Mousley per Q^r		169		186
	1 July to 24 July Premium	50	1	51
	24 July to 30 Sepr Piecework	180	30	210
		230		261

Observations
Mousley had during the Premium Quarters, the assistance of Chas Betts the App^{ce} & of his own son Wm Mousley, with occasional aid from the labourers in the foundry.
At Piecework he has been Assisted by Chas Betts, and Jno Mousley with occasional aid from Wm Mousley, Wm Turton & S. Pritchard. He has done the greatest part of his chipping with his own men, but has had some done by others.
Thos Mousley has drawn various sums weekly according to the number of hands employed under him. He says that after he has paid his Men & paid for chipping he has not had more than 4/- per Week for himself.
The Ballance coming to him for 9 Weeks Piecework is £1. 16. 8 = 3/- p^r Week which added to 14/- would not make his wages — — which are 19/-. His Wasters amount to £4.

6. James Middleton

		Goods	Wasters	Total
1802	1 Jany to 30 March	211	—	211
	1 Apl to 30 June	207	12	219
	divide by 2)	418		2) 430
		209		215

Average Weight of Pipes of various decriptions but chiefly side pipes & Sockets made per Q[r].

1 July to 17 July	Premium	43		43
17 July to 30 Sepr	Piecework	355	10	374
		398		417

230 Cwt of this consist of Shafts, of which few or none were made in the preceding quarters.

Observations

James Middleton had only one Assistant when at Premium work viz, W[m] Dag who has since left.

At Piecework, he has had 2 Assistants Tho[s] Wells & Rich[d] Webster and during a part of the time, a boy W[m] Middleton.

His Assistants have done all the Chipping for which he has paid them extra.

James Middleton has worked exceeding hard during the Quarter & done his work very well. He states that the weekly sum he has drawn has not been sufficient to enable him to pay his Men & chipping & to keep Wages for himself, upon which Account he is in advance.

The Ballance coming to him is £6. 19. 5½ made during 10 Weeks. It is supposed that not more than £4 of this may be clear gain = 8/- per Week.

7. Vacant, formerly Marsh.

8. Daniell Pearsoll

NB No Comparison can be formed with the first quarters of G. Halligan, as he was much employed at that time upon Nozzles & shafts in Dry Sand as well as upon Open Sand Casting.

	G. Halligan		Goods	Wasters	Total Cwts
1801	1 July to 30 Sep[r]	92 Nozzles &c			
		10 Gudgeon			
		364 Fly Wheels			
		791 Grate Bars &c			
		84 Plate Beams			
			1341	2	1343

	1 Octr to 31 Dec^r	5 Trough 255 Beams 36 Crane 225 Fly Wheels 613 Grate Bars &c			
			1134	4	1138
	Halligan to 26 Feby Pearsol from thence to the end of the Month				
1802	1 Jany to 30 March	172 Plate Beams 231 Fly Wheels 203 Bearer Bars &c 606 348 Fly Wheels 204 Bearer Bars			
			1158	—	1158
	Pearsol 1 Ap^l to 30 June	501 Fly Wheels 6 Small Plates 528 Bearer Bars &c 143 Plate Beams			
			1178	—	1178
		divide by 4)	4811	4)	4817
Average Weight of Open Sand & Castings with Loam Cakes per Q^r			1205	Cwt	1204
1802	1 July to 1 Aug Premium	286 Fly Wheels 67 Open Sand			
			353		
	1 Aug to 30 Sepr Piecework	528 Fly Wheels 8 Stand plates 80 D & L plates 22 Sundries — Do. 372			
			1010		1363

9. George Croft

		Goods	Wasters	Total
1800	1 July to 30 Sep^r^	270	10	280
	1 Octr to 31 Dec^r^	263	23	286
1801	1 Jany to 30 March	325	13	338
	1 Ap^l^ to 30 June	360	13	373
	1 July to 30 Sep^r^	478	45	523
	1 Octr to 31 Dec^r^	392	10	402
1802	1 Jany to 30 March	457	15	472
	1 Ap^l^ to 30 June	434	15	449
	divide by 4)	1761		4) 1846
	Average weight of the 4 last Quarters	440		461
	1 July to 7 August Premium	145	6	151
	7 Aug to 30 Sepr Piecework	386	suppose 15	401
		531		552

Observations
George Croft whilst at Premium Work had the Assistance of his two brothers Alex. Croft Barn[s] Croft and during part of the time of Tho[a] Philips. Since the Piece Work, he has been assisted by Barn[s] Croft, Tho[s] Phillips and during part of the time by the boy John Phillips. He pays for his chipping, which is done by John Scott chiefly.
It is supposed that G. Croft's weekly drafts have been scarcely sufficient to allow him Wages for himself & pay his Men & Chipping. If he stood clear of Debt to his Assistants at the end of the quarter, the ballance of £6. 6 which has accrued in 7 Weeks would make his gain = 18/- p[r] Week but it probably has not been more than 12/- to 15/-.

10 & 11. Collier & Young

NB These two have agreed to take the Piecework together.

			Goods	Wasters Cwts	Collier Cwts	Total of Young Cwts	Joint Amount
1800	1 July to 30 Sep^r	Collier	359	14	373	—	765
	Do. Do.	Young	364	28	—	392	
	1 Octr to 31 Dec^r	Collier	309	7	316	—	726
	Do. Do.	Young	390	20	—	410	
1801	1 Jany to 30 Mar	Collier	393	6	399	—	808
	Do. Do.	Young	393	16	—	409	
	1 Ap^l to 30 June	Collier	405	12	417	—	819
	Do. Do.	Young	377	25	—	402	
	1 July to 30 Sep^r	Collier	350	13	363	—	715
	Do. Do.	Young	322	30	—	352	
	1 Octr to 31 Dec^r	Collier	175	15	190	—	548
	Do. Do.	Young	346	12	—	358	
1802	1 Jany to 30 Mar	Collier	184	10	194	—	730
	Do. Do.	Young	521	15	—	536	
	1 Ap^l to 30 June	Collier	225	3	228	—	638
	Do. Do.	Young	414	46	—	440	
		divide by 8	5527		2480	3319	5799
Average Weight P^r Q^r at Prem^m			691		309	415	725

NB It will be observed that during the 3 last quarters Collier has done a very small proportion of work to what he did in some of the first, having been absent during great part of them.

		Goods	Wasters	Collier	Young	Joint Amount
1802	1 July to 7 Aug. Prem. Collier	87	—	87	—	282
	1 July to 7 Aug. Prem. Young	170	25	—	195	
	7 Aug to 30 Septr Joint Piecework of Collier & Young	473	30	—	—	503
		730			Cwts	785

Observations

Collier whilst at Premium work had the Assistance of James Pearsol & Samuel Walford.

B. Young was assisted whilst at Premium work by Frank Walker, who has since gone to the Smithy. Collier & Young since at Piecework jointly have been assisted by James Pearsol, Samuel Walford & John Young. Part of the Chipping has been done by themselves & part hired out to the Air furnace Man.

Comparative View of the Quantity, say of the Weight of Good Castings done in the Various Quarters since the Institution of Premium, including the first Quarter of Piecework

		1800 1 July to 30 Sepr	1800 1 Octr to 31 Decr	1801 1 Jany to 31 March	1801 1 Apl to 30 June	1801 1 July to 30 Sepr	1801 1 Octr to 31 Decr	1802 1 Jany to 31 March	1802 1 Apl to 30 June	Premium & Piecework 1802 1 July to 30 Sepr
Wm Gilpin		229	—	—	—	—	—	—	—	—
Joseph Croft		383	230	468	473	338	253	588	359	937
William Williams		288	257	277	347	354	246	214	197	—
Halligan & Owen		—	—	—	—	—	—	—	—	331
Isaac Croft		222	258	198	240	243	221	239	191	346
Thos Mousley		205	165	183	164	160	97	158	150	230
Jas Middleton		—	—	—	—	—	125	211	207	407
Wm Meadows		144	231	212	269	152	98	362	217	96
Jos Marsh		251	219	287	206	152	99	140	120	59
Saml Williams		—	—	—	—	—	159	179	191	396
G. Halligan		579	509	627	657	1331	1135	607	—	—
D. Pearsol		—	—	—	—	—	—	552	1180	1363
G. Croft		270	263	325	360	478	392	457	434	531
E. Collier		359	309	393	405	350	175	184	225	730
B. Young		364	390	393	377	322	346	524	414	
Weight Pr Qr	Cwts	3294	2831	3363	3498	3880	3346	4412	3885	5426
	Tons	164.14	141.11	168.3	174.18	194.-	167.6	220.12	194.5	271.6

DOCUMENT 121

Memorandum of Conversation with Samuel Richards relative to Boiler making 20 March 1809 and settlement 29 April 1809.

Samuel Richards says he has been 13 years with Thomas Horton, who allowed him 21/- per week with some perquisites of piecework & overtime; and that when he was sent from home, he had his expences of living allowed him in addition to his wages.

Says that he takes the boilers under Baker by the piece, in paying the men and doing the whole work to the boilers which he undertakes, says that Baker allows him £3 per ton for the Boilers made in pieces and £4 for those made whole. That the £3 per ton is too little and he cannot undertake to do them properly for less than £3.10 per ton. Does not object to the price of £4 pr ton for those made whole. Thinks some extra allowances should be made upon the Tube boilers, as being more difficult, but does not urge this strongly: and I suppose would be satisfied to undertake the boilers as they come, at £3.10 for those in pieces & £4 for the others.

Is willing to article with us for 27/- per week, in which he includes all the overtime he makes; and it is to be observed that he is generally there by 4 o Clock in the Morning & remains late in the Evening.

One of the men who works with him will engage for 21/- pr week and I have given Mr. Hart directions to make out his Agreement.

The other man will not engage under 24/- which Richards thinks more than he is worth and advises to let him go and get another in his place, which he will have no difficulty in doing. The price which we allow Baker is £6.10 per Ton — upon all sorts and sizes.

29th April 1809

Upon further explanation with Richards, he says that he omitted to mention that in the prices stated by him he did not include the cost of making rivets, those being found by Baker as well as a boy to attend him.

Richards thinks there are 2 Cwt of Rivets in the Ton of finished Boilers. Horton allowed 5/- per Cwt of iron delivered for making the Rivets out of all sorts of Iron, which allowing 1/10 for waste would be equal to 5/6 pr Cwt of the finished Rivets =

11/-per ton of Boilers. A Rivet carrier's wages may be stated at 3/-per week.

Suppose price for Boiler in pieces not including Rivets or Rivetter	£3.10
Add for Rivets	11
Rivetter	3
	£4.4

Suppose price for Whole Boilers	£4
Add for Rivets & Rivetter	.14
	£4.14

Richard asks £4.10 for doing the boilers in pieces & £5.10 for those made whole. After much discussion he agrees to take the Piece Boilers at £4.4 per ton finished weight & the whole Boilers at £5 per ton.

Thinks that with 2 men & a boy he could finish a Ton of Boilers whole, per week.

His own wages	£1. 7
Two men at £1.1 each	2. 2
2 Cwt Rivets	11
Boy for Rivetting	3
	£4. 3

Box 45

DOCUMENT 122

Accounting guidelines

1816

In course all articles which have been delivered from the Manufactured Goods account to the different Departments will be taken in the Inventory as Goods.

Those manufactured goods remaining in the Store house will be taken as the Stock of manufactured goods from their not having been delivered.

The Rough brasses when once delivered become goods & in course are taken as such in the department to which they have been charged in the delivery.

Box 46

DOCUMENT 123

To James Brown, Soho, 16 Oct. 1824

The fact is there is more work than can be done, 7 or 8 founderies have now orders from thence but castings can scarcely be had for money — at SF[1] they are preparing for filling another Foundry with cupolas but cannot get on for want of pipes for blast which are necessarily procured elsewhere.

At present the Manchester foundries work only 3 days pr week, for the men get such wages it takes the remainder to squander the money away.

W. Creighton

Box 45/1

[1]Soho Foundry.

Section J

The engine business – engine sales

The engine business – engine sales

Documents 124-213

The domestic market was, both in terms of value and the number of orders obtained, of greater significance than the overseas one. Cornwall was, from the outset, of major importance since the premium system of payment operated here to Boulton and Watt's advantage. It was partly for this reason that the partners were particularly keen to stamp out piracy in Cornwall. For their first Cornish engines, Boulton and Watt carried out careful trials in order to ascertain the potential coal savings to be made by their engines compared with atmospheric ones[127].

Nevertheless they were prepared to sell engines in less profitable regions where they were hopeful of further orders because the entrepreneur in question undertook to advertise them[132]. And the partners deliberately sought out leading innovating entrepreneurs in the hope that if they could be persuaded to purchase a Boulton and Watt engine, others in the same industrial sector would follow suit[142 & 144].

But Boulton and Watt failed to make much impact in some markets where the demand for power was high as McTaggart's list of western Scotland engines shows[147] for Glasgow had become a noted centre of steam engineering and manufacturers preferred to buy locally.

Some of the most interesting information was furnished by those entrepreneurs who made a negative decision, either because they felt that Boulton and Watt's claims to coal saving were misleading[160] or because they wished to use the engine as a supplement to water power[161], in which case the premium acted as a deterrent.

The overseas market appealed to both Boulton and Watt. Boulton was an extensive exporter of hardware before the commencement of his partnership with Watt and he employed his well tried consumer goods marketing techniques with this new capital good. Since many of the overseas customers were monarchs, noblemen or members of governments, cost effectiveness was not of first importance and Boulton's marketing methods were in some measure appropriate. For certain overseas customers a total package of engine, millwork and machinery was required[176].

Watt Jnr. recognised the importance of the expanding marine engine market, but the adaptation of boat designs to accommodate engines was very much a case of trial and error. As with late 18th century stationary engines the firm's prices were consistently undercut by other engineering firms. With the expiry of Watt's patent looming, the firm became more competitive in its attitude to marketing. Although the prices of Boulton and Watt engines fell very little after 1800, the partners introduced a system of after-sales service which operated as an effective sales technique. Partners and senior engineers visited engine owners in different regions recommending maintenenance work and seeking new orders. Much of this work was undertaken by Watt Jnr.[209-213]

Domestic Market – Cornwall

Documents 124-131

DOCUMENT 124

Proposal to Poldice n.d.[1]

The Engine to be erected at charge of adventurers B. & W. to furnish drawings and directions of all sorts for ye use of ye Companys Engineers, who are to adhere to them –
The Coals now burnt by the Engine in keeping ye water in fork to be compared with those wch shall be burnt by the new Engines when in fork. The savings to be divided into 3 parts of wch one part to be paid to B & W & the other two parts to remain with the Adventurers. If the present Boilers should not be found to perform as well as that at Chasewater Others to be furnished by Adventurers. Each Engine to have 2 Boilers –
B & W. To have power of rejecting any improper Coals & to displace any Engine men.
The number of Strokes p^r minute necessary to keep ye water in to be agreed upon now –
Engines to be made capable of going ten fathoms deeper –
premium to be paid monthly
Charge of Drawings 25 Guineas p^r Engine
The Engines not be sold off the premises without Consent of B & W which they shall grant upon ye purchaser entering into agreement to pay their premium in proportion to savings in new situation.
B & W. to have liberty of visiting the engine and directing repairs when any complaints are made of extra Consumption. premium to commence from day Engine is sett to work & to continue while ye Engine is used No extra premium to be paid during time of forking y pool

(J. Watt's hand)

Box 9

[1]This is a confusing document. The style suggests an early date. And it is not clear whether one or two engines are intended. The most likely date is 1778 when a 63 in. cyl. 9 ft. stroke engine was ordered.

DOCUMENT 125

Poldice Mine October 30th 1778

We the under Subscribers having carefully examined the Books of this Mine touching the Consumption of Coals by the two Eastern Engines[1] during the Months of August and September of this present year and attended to the Working of the Engines during these Months do hereby certify that the said Two Eastern Engines did consume in Sixty one Days of these two Months $\overset{W}{220}$. $\overset{Q^r}{0}$. $\overset{B}{0}$ Ways of Coals each Way being 64 Winchester Coal Bushells which amounts to 14080 Bushels in Sixty one Days.

We do also certify that the said two Engines together do Work pumps in 4 lifts which are Seventeen Inches Diameter in the Working Barrels and the whole Depth from which these pumps draw the Water to the Addit is Fifty eight Fathoms.

We further certify that the said Engines did during these Months of August and September last Work the said Pumps at the Rate of 6 Strokes of 5½ Feet long each in every Minute which amounts to 8640 Strokes p^r 24 Hours.

We have also made an accurate calculation by which it appears that when the new Fire Engine[2] to be erected by Messrs Boulton and Watt is completted and actually Works a pump of the same Depth of Fifty eight Fathoms and Seventeen Inches in Diameter at the Rate of 6 Strokes of 5½ Feetlong each in every Minute and Consequently makes 8640 Strokes p^r Hours it will draw a Quantity of Water equal to the Quantity now drawn by both the Present Engines and Consequently whatever smaller Quantity of Coals it uses than 14080 Bushels for Sixty one Days when going at the Rate of 6 Strokes p^r Minute will be the Real savings in fuel Occasioned by the use of the said new Engine at that Rate of going —

James Watt
Matt^w Boulton
H Hawkins Tremayne
Richard Williams
John Williams
Thomas Brown

Box 9

[1]These were Newcomen engines.
[2]A 63 in. cyl. 9 ft. stroke engine with a horse power equivalent of 95 h.p.

DOCUMENT 126

To Mr Daniel

Soho Dec 1 1779

Sir

Since we had the Pleasure of meeting you at Truro we have again considered the Standard for common Engines where with ours ought to be compared and beg leave to submit the following articles to your deliberation.

1st When we proposed to the Public at large to accept of ⅓ of the savings as a Compensation for the use of our Engines, we did not believe that there were any Common engines which produced so great Effects with the fuel as are attributed to the present engines at Dolcoath, and had we been desired to compare with any such engines we should have demanded a much greater Share of the savings, as we believe if any such performances take place that it is owing to some particular construction of a boiler, perhaps accidental, to superior skill in the fireman, to better coals, larger measure or some other local advantage which we cannot command in every situation nor perhaps in any situation.

2^{d} It has always been our custom where there were engines on the spot to compare with them & no other; and if there were no engines on the spot, to compare with the nearest as those which were most likely to be affected by the same accidents of bad water, or coals &c.

3^{d} In all kinds of fire Engines that have hitherto been used there are some which produce much greater effects with the same nominal quantity of fuel than others do, which is frequently owing to accidental circumstances which cannot be discovered and consequently cannot with any degree of certainty be corrected. They are also more or less perfect according to the skill of the engine men the goodness of the water The quality of the coals, the measure of the Coals, with other circumstances such as the opportunitys which can be allowed for repairs &c. Our engines are subject to be affected by all these Circumstances in a greater or lesser degree; And surely the Proprietors of Wheal Virgin Mines would reckon it very unfair if we told them that we should not calculate the savings

from the actual Consumption of any engines we might erect for them but from such other of our engines in this county as we found to produce the greatest effect. If therefore this would justly be esteemed a grievance when imposed on the proprietors of Wheal Virgin It cannot be esteemed justice to require it us to compare our engines with the very best of the common sort that can be found.
4th As we must adhere to the effects of our engines under all such local disadvantages as cannot be remedied An average of the common engines should be taken as they realy exist; or if the proprietors of Wheal Virgin should claim a right to strike out of the list such engines as they think defective we should on the other hand beg leave to reject an equal number of those we think exaggerated or not applicable to the present case.
5. In our agreement with the proprietors of Poldice mine we made a very considerable deduction from the quantity of Coals really consumed and settled the other matters of fact in such a manner as appeared to us to throw the advantage on the side of the proprietors, and we still view it in the same light. We have offered the facts then agreed upon to be fixed as a Standard for the whole county which will appear from the accounts of Engines herewith sent you to be considerably within the average effects of the engines in the County taken at large even allowing the Assertions concerning some of the common engines to be true, which we cannot believe do not admitt of except for the sake of argument in this case.
6. By the accounts we are furnished with it appears that in the month of Sepr 1778 The seven engines on the combined mines consumed 34,454 bushels of Coals, and we are told that the 5 engines of our construction now proposed to be erected would draw the water of that month at 6 Strokes per minute. At that rate of going and at the loads they would sett and with the tables made from the two Standard engines at Poldice would rate the consumption of 5 Common engines of equal power with those we propose to erect, at —
26,105 Bushels monthly and our engines would probably consume
10,110 bushels which being deducted from the estimated consumption of the Common engines would leave 15995 Bushels the total savings, one third of which (= 5331 bushels which we claim for our share) added to the actual Consumption gives 15441 bushels for the monthly expence in coals to the proprietors when the engines went at the rate of 6 strokes pr minute. Hence it appears that the proprietors would save 19013 bushels from their former Consumption even in the dryest months and consequently much more in those months when the water was quickest and still more as the mines become deeper.
7. The water which must be used for the Boilers and injection water

at Wheal Virgin is exceedingly is exceedingly acid and corrosive and must sooner or later destroy any iron Cylinders of Common engines which could be employed and consequently would render them worse than other engines equally well constructed which were supplied with better water. From this Inconvenience our Engines would be freed except in so far as regards the iron work of the Condenser, and there the effects would not prove so bad as they do in Common engines; But these corrosive waters produce very bad effects in the Boilers and would thereby hurt our engines as well as the common ones, though not in so great a degree; for as the destruction of the boilers is proportioned to the quantities of coals used and as our engines use smaller quantities than the common ones, they would suffer less injury in the same proportion, though the effects of the engines would still be sensibly affected by it.

(James Watt)

Box 9

DOCUMENT 127

An account of the Consumption of Coals by the two Eastern and the Middle Fire Engines[1] at Poldice Mine, and Calculations therefrom of the Coals which would be consumed by a Common fire Engine equal in power and length of Stroke to the new Engine erected at Hallamanin Mine[2] according to the invention of James Watt —

Poldice	Diameter of Cylinder	Depth in Fathoms	Box	Strokes p^r Minute	Length of Stroke		Load p^r Sq^r inch	Bushels of Coals Consumed p^r month of 30 days
					ft.	ins.	lb.	
N° 1 Eastern Eng^ne	66 inch	33	17 inch	6.	5	6	$5\frac{7}{10}$	3945
2 East^n Engine	60 inch	25	17 inch	6.	5	6	$5\frac{2}{10}$	2979
3 Middle -	66 inch	63	12 inch	5.	5	6	$5\frac{4}{10}$	2864

If a Cylinder of 66 inches dia^r., Loaded to $5\frac{7}{10}$ pounds p^r square inch of the area of such Cylinder, require 3945 bushels of Coals in 30 days what will a Cylinder of 60 inches dia^r require (length & number of Strokes equal). Say as the square of 66 (= 4356) is to the square of 60 (= 3600) so is 3945 Bushels of Coals to 3262 bushels the Coals necessary for a 60 inch Cylinder.

The Engine N° 3 being also a 66 being calculated in y^e^ same manner gives 2366 for the consumption of a 60 inch Cyl^r^ going the like length & number of Strokes, but as N° 3 goes only 6 Strokes p^r^ Minute ⅕ must be added to give the Consumption at 6 Strokes

2366 ÷ by 5 = 473
2366
2839

the Consumption at 6 Strokes p^r^ minute

N° 1 Loaded to lb. 5 7/10 will consume 3262

N° 2 Loaded to 5 2/10 will consume 2979

N° 3 Loaded to 5 4/10 will consume 2839

divide by 3/9080

Hence the average Load & Consum^n^ p^r^ monthly are 3027 bushels

Now it being agreed that a Cylinder 60 inches dia^r^ with a Stroke of 5½ feet long and making 6 Strokes p^r^ minute when loaded only with a Column of water equal to one pound p^r^ inch would consume in 30 days 1920 bushels or 64 bushels p^r^ day, deduct 1920 bushels from ye average burning of the 3 Engines (3027) remains 1107 which is ye monthly additional quantity of Bushels necessary to increase the power from one pound to 5 4/10 pounds p^r^ inch but, in order to take away fractions, say if 5 pounds 4 tenths require 1107 what will 5 pounds require. Answer 1025 bushels wch added to 1920 the Coals necessary for one pound p^r^ inch Cylinder going 6 Strokes of 5½ feet long each minute and loaded to 5 pounds p^r^ inch — Then dividing the aforesaid number 1025 by 4 the number of pounds intermediate between 5 pounds & one pound, gives 256¼ the number of bushels necessary for every pound in y^e^ inch increase of load —

As Hallamanin Engine will raise a Column of water in y^e^ pump equal to 10 pounds & a half on every square inch of the piston and the Common Engine only a Column of seven pounds p^r^ inch Square 40 = 1600 multiply by 10½ = 16800 divide by 7 = 2400 which is equal to the square of y^e^ Cylinder of Common Engine = in power to Hallamanin (about 49 inches dia^r^) — then by the rule of three say if 3600 (sq^r^ of 60) require 1920 what will 2400 require ans^r^ 1280 bushels p^r^ month for 6 Strokes of 5½ feet long at one pound p^r^ inch. And if 3600 require 2945 what will 2400 ans^r^ = 1963.33 bushels the consumption at 5 pounds. But these numbers corresponding to 5½ feet stroke must be further increased in y^e^ proportion of 5½ feet to 8 feet the Stroke of Hallamanin Engine. From these reasonings Facts & calculations the under table is formed —

60 Inch Cylinder 5½ feet Stroke 6 Strokes p^r minute 1^lb to 7 on sq.inch Consumes in 30 days		49 inch Cylinder 8 feet Stroke 6 Strokes p^r minute 1 pounds to 7 pounds p^r sq inch would consume in 30 days	
Load p^r inch	Bushels of Coals p^r month	Total Load in pounds weigh	Bushels p^r month
1	1920	1884.96	1861.8
2	2176.25	3769.92	2110.3
3	2432.50	5654.88	2358.8
4	2688.75	7539.84	2607.3
5	2945.00	9424.80	2355.8
6	3201.25	11309.76	3104.3
7	3457.50	13194.72	3352.8

In order to find the Savings according to the preceding Table It (is) necessary in ye first place to find the weight of the Column of water in ye pumps — to do which Sqr dia^r Working Barrel and add one fortieth of s^d square (and) multiply it by the yards the pumps are deep Example 12 inch box Sq^r 144

add 1/40 3.6

147.6

46 yards high

6789.6 pounds weight the load of the engine.

In order to find the monthly bushels of Coals corresponding to the given load — It is necessary to establish some Speedy rule which may be thus formed. Say as $\frac{\text{pounds}}{1885}$ is to 248.5 so is 10,000 pounds to 132 or rather 131.8 very nearly therefore the given number of pounds 6789 being multipled by, $\frac{132}{1000}$ will produce 896.148 wch being added to 1613 (the bushels which would be consumed monthly if the Engine was loaded to 0. pounds p^r inch) will give the month Consumption at the given load when going 6 Strokes p^r minute. But as the Engine may not go exactly 6 Strokes in a minute Divide the number of Strokes shown by the Counter by 8640 (The Strokes in a day when the Engine gives 6 Strokes p^r minute) the quotient will (be) the days it would have required at 6 Strokes p^r minute to have made equivalent the same number of Strokes from which deducting the actual consumption the savings will remain.

Box 9

[1]Newcomen engines.
[2]A 40 in. cyl. 8 ft. stroke engine.

DOCUMENT 128

To Ann Watt (his wife) Chacewater Jany 3^{d} 1780

Dear Annie

Since I wrote you last I have the pleasure of one of yours and am very Glad to hear that you and the Children continue well — You complain of bad weather it has been very severe here but no snow, it is now become fine mild weather again and we have the advantage of being able to read without candle to near five o clock after noon — Whele Chance[1] and Poldice[2] engines are both gone to work tolerably prosperously hitherto only Whele Chance has broke a rod and Poldice has dropt the ballance Bobb last night which will soon be repaired. Dolcoath[3] Engine has also broke the main rod and done much mischef on Friday last. —

We had a meeting with Hallamanin[4] last friday and have come to a sort of conclusion with them if they can be got to Stand to it for a week untill the agreement be executed — It was 12 oclock at night before we concluded any thing with them after wch we went to bed but Mr Wilson & others kept it up till 5 in the morning, by which he got a fever and a sore throat but is now better. Blewett & Bevan have never spoke together since the former corrected the latter — It is said that company will be disolved we have now only the United Mines & Wheal Treasury[5] agreements to execute both which will be done the next week — Poldice will be long in forking the water being very heavy there at present & 27 fathoms in — both the old Engines disabled so as to do us little good — Whele Chance has got down 24 fathoms out of 104 — There is ½ a dozen people speaking to me at present and I am obliged to go over to Poldice — so good night.

Compliments to all friends

Sincerely Yours

James Watt

MIV Watt (1)

[1]A 63 in. cyl. 9 ft. stroke engine ordered in 1779.
[2]a 63 in. cyl. 9 ft. stroke engine ordered in 1778.
[3]a 63 in. cyl. 9 ft. stroke engine ordered in 1779.
[4]a 40 in. cyl. 8 ft. stroke engine ordered in 1778.
[5]a 36 in. cyl. engine ordered in 1780.

DOCUMENT 129

Engines' Savings at Dolcoath Mine.

April 1782

Dolcoath Engine, Erected by Mr John Budge on Mr Newcommons Construction, Lifted a Column of water 21088 pounds Six feet. Stroke Six Strokes p^r^ minute — Consumption of Coals for Thirty days — 3240 Bushels — and if you refer to Messrs Boultons & Watts Table, for the Engine in the Said Mine it will appear they have allowed 6634 Bushels of Coals for Thirty days for an old Engine, to Lift 21009 pounds Six Strokes p^r^ Minute Six feet Stroke.

Box 9

June 1782

Consumption of Coals With Dolcoath Fire Engines from the 30th May 1782 To the end of June being 31 days

	Wey	Q^r^	B	
Boultons Engine	46.	2.	0	Weight Lifted 37381 pounds
Budges Engine	48.	1.	0	Weight Lifted 21088 pounds

Then if 21088 pounds require 48. 1. 0 (Wey Q^r^ B) what will 37381 require

	Wey	Q^r^	lb
in going Equil Strokes — the ans^wr^ is	85.	2.	2
and Boultons Table give	138.	3.	5
an over Charge in the Savings	53.	1.	3

	Wey	Q^r	lb
The Consumption with Boultons as above	46.	2.	0
one third of the Savings Charged To Boulton	30.	3.	1
Left for the adventurers Two thirds Saving	8.	1.	1
	85.	2.	2

The Real Savings that month 39. 0. 2 (Wey Q^r lb) at 52/-	£101.	9.	4
Savings — ⅓	£33.	16.	5
And Boulton was Charged for that Month	£79.	19.	8
an over Charge in one Month for one Engine	£46.	3.	3

Box 9

DOCUMENT 130

Captn E. Rogers Observations and objections ag^t the comparative trial made by Capt Trevithick & Budge at Dolcoath. n.d.[1]

Reasons to Show that Messrs Boultons and Watts and Mr Budges Engins on Dolcoath Mine were not Situated with Equal advantages to make a fair Experiment —

First Mr Budges was the Top Engine and Drawd the water from Mr Boultons Engine So that Mr Budges was put to her Load and Mr Boultons not half Loaded which must be in favour of Mr Budges Engine —

Secondly Mr Boultons being the Bottom Engin Drawd 100 fathom of Large Rods deeper than Mr Budges Engine, and to put them in Motion Eight Strokes a Minute, must be against Mr Boultons Engine —

3^ly As there were no Counters on the Beam to Count the Number of Strokes, and Mr Budges Drawd from a Cistern with a piece of wood Swiming to Regulate her Strokes, and Mr Boultons Drawd from the Bottom, all Miners Must know that to keep the water out of the Sump, the Engine is great part of her time working in fork to keep they sump Men to work, and of course a greater Number of Strokes —

4^thly if the water run^d over the Cistern of Mr Budges Engine or if Shee Slop^d the water must run down again and as there were no Counter the Bottom Engine must Draw it for Nothing with all the water that the pumps of Both Engines lost which must be Greatly (against) Mr Boultons Engine.

Box 9

---

[1]Probably 1782.

DOCUMENT 131

To James Watt Cosgarne Augt 19th 1784

Dear Sir

I wrote to you ye 12th In^t since wch I have recv^d yours of ye 7th & 10th In^t — I have carefully attended to all your remarks respecting the Loads & prices of Engines which I shall avail my self of when the time comes for closeing agreements, but not make any absolute bargins untill I have your previous concurance. The best times for makeing agreements are when Mines are getting money, which is not ye case at present with ye United Mines[1] & Poldice however they just save themselves. These circumstances added to the hope of ye Trump^{trs}[2] coming to their relief puts me in a critical situation for makeing bargins pray what is the tenor of Trumpets advertisement?

We all paid a morning Visit a few days ago to the Sheriff & his Lady. I at ye same time applyd for permission for Rob Moor to return & finish poldory Engine, the Squire promised to use his influence with Cap^{tn} George Symons (who is an overseer of ye poor) but Cap^n G: refuses to permit him to return unpunished as he righteously says it will be a bad example to the parish; the Wom^n hath had 6 or 7 Bastards & was tryed last year for the Murder of one & narrowly escaped ye halter she tells ye overseers she will not lye fallow if she can find a Man in the Country. As she is a com^n wom^n & as Robt swears that she hath foreswore herself & as he has already been much punished I think I should not give him up but let him return to Soho. He can erect an Engine & therefore I doubt not but you may find him some employment & send us another or two in his room.

Tomorrow is N Down[3] Acc^t when I shall attend & if any thing new occurs I will write to you. None of the Nossells &c for Crane[4] & W^1 Towen[5] are arived wch is vexatious as I wished much to see Crane work two pumps of $\frac{\text{In}}{6\frac{1}{4}}$ ballanced in y^e new way, the working geer is in good forwardness the ∇ bobs are made & fixed & all wants for the s^{d} nossells &c.

The new adventurers for Chase ∇ are Jenkins, Mr John Vivian, Mr Williams alias Col^{l} Williams, Mr Danbos of Falmouth, Mr Nankivell, Mr James of St Anns, who are all 1st rate men i.e. men of Sense & clever in business, besides wch are Cap^{tn} Paul, Moyle of Chase ∇, Boden, Plummer of Truro one or two other small adv^{trs} Wilkinson & our Selves & yet after all I begin to fear the whole will fall to the ground for after much Negotiation between the afores^{d} & Lord Falm^{h} he (Lord F) hath now sent us word he will not take even $\frac{1}{16}$ as Cap^{n} Magor hath told him there is no necessity for his takeing any share as he Magor can make up a C^{o} without L^{d} F takeing any part, but that is a Lye for I dont believe he could raise Advent^{rs} for $\frac{1}{4}$ part in Cornwall, he is too well known here & is disliked by every body. Mr Wilson hath made an offer to the Old C^{o} to induce them to take one Eigth but if they refuse I know not how y^{e} C^{o} will be filld up, as we can't prevail on y^{e} Truro Gen^{tn} to take any more. Mr Wilson hath taken much pains but now seems tired out with disapointment.

I presume you will have finished ye Specification by the time this reaches you as ye 4 M^{o} will be then expired within one or two Days. Have you had any further talk wth Jno Stead about y^{e} (crank.) I think it ye only good means left for Rotative motions besides those you have already secured. Mr Edwards sets out this day to attend the ♀ Meeting in London & to settle ye offers to ye East India C^{o} — He thinks with me that certain persons interested in Cornish Mines should endeavour to take a new set of ye expireing Lease in Anglesey. The Load of that Mine is now distinctly discoverd & defined w^{h} ye underlay of it. You may order one or two more drawing Wim Engines as they will shortly be wanted. I think Brown is getting better into tune but I must say something more to him before I leave upon ye Car^{g}[6] subject, or he will relaps. We must make the best of Mankind as we find them.

Mrs Phillips desired I would talk to Molly Williams about Lawson but as I know the progress of <u>advice</u> in matters of that kind I declined it. She came over to Cosgarn herself a few days ago to ask my opinion & speak in his favour. I find the match is agreed & now they want to find out how they shall live. He says he can erect an Engine himself & that he did most of the work at Trevascus.[7] He doth very little that benefits us, & as to the mine drawings it may pay him for once but I have no hopes of the Mines establishing a new office (draughtsman). I beg Pierson would send to Mr Thos Wilson by the return of the post the state of Lawsons acc^{t} in his Books for I perceive it is necessary to settle it directly. We are all

well & joyn in respectfull Compl^{ts} to Mrs Watt remaining always
Dear S^{r}

Yours most sincerely
M: Boulton

Box 20

[1]They eventually ordered a 45 in. engine in 1786.
[2]The Hornblower family.
[3]They ordered a 48 in. cyl. 8 ft. engine in 1784.
[4]for a 15 in. cyl. engine.
[5]for an 18 in. cyl. 4 ft. stroke engine.
[6]Wheel carriages.
[7]for a 45 in. cyl. 8 ft. stroke engine ordered in 1781.

The domestic market – elsewhere (excluding Cornwall)

Documents 132-167

DOCUMENT 132

To Joseph Crouchley,
Leigh, Lancashire

Birmingham Aug 11 1780.

Sir,

It has long been a resolve of our partnership not to erect any of our new invented steam engines or to consent to their erection & give the necessary directions in any case where the consideration money did not amount to £300 at least.

But as you propose to take some pains to introduce them into your neighbourhood, and as according to the small value you sett upon your coals, it would not be to your interest to give so much money; we have agreed to depart from our former resolution, and hereby offer you our Licence, to erect and use at your Colliery near Leigh in Lancashire one of our new Steam Engines with a Cylinder 28 inches dia^{r} and 9 feet 7 inches long[1]; provided that you pay us the sum of one hundred and fifty pounds for such licence and for our trouble in making drawings, and giving directions. You are also to pay a workman whom we shall send to assit you, to complete the engine, for his trouble and travelling charges. That part of the materials which we comm^{ly} furnish shall be charged at reasonable rates; and the materials furnished by Mr. Wilkinson you will pay to himself.

If this offer should meet your approbation you will please advise us, and we shall give immediate orders to cast the cylinder etc. If it should not prove acceptable we thank you for the trouble you have taken and wish you success with a common Engine.

We are Sir
Your Obedient
humble servants
Boulton & Watt
(hand of James Watt)

Office Letter Book 5 p. 19.

[1]In fact no order was placed until 3 years later. This was for a 34 in. cyl. 8 ft. stroke engine.

DOCUMENT 133

To James Watt May 1 1777

Dear S^r^

I had the pleasure of writing to you last night. This morn^g^ I visited Sr Char^s^ Whitworth at ye Foundling Hospital. He said I was just in time to accompany him to the Comm^tee^[1] in Arlington Building, where we accordingly went in Sr Char^s^ Chariott & where I underwent a long examination for they had been falsely inform'd that Stratford Engine[2] could not be worked. They dwelled much upon Mr Chaponeris's argum^t^ but upon the reverse principle. Each party chops their Logick to their own palit. However the result was that I was requested to examine their Engine at Chelsey[3] & to form some Estimate of the expence of converting it into one of ours. They wish to raise more Water & I find their present Engine makes 15 Strokes p^r^ M^t^ — Experience hath not yet taught me to estimate such an alteration & as they have adjourned their Committee till tuesday morn^g^ next at Ten oClock & then I am to attend 'em at the Fire Engine where breakfast is orderd for all the Comm^tee^. It becomes absolutely necessary for me to be as well prepared as possible wherefore I must beg that you will either by the return of post or by the dilligence or Fly on Sunday afternoon send me such estimates, hints, directions, or orders as you may think necessary, for when one is dealing with Committees it is necessary to speak by ye Card. Nothing but precision & firmness will do with Committees & I shall be diffident if I dont hear from you by Monday. Mr Henderson hath not yet accomplished his wishes but nevertheless proposes returning in a few days. We have an idea that your contrivance for pleasing ye Curious is puting Glass Covers over the Valves in ye new Nossell as ye red ink by wch one may see the working of ye Valves. I should be glad also of your remarks upon the inclosed Copy of Wilbies acc^t^. I think Webb should not go away untill we have obtaind our Wishes abroad I am surprised at hearing nothing from Holland nor Paris[4] remember me kindly to Mrs Watt

farewell

God bless you

(Matthew Boulton)

Box 20

[1]Of Chelsea Waterworks Co.
[2]an 18 in. cyl. 3 ft. stroke engine.
[3]The engine was altered to a 30 in. cyl. 8 ft. stroke in 1778.
[4]Tentative enquiries had been received.

DOCUMENT 134

To John Southern. London 26th Feby 1795.
Thrales Brew house.

The high opinion I had form'd of your Steam Engine being confirm'd by so many years successful experience of our own, erected upon your improv'd principle, has induced me to recommend the use of your Engines upon all occasions that require such powers.
Being one of the Proprietors of the Borough Waterworks, work'd upon the old expensive principle, I have lately spoken so much in favor of your improvements as to obtain their consent to my requesting one of you to examine our Engine the first time you come to town & give us your opinion of the estimate necessary either for such alterations as you see fit to recommend, or the whole expence of a new Engine, remembering that we cannot discontinue our wanted supply — our hours of working are 6 in winter & 10 in summer, or 8 hours on an average, raising 1200 barrels p^{r} hour (I think about 50 feet) in which work we consume full 3 buckets p^{r} hour — if you have other business which calls you to town at the present time I could wish for the pleasure of seeing you here soon after your arrival to confer on this business, as from the excessive high price of coals even at this day, as well as thro the late severe weather, I persuade myself our Company would be more likely to adopt my advice of introducing your engine than if it defer'd to a more distant period.
The increasing grievance of bad half pence will I hope revive the consideration of your M^{r} B's offer to coin that money upon the wise plan & beautiful model he offer'd to our Trade, who at a meeting held a few days since (for the purposes of our comon interests) receiv'd a proposal from the Agent of a Copper company to supply us with tokens in the manner of those now so current, but I persuaded the Gentlemen to defer the consideration of that offer to our next meeting (the 25 March) that I might be able to lay before them your M^{r} B's proposals to whom I consider'd ourselves pledg'd

to give a preference on equal terms. I therefore wish he would send me up a few (say 3 dozen) of specimens of his halfpenny offe'd to government to distribute among our trade to influence them in its favor, as unless we can obtain redress from government in removing this grievance[1] there are several gentlemen who appear inclin'd to join in my plan of issuing a token for our Trade upon the former principle of issuing the value in the weight — but as this is mere matter of opinion I only offer it to prove my readiness to co-operate in so necessary a measure if it may legally be carried into execution. I am told there are many tokens current in your neighbourhood of wch I should be glad to receive a few specimens. I am respectfully &c

Robert Barclay.

Box 6 (Water Works)

[1]The shortage of small coin.

DOCUMENT 135

To Boulton and Watt

Princess st London
28 Nov 1789

D[r] Sirs

When Mr. Thompson[1] was at Soham he attended the Commiss[rs] and offered to erect an Engine on the common construction which sho[d] with pumps throw out 220 Barrels of Water p[r] Minute no mention was made what measure nor was Hogsheads mentioned. Why Mr. Wilkin uses Hogsheads instead of Barrels I cannot guess unless he supposed that 160 Hhds were equal to 220 Barrels but I cannot make this out in proof by any figures in use here altho so to do I have tried as many figures as if words wo[d] constitute a moderate Bill in Chancery for a discovery —

We will reckon upon and abide by the Calculations you have made of an Engine to thro out 160 Hhds Wine Measure per Minute will cost 723£ or thereabouts.

The Pump work you have not mentioned the price but supposed you have informed Mr. Wilkin —

The Article of your Premium I think cannot be ascertained more justly than by the difference of savings between your Engine and Mr. Thompsons every 24 hours laying his at 3 Chaldron for that time of 36 Bushels to the Chaldron each B^{l} 8 Gallons — we have Newcastle Coals from Lynn in Norfolk our nearest Sea port which cost from $\overset{s}{26}$/ to a Guinea and a half per Chaldron — Mr. Thompson offered to give Security that his Engine wo^{d} perform the Work and not exceed 3 Chaldron in 24 hours expenditure which so far establishes a certainty from which to stipulate tis generally understood that you have hitherto stipulated from the savings and the Quantum ⅓ and so I informed the Commissioners from the first.

The Fens has been in great danger of being under Water all the Summer from the heavy Rains the Ditches were quite full and the Water kept of(f) the Surface of the Land by the exertions of our two Mills which have been employed all the Summer aided by strong Winds which are very unusual with us in Summer — We have now our Ditches quite full and the furrows of the Land nearly full so that we daily expect to have the Surface covered with Water — Our Mills can only go when there is Wind and I have remarked that in a stiff Wind they have lowered the Water in the Main drain near my house two Inches in Eight hours.

According to the Quantity of Rain and Wind we have more or less work for Engines. Our Mills were employed the greatest part of last year seldom less than 6 Mo^{s} in a year often eight or nine so that it will be very difficult to ascertain any precise time that the Engine must be worked — By whatever measure you fix our premium we must rely much more on your honour and Justice than upon our Judgments hope we shall pay no more than others — I beg leave to recommend you to settle the Premium per savings it appearing to me the most Equitable because in any other way it may happen in particular circumstanced years you might get the advantage of the $Comm^{rs}$ in others the $Comm^{rs}$ might gain the advantage. I am sure neither party wish for an advantage hurtfull to the other.

I consider this treaty as a business of the first magnitude which perhaps has occurred in your business for the object is extensive indeed every particular District in the Bedford Level (being upwards of 400'000 Acres) must apply to us and from us to you and so wet a Summer and Autumn having happened at this time is very apropos for your purpose as it will remind all Fen Men of the necessity of Steam Engines and a speedy Erection of ours will afford them an early proof of the Utility.

The paper I had from Mr. Bolton I gave to the $Comm^{rs}$ which they now have and I have no copy. One Article of that paper as well as the first I had from Mr. Bolton was ⅓ of the savings.[2]

I will assuredly keep up the honour of attending Mr. Bolton & will so soon as he comes to Town and perhaps may bring one or two of our Commissioners with me so pray tell Mr. Bolton he must make up his mind with you in all points before he comes to Town.

I am

D Sirs

Y^{r} m^{t} Obed Servt

Robt. Wild

Box 4.

[1]Probably Francis Thompson.

[2]The Commissioners of the Fens purchased a second hand B & W pumping engine of 36 in. cyl. and 8 ft. stroke in 1803 and a new 6 h.p. rotative engine in 1804.

DOCUMENT 136

To James Watt

Newark
25 Jan. 1796

Dear Sir

finding considerable difficulty and a certainty of great Expence in the execution of a Tunnel at Blisworth on the Grand Junction Canal, I have recommended to the Committee of the Grand Junction Canal to desert it and pass over the Hillby Locks and to supply it with water by Steam Engines; by so doing I have raised a nest of Hornets about me, and particularly among some Proprietors in the Neighbourhood of Birmingham who have consulted Mr Houghton the Cl^{k} to the Birmingham Canal C^{o} & Mr Bull who referring to their own Experience tell them that the adoption of Steam Engines will be fatal to the Existence of the Canal.[1] In vain have I urged that there is no uncertainty in a supply by Steam Engines — that there will be less delay in passing by Locks than by a Tunnel, that the execution of the former is easy & certain, the latter expensive and hazardous, and that a year or two will be saved in the term of completion, the saving of Expence will be more than £20,000 and I flattered myself I could convince them that this Sum would build Steam Engines & leave a Sum the Interest of which would maintain them; but they have got from Mr Hunt — Mr Houghton & Mr Bull, such an unfavourable account of Steam Engines, that I shall have much difficulty in combating their Prejudices; Old Bull goes so far as to say that we had better Cut the Hill Open be the Expence what it may than resort to Steam Engines — so far entre nous — to the following I request your answer which I may produce to the Committee.

Our Summit will be about 1300 Yards in length — the height to lift the water will be 94 feet the quantity wanted may be 40 Locks p^r Day — each Lock containing 290 Cubic yards — we can have a small Reservoir that will hold about 500 Locks full, which may probably be so often replenished as of itself to supply the Canal two or three Winter months, and may be resorted to for a Day or two occasionally, when the Engine may want Repair — from this Circumstance one Engine may probably be sufficient. Now Sir I will request you to inform me what will be the cost of such an Engine, what quantity of Wednesbury Coal p^r Hour will work it, and what may be the other annual Expences attending it.

I know I can maintain my Ground on every other Position and have little doubt that you will enable me to convince them that a supply by Steam Engine is neither impracticable nor uncertain — be the future increase of trade what it may, for from the experience we have had in the tunnelling we have already executed in another part, it is probable that the casualties and loss of time by tunnelling[2] may make the difference in favour of Locking nearer £60,000 — than 20.[3]

Your answer as soon as convenient will much oblige.

Your very humble Servant

W. Jessop.

Box 6 J

---

[1]Yet the Birmingham Canal Co. was amongst Boulton and Watt's best customers. Eighteen engines were purchased between 1778 and 1818.

[2]Jessop was over-ruled. A slightly different route for the Blisworth tunnel was proposed but it was not executed until 1804/5. In the meantime, the two halves of the Grand Junction canal were linked first by a toll-road and then by a horse tramroad.

[3]In fact the Grand Junction Canal Co. did not order a B & W engine until 1805. This was a 20 in. cyl. 5 ft. stroke pumping engine. Two further engines were purchased in 1810 and 1817.

DOCUMENT 137

To Boulton & Watt — Saxthorpe July 7 1795.

Gentlemen

Having had some experience in erecting of Steam Engines upon the plan known by the Name of Savorys Engine for Manufactorys where the fire is apply'd to heating of Irons as well as produceing of Steam so that the Quantity is Consider'd as no object — & my work has given in general very good Satisfaction. I have therefore been apply'd too by Gentlemen in and about Norwich who wish me to Erect them Engines of the Lever kind, but as I have for some few years Spared no pains in informing myself the exact nature of Engines in General and the principal upon which they must work, either to give Credit to the builder or Satisfaction to the employer, I therefore find it impossible for any person to Erect an Engine to answer a Good purpose except they make use of your principal. Now Gentlemen I shou'd be Glad if youd be so far obligeing as to inform me what Sum of Money I must give you for every Horse power to let me make use of any part of your principal I please for Building Engines some so Small as 2 horse power and very few to exceed 4 as our Manufactorys are small. I'd produce you a Certificate of the exact Diameter of each Cylinder & either pay you the Money Annually or a Certain Sum when every Engine is finished. If therefore you think I'd better come over to Birmingham to agree upon a Business like this I'll certainly wait upon you if you'l first be so Obligeing as to Send a few lines to inform me upon What terms[1] you can do it that I may be able the better to judge wither it will answer my employers purpose or not as they Cant go to a very great expence — the Manufactorys whon't Answer it.

I am Gentlemen

Your very Hb^{e} Servt
W. Hase

pray direct for Willm Hase Saxthorpe by way of Aylsham

Saxthorpe near Aylsham
in Norfolk

Box 6 H

[1]B & W refused licences to other engine erectors. Hase eventually patented an engine of his own in 1801 (pat. no. 2501).

DOCUMENT 138

To Boulton & Watt

Hunslet Near Leeds April 13th 1790.

Sirs if you please to send me the Expence of Smaul Steam patten fire ingon a Compleate wone; to Lift a piller of Water 30 yards and to have a Six inch Diameter Working Barrel and the same to be fixed Whithin 6 Mile of Shefield and if there by any aftter Charge youl mention I sepose the frame and Working beam you omitt, butt youl please to mention Whatt you so find and writte by return of Post.[1]

I am Sirs

Y^{s} Robert Wood
Milwright.

Box 4.

[1]No order materialised.

DOCUMENT 139

To James Watt London Augt 27 1781

Dear Sir

I wrote to you p^{r} last post since which I am fav^{d} wth yours of the 20th — I bought some sheets of large post paper in order to have wrote you a long letter upon various things but I have been extreamly hurryed all day & distressed by my Wifes illness. We are in our old Lodgings at Birds in Bridge Street but the poor Wom^{n} is dying, the House full of Buggs & nastiness, the Weather extreamly hot & my Wife in a fever & quite outragious at London insomuch that if I had not promised to return home tomorrow she would have gone by herself to day. I therefore must quit London before I have finished my business which will oblige me to return again soon. The only thing that I was anxious about was to have obtained from L V W[1] your release but from Lows absence & Veres being at this time not in very good humour & from sundry other circumstances I think I had better postpone that business for a few weeks. I therefore beg you will not vex yourself about it for if you do you will vex me. My life on't you shall shortly have a full discharge. It is impossable that you can judge of ye propriety of my Conduct unless you had been here & saw all ye circumstances. A Gentleman who hath got from Goverm^{t} some good contracts one of wch is supplying the Navy with Bread wants 2 Corn Mills, one at this place & one at Southampton, at each of wch he has built several great Ovens. He would give £1000 more than common cost to have them erected in 1 M^{o} or 2 but I could not undertake to be so expeditious. He then sent for Jno Stead,[2] who is in Town, & Jno offers to contract to build one in 3 M^{o} with ye Mill & has now accordingly contracted. This Gent^{n} then returned to me & told me what he had done wth Jno & said he w^{d}

also have one of ours to be erected as soon as we could, beging I w^d take my choice as to situation either at Southampton or London — I have chose ye Latter as y^e ∇ is bad at Southampton & more out of the reach of good Workmen. The(y) are both to be 36 In Cylinders with 8 feet Strokes. I told him y^e nature of our agreements to wch he readily consented & to pay for everything as soon as deliver'd.[3] He is a Man of Spirit & honor & not a difficult man to deal with. George Matthews is going on wth a $\overset{In}{30}$ Cylinder for a Corn Mill upon y^e Thames. Thus the publick will now have their choice of 3 different Mills. The greatest defect of our eliptical horozontal Wheel is the quantity of Vis inertia for a $\overset{In}{36}$ Cylinder will be equal to $\overset{lb}{12000}$ & if we apply ½ y^e weight viz 6000 & that weight is be raised up 8 ft & let down 8 feet twelve times p^r Min^t. you will find y^e Vis inertia amounts to a great deal of y^e whole power wch w^d not be y^e case if y^e weight was applyd to a Wheel or if you could overcome the inconvenience of a piston working up & down by wch means a weight may be saved & friction also.

As this Engine will be in conspicuous point of light I wish it to be as free from all taxes and imperfections as possable for on it will depend the application of many of our Engine(s) & in proportion as we get forward other Engineers will get backward & Visa Versa.

Dont forget Walkers drawing but send the plan to me directly. I am sorry to hurry you there is a Tyde in ye affairs of man that must not be lost.

I will put the equalising Machine in hand so soon as I get home. I shall lose ye post if w^t another word

adieu God bless you all

M B

Box 36

---

[1]Lowe, Vere, Williams, the bankers from whom Boulton had borrowed a large sum.

[2]Who was connected with Matthew Wasborough of Bristol and James Pickard of Birmingham.

[3]See [45] B & W did not obtain this contract.

DOCUMENT 140

Minute of a Conversation with Mr Arkwright. 30th July 1782

Mr Southern was with Mr Arkwright this morning & had conversation with him a considerable time — introduced himself to M^{r} A — on visiting Soho with Mr Norcott from Manch^{r} who s^{d} (amongst other conversation) that M^{r} A was agoing on afresh with ye Cotton construction there and much wondered that Mr Arkwright did not employ Mr Boulton & Watt's principle in ye great work;[1] however the movement might be intended he, Mr Norcott, being convinc'd of the great Utility of the patent priviledge if imploy'd by M^{r} A, that he had seen a model just finish'd and secured by Patent[2] to B-n & Watt of amazing rectitude of motion and convertable to Rolling, slitting, Tilting or battering of Iron, Copper, Lead, as well as Grinding of Corn &c, &c, by steam annex'd to a very trifle of water wanted for condensation, so fortunatly adopted might very safely be erected on almost any situation whatever, where water to turn a wheel cou'd not be had; with many other valuable observations, caus'd Mr Southern to think it his duty to wait on M^{r} A — as a neighbour to inform him in time yf he might not slip an opertunity of benefiting himself thereby — same time told M^{r} A that J S had seen his fire Engine at Wirksworth[3] and was sorry that he, Mr Ark^{t}, had not incouraged the Patent Steam Eng^{e} there and durst engage that $^{1}{}_{5}{}^{th}$ of the Coal wou'd raise an equal quan^{ty} of water in same time on the patent construction if made use on there. After a good deal more M^{r} A—t say'd he w^{d} show me a model of construction which he s^{d} farr exceeded Mr Boulton &c or Pickards or any other — so was introduced wth 4 other Gentl^{n} from Manch^{r} to see this wonder — I aplauded the invention for simplicity and (suspend saying further here) ended with saying yf Mr Boulton woud be at Matlock Bath I believed this afternoon & evening and wished M^{r} A wou'd see him so conveniently for both — he seem'd startled at this & answ^{d} he shou'd be glad to see Mr Boulton — I then said M^{r} B-n intended himself the pleasure of calling on M^{r} A last summer when at Bath but was told M^{r} A was very ill. M^{r} A remembered this but did not say further on seeing Mr Boulton — but thank'd Mr Southern very kindly for calling on him.

In the conversation M^{r} A-t say'd he had been made to understand (rather believe) that M^{r} B-s Engine was so subject to disorder & so complex prevented him from calling in M^{r} B & W assistance. This I told him was from Ignorant being Tompson;[4] he reply'd it was he. On this I cou'd not contain — told him Tompson had given out publickly that M^{r} A had thefted from him great part of his mechanism and obtain'd a patent of such — here he laughed at ye ignorance of T- and concurr'd in that he knew nothing.[5]

MB — Southern/1

---

[1]Shudehill Mill, Manchester, powered by an engine designed by Hunt.
[2]Patent 1321.
[3]An atmospheric pumping engine.
[4]Francis Thompson, engineer of Ashover.
[5]In fact Arkwright did not order a B & W engine until 1790 when he requested one for his Nottingham mill.

DOCUMENT 141

To J. Watt Dundee 22 Feb 1788

Dear Sir

I wrote Mr Hamilton some time ago, to learn when you should be at Glasgow, as I heard you were expected there — I wished to have seen you, and would have come so far for that purpose. If you came however, he had forgotten to advise me.

There are many occasions in this County when your Engine is spoken of as applicable to Mills of various kinds — not, as you know, because we want Water in general in Scotland but it is often so far from the towns in which our Manufactures are, as to be unserviceable for that purpose; at this very town for example, we have no Stream nearer than two or three Miles from us, in which moreover their is too little Water for our purposes — but running past our doors, is the firth of River Tay, with a pretty high ebb & flow — several Methods have been talked of to make it useful in driving the machinery neccessary to our manufactures M, one of which is tide Mills, another raising the water from the River by pumps into Reservoirs, for which purpose the Perth Machine described in the inclosed[1], has been much spoken of, I shall not say with what propriety, most likely it may appear futile at once with you — tho' there are Gentlemen here who think so well of it, that they are going to try it.

I want very much to have a descriptive Account and common measure of the powers of your Engine and its Expence, — both as applicable to smaller Mills and Machinery, and greater; — the knowledge of which in this Country I really think would be of Service — You once told me this in a certain degree, but to my shame be it said, I have forgot it. Horses, and Weight are to be sure good popular methods of defining the powers — but a comparison with a power necessary to drive a Flour Mill, or a Cotton Mill of 1500 or 2000 Spindles would still be more so in this Country. — Best coals cost us here, about 10/- a Ton of 20 Cw^{t}.[2]

You'l do me a favour to answer this particularly, and as speedily as you can.

By the bye, can you recommend to me any treatises on Mechanics and Hydraulics, teaching in the most practicable and intelligible manner the construction of Mills, the calculation of the necessary powers, and how to make an estimate of the quantity or fall of Water necessary to a required effect — I presume these are better than Belidor's.

I write this by assistance of an Amanuensis — my apology for which is, that you may stand on no ceremony with me, but use the same conveniency — for I know that you are still more hurried than I am, and your time much more precious — upon that I tresspass too long — I shall only add, that I should be glad to know if it be true that you have made great late improvements on your Engine, and that a person in London makes great saving in fabricating the Boilers. If you have as much time, favour me with a Sentence of Birmingham News. —

Present my best respects if you please to Mrs Watt and the family and to M^{r} & Miss Boulton & believe me sincerely your most faithful

obliged and Humb^{e} Ser^{t}

Richd Neilson

Box 4 N

---

[1]Enclosure missing.

[2]No order materialised.

DOCUMENT 142

To Boulton & Watt Manchester 3rd Ap. 1789.

Gentlemen

Your Assistant Mr. Southern possibly woud mention to you a conversation I had with him respecting a fire (or Steam) Engine which I wish you to furnish me to use in the Cotton Manufactory for the purpose of preparing & Carding (not spinning) cotton[1] – in a factory or Build^{g} conformable, or nearly comformable, to the enclosed Scetch.

As on one hand I am almost an entire stranger to the Nature of your, or any other, Engine of that sort – & on the other the intended establishment taken altogether will be rather expensive – you will excuse me being a little tedious and inquisitive at the outset dispensing with this; you will I hope not find me more troublesome than I ought to be afterwards.

The Engine I would set out with – is one with a Power equal to six Horses – & of the double kind – & the first & immediate object of my application is to request you will send me as soon as possible a scetch for my regulation in diging out the foundation for the walls of the whole build^{g} for which purpose I have now a considerable number of hands in waiting – a more correct plan & elevation with such other Instructions as you think necessary may follow afterwards at your convenience.

Of the inclos'd plan I propose to build no more at present than the front part with the two stair Cases – the most proper Place for introducing your engine will, I presume, be about the Center in the back part as dotted with red – & so as the Axis of the fly wheel may pass directly thro' the Middle to give motion to the Horizontal wheel placed upon an upright Shaft in the Center of the great room – But please to observe that what respects your work, I mean only to give mere conjectural outline, leaving the Place – Form – Width – Height &c &c of the Brick & other work to your fixing & direction – in doing which however I wish you to block up no more of my intended windows than you can help by either ye width or the Height of your Brickwork or Roof – for to give the build^{g} (I mean the Factory part) its greatest possible convenience it should have a continued series of Windows one & a pier introduc'd – in the Distance of every 8 or 9 foot or thereabouts. I woud likewise wish you to introduce room for 4 small single Necessaries – one for each Room to be placed nearly one over another & so if possible to be so managed as neither to be offensive or endanger the Health of the People at work — To which purpose I shall be thankfull for your advice — whether it will be proper to pass a current of water thro' them to a common sewer by help of the Engine pump, at Nights & Mornings & before the Works are set a going, or whether make^{g} air pipes or vents in the Brickwork — from the excrement in the Bottom to the top of the Roof will not answer the same end — or whether you think both or any other means preferable.

In respect to the place & the width of the Well for pumping up the Condensing water — you no doubt will mark it out, — the depth before we come to a body of water suff^{t} to depend upon in all seasons will I believe be about 20 or 30 yards — referring you for other particulars to the scetch & beging leave to repeat my entreaties that you will favour me with a ruff outline for Proceeding to the ground works as soon as possible. I am very respectfully Gent^{n} y^{r} most Obed^{t} Servt

Peter Drinkwater

I should have wrote you on this subject much sooner had I not been incommoded on all sides by threats of a prosecution for erecting a nuisance — & indeed the prejudice is not yet much — if at all abated — the fact is we have already a great Number of the common old smoaking Engines in & about the Town which I can see are far from being agreeable — & the public yet are not all inclin'd to believe otherwise than that a steam Engine of any sort must be highly offensive.

My near Neighbour young Mr. Watt informs me that you are now about making or putting up an Engine or two not far from Manchester[2] — It is possible I may not have been sufficiently explicit — or that further doubts or difficulties may arise respecting the foundations or some other Matter — if therefore any of your People should soon come into this part of the Country you possibly woud do me the favour to let 'em call upon me.

Box 6 D

[1]This was the first B & W engine to be ordered by a Manchester cotton spinner.

[2]Possibly the 14 h.p. engine ordered by the Royal Plate Glass Works, Ravenshead in 1788.

DOCUMENT 143

To Boulton & Watt   Manchester Jany 17th 1791

Gent^n

I duly receiv'd your esteemd favor of the 14th Ins with particulars of ye Engines as therein mentiond: With respect to the mode of payment for the materials, woud it be agreable for you to take it at four instalments, say one fourth on delivery, and the remainder at the expiration of two, three & four years. I do not enquire for myself, a Relative[1] of mine of undoubted Stability has thoughts of erecting one — and as the expence of Building the factory, the Engine and Machinery are all together very great it may be more convenient to him to order the mode of payment as above.

With regard to an Engine of Twenty Horses requiring 60 Gallons of Water p^r Minute, is it requisite there should be a Stream which constantly supplys that q^ty or you can work the same water over again. The Stream he wishes to erect his building upon does not run more than woud pass through an Inch bore in a dry season, you will therefore judge whether that woud be Sufficient. likewise desire to know if heat can be carried into a Stove House from the same fire that heats the Boiler, so as to be sufficient to dry Goods in a Room of abo^t 40 yds long and 10 yds wide.

A great number of factorys are now erecting in this Town, the greatest part of which are intended for Fire Engines of the old construction[2] it is quite evident to me yours are very much preferable to them, but I think they are not generally known here. I take the liberty of mentioning for your government, whether it would not be proper for you to advertise them in the Manchester papers, mentioning some particulars, and if you was to refer the publick to look at that of Mr Drinkwaters (if this woud be agreable to him) it might bring them to be generally used in the Cotton & Woollen branches. It is truly astonishing the money that is at this time expending in this place in erections of that nature. I have taken sevral good mechanicks to look at Mr D. engine who all agree that one of them is worth all the rest put together on the old principle.

I have receiv'd the Articles from Mr Thomas Gill, which are executed in an astonishing masterly manner, indeed much better than I, or the Mechanick employd ever expected they could have been. I return you thanks for recommeding this Gentleman, and shall be happy to return the Obligation whenever it lays in my power

I am

Gent

Your mo Obed^t Servt

Benj. Lees

Box 3L

[1] Possibly Charles Lees who ordered an 8 h.p. engine for his Stockport mill in 1796.
[2] See also Glasgow [148].

DOCUMENT 144

To Boulton & Watt

Holt town Mill near
Manchester Jany 9th 1792

Respected Friends

We presume you will ere this be informed by your Mr Howard of our determination to have one of your Fire Engines, of a power equal to that of thirty Horses;[1] which we have particularly to request, you will use all possible expedition in the Completion of; Our situation being particularly unfavourable, as during the summer Season, & in an especial Manner when a drought prevails, we can place no certain dependance upon a supply of Water in any measure adequate to our Wants. From the River near which we have Built even with the number of spindles we have at pres^t in Motion, 'tis only when a fresh comes, that we are enabled to work them, without the aid of an Engine; So circumstanced, having no other resource, we are either under the necessity of discharging our Hands, or keeping them unemploy'd, at an immence Loss, from an encouraging hope that a change in the weather may come & bring with it the assistance (tho' but momentary) we may stand in need of.

Your Mr Howard has inspected the situation and informs us, you have frequently so constructed your Engines, as to Work them in Conjunction with a Water Wheel, which we particularly desire may be the case with ours; Or rather, that the Gear which apertains to the Engine, and which communicates Motion to the Machinery, may be so constructed, as either to work the same, without affecting

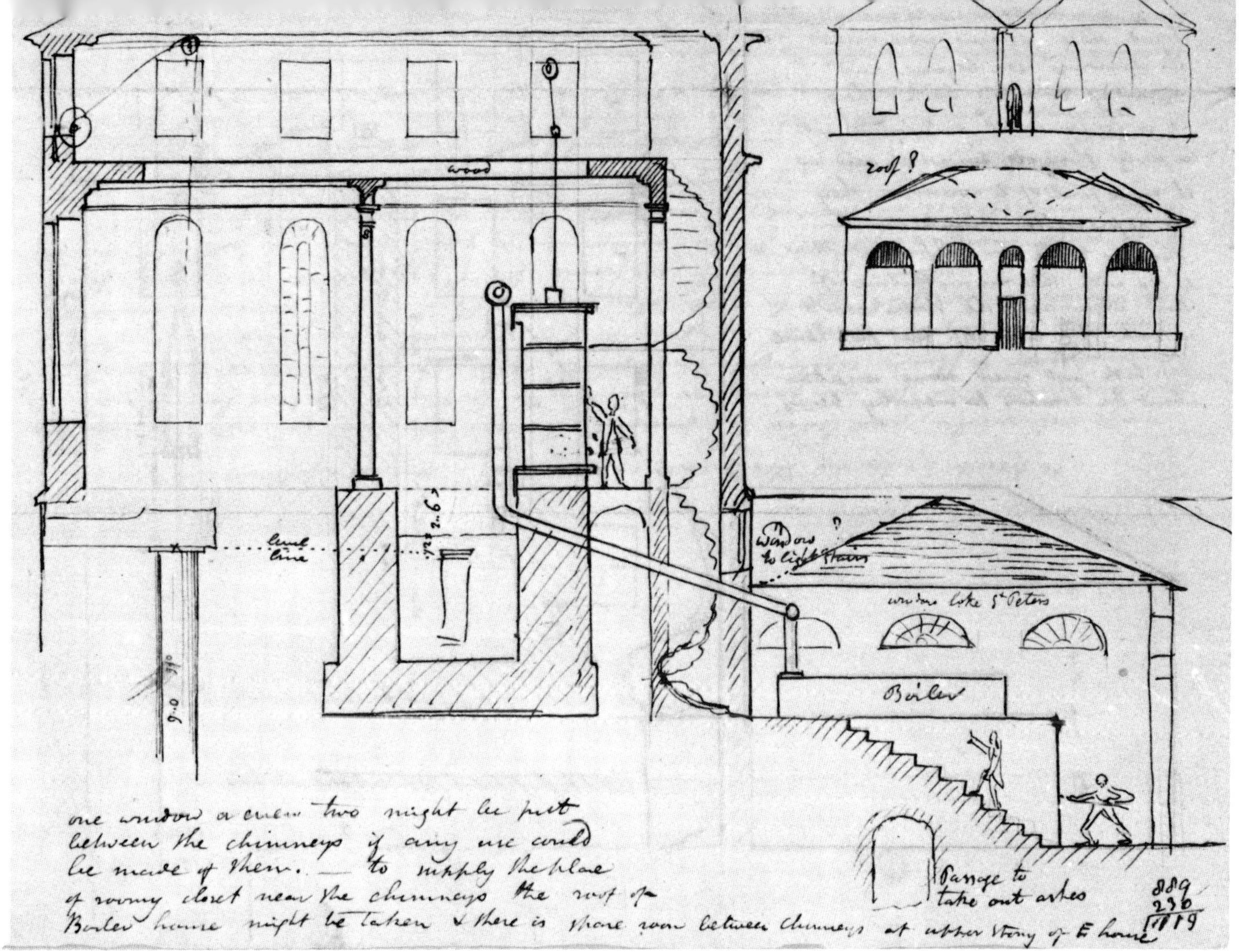
wood
roof ?
level line
Window to light stairs
windows like St Peters
Boiler
Passage to take out ashes
one window or even two might be put
between the chimneys if any use could
be made of them. — to supply the place
of roomy closet near the chimneys the roof of
Boiler house might be taken & there is spare room between chimneys at upper story of E house
889
230
1119

Plate [8]

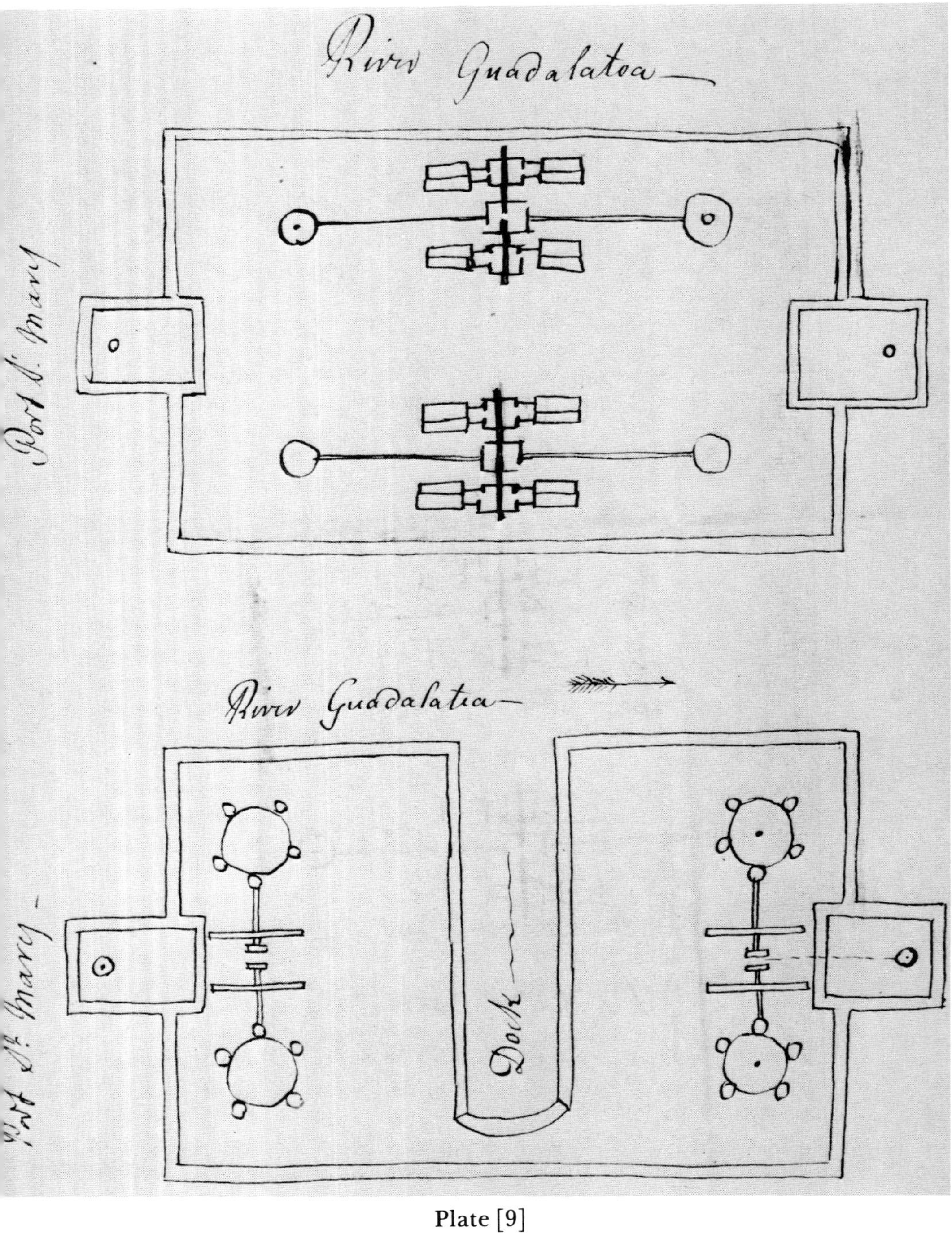
River Guadalatoa
Port St. Mary
River Guadalatia
Port St. Mary
Dock

Plate [9]

our present Water Wheel & Upright Shafts, or in conjunction with them, as occasion may require, as at some times, I believe we may say during 6 mos out of the 12, we raise a supply of water amply sufficient, without any other Aid, to work double the number of spindles we have at present. But when that may cease to be the case, we are fearful, the very Irregular & Small quantitys of water coming at such times, may rather prove prejudicial to us than otherwise, by destroying that Harmony & regularity of Motion, The Engine alone might produce, and by giving motion to the Water Wheel with its attached Gear, we should be deprived of a proportion of Power, that might be applied More profitably. You will confer a favour by communicating your Sentiments upon the Business, and by the very shortest time you would undertake to furnish us with an Engine equal to the power before quoted, with all its apendages, and prepare the same for work; and what will be the am^{t} (or near it) of the expence attending the same —

Your Compliance will oblige

Yours mo Respectfully
David Holt & Co

Box 6 H

[1]See next document.

DOCUMENT 145

To Boulton & Watt

Manchester 14th Jany 1792

Gent^{n}

We are favor'd with yours of the 12th & are very sorry to find that you cannot undertake to furnish us with an Engine the beginning of next Summer as the Consequence of our not having an Engine to work in the Summer Months would be a certain heavy Loss to us. We should have given a decided Preference to one of your Engines — if it could be procured in Time but as the prior Orders you have on hand preclude us from that Advantage we have no other Resource than that of ordering a common Beam Engine which may be compleated by the time we shall have Occasion for it.

Though Necessity compels us to countermand our Order, we hope it will soon be in our Power to give you another — having it in

Contemplation to erect in the Course of next Summer — a Building to be worked entirely by one of your Engines but we shall write you more particularly as soon as our Plan is fixed — in the mean Time

We are, with respect

Gent^{n}

Your Obed. Hum^{e} Servt

David Holt & Co

Box 6 H

DOCUMENT 146

To Boulton & Watt

Holt town Mill April 3^{d} 1792

Respected Friends

Pursuant to our promise, we now request you will furnish us with one your Patent Engines, equal to the Power of 10 Horses[1], as speedily as the Nature of Your prior engagements will admit of, we are in expectation that the Building for the Reception of the Machinery, the proposed Engine is intended to work, will be compleated in the course of 6 mos hence, but as we purpose soon to make a beginning, will thank you for your instructions in order properly to prepare the Engine House.

We remain with due Respect

Your ass^{d} fr^{d}

David Holt & Co

Box 6 H

[1]In fact they settled for a 14 h.p. engine, purchasing another of the same power in 1804.

DOCUMENT 147

To Boulton & Watt

Mansfield the 19 Feb 1793

Sirs

We again solicit you to assist us in the Disposal of our Steam Engine[1] the Excessive Scarcity of Coal in our Neighbourhood damps every Idea of Benefit which it ought to participate to us. In your Last Acc^{t} you charge us with ¼ Box of Cement 10/6 which never came to hand and for which pray give us credit. We are

Sirs

Your most humble Servt^{s}
for Gorton & Thompson
Thos Green

We think it best to keep the Governor a while Longer As it may be wanted with the Engine.

Box 6G

[1]This was a 14 h.p. engine ordered in 1787.

DOCUMENT 148

McTaggart's List of Steam Engins at present in use for Cotton Machinery. Aug 1798[1]

Glasgow	Scott Stevenson & Co. Coal quay	8 Horse power
”	Alex^{r} Henery Greenhead	4 do.
”	John Brand & Co. Calton	4 do.
”	Jno Pattison[2] & Co. Brigton	8 Suppos'd
Paisley	W. Leach Twigg & Co.	8 do.
”	Mr. Twigg & Co.	24 do.
	The Underwood Co.[3]	
Johnston	Campbell Spier & Co. 38^{inch} Cyl^{r}	10 do.

Stravethen	& Smith & Currie	
	N^b the two last are us'd about 3 months in the year having water the remainder	
	Horse gins us'd by	
Glasgow	Chadwick & Whyte Grahams Sq^r	24 Mules
"	McTierlie & McTaggart do.	14
"	John McTaggart Ruthglen	14
"	James Monach Canon Street	Common (Jennys) & Mule
"	Peter Buchanan George Street	Common (Jennys)
Paisley	Matthew Corse & Co. Newtown	20 Mules
"	David Finlay & Co.	14
"	Wm Clarke Snedden	20
"	John Davidson	14
"	Jno Dalyel & Co.	14
"	A. Pollock & Co.	24

Box 3

[1]None of these is a B & W engine.
[2]Pattison ordered a 32 h.p. B & W engine in 1800.
[3]The Underwood Co. ordered a 16 h.p. B & W engine in 1798.

DOCUMENT 149

To J. Watt Junr. Soho 22 April 1799.

Dear Sir,

On Saturday evening Mr. Jewsbury from Joseph Wilkes Esq & Co Measham nr Burton on Trent came here to enquire the price of an engine of the most approved construction for a cotton mill containing 3500 spindles. He did not dissemble that they were making enquiries of other engineers, as they had determined to have that engine which under all consideration should appear to be the best, and had been to look at one of Bateman & Sherratts at Alrewas near Lichfield — I said "they use an air pmp there I believe" he replied "they say not, but both myself and Mr. Woodhouse who was with me thought so" — (Mr. Woodhouse is a common engine maker, and has erected several for Mr. Wilkes in his collieries, as well as for many others about Nuneaton & that district of coal country). Jewsbury said that Mr. Wilkes own opinion was in favour of the common engine, but the company had nevertheless determined to have the best, for which purpose they were making all the enquiries they could ye way of inabling them to judge. It is to be erected for an old mill at Measham which at present is worked by a water wheel supplied by water raised by a common steam engine; but as they are going to enlarge the mill they will be deficient in power and have now an opportunity of disposing of their present engine to Mr. Wilkes for a new coal pit, and therefore this is the time to think about another. They seem to think that converting the present engine into a crank engine instead of a pumping one will serve them if they determine on a common engine — They had heard, when the patent was out you had a great improvement to bring forward — This has lately been mentioned by several others, & I find myself always at a loss to reply — to Jewsbury I said that the report was totally without authority. This did not come up to his wishes I plainly saw, but I stuck to the word — adding that whatever improvement might be made, it would be applicable to our engines as at present constructed. He did not press for an estimate then & I promised him one in the course of the present week. He said their machinery would be entirely new and that if their present was not of the best kind, new would be substituted, as they want to proceed on the best plans (cotton machinery for water spinning).

I am satisfied to recommend a 32 horse engine with 2 boilers — this with a steam case will inquestionably turn 3500 spindles. I might have given them the estimate directly but if you should think it necessary to describe to this company either the advantages of your engines, the defects of the old ones, or the probable bad execution of your engines by others after the expiry of y[r] term, I rather wish you to do it, as you will do it better than I.

If you think, on the contrary that simply the estimate is enough for them, one of whom has had 2 of yours already, but who has never shewn himself very hearty in their adoption (I think more from ignorance of them than want of good will) I will if you wish it send them the follow[g].

MM with two ample iron boilers & grates, feed pipes, steam pipes, safety valves etc for the same, £1357 made out as follows.

32 per new list	1109
Forman's price of said Extra articles for Marsland without boiler	76
Extra boiler at £2.2s	172
	1357

On Forman's price for extra articles I have put no pr centage.

Coals such as our Bilston about 3 cw^{t} pr hour. Framing, cisterns & their bolts nuts etc., also working beam may amount to £160 Putt^{g} together about £75.

Do you remember whether Bateman & Sherrat have given you the names of Dicken & Finlow of Alrewas near Wicknor Bridge between Burton & Lichfield. I think they should be enquired into. In May 95 they had estimates of 8 & 12 horse engines.

I forgot to mention that Wilkes will want their engine in the course of the summer if they get one at all.

I hope you and Mr. W had a safe journey & that you found Mr. B etc. well.

I remain Dr Sir

Yours Truly,
John Southern.

Foundry Letter Book 68, p. 126.

[1]They ordered a 36 h.p. engine in 1802. This was not their first, however, for Wilkes had ordered an 8 h.p. engine for a flour mill in 1786 and a 16 in. cyl. 4 ft. stroke winding engine for a coal pit in 1787.

DOCUMENT 150

To J. Southern
Soho

London 24th April 1799

Dear Sir

I have to reply to your two favours of the 22nd Inst^{t} — but as time is precious here, I shall be brief. I have been writing constantly since Seven oClock this morning, till now Six oClock in the evening and feel therefore little disposition to be tedious.

I think the estimate you propose sending to Mr Jewsbury very proper, and I also think it very right that the topics you state should be urged at some length, particularly that our prices now, comprehend nothing but a manufacturing profit & will not be affected by the expiration of our exclusive privilege. That it is not our wish to vie with others in lowness of estimate, but in goodness of workmanship, being well convinced by long experience that the best Engines are the cheapest in the end. These & other points which will occur to you, I shall thank you to state, for I know not when I shall have leisure here.

Bateman & Sherrat never gave us the names of Dickens & Finlow & they are quite strangers to me. — A 32 Horse might be prepared in 3 or 4 Mos from receipt of Order.

Your answer to Mr Jewsbury respecting pretended new Improvements was very proper. As such an idea has got abroad, I think it would be impolitic to discourage it altogether, but very necessary to qualify it as you have done. We are obliged to the public at least for this good opinion of us.

My father's opinion of the sliding Valves & Beam less Engine agrees pretty much with yours; as to myself, I cannot say that I have any prepossession either for or against one or the other, but am clearly for having them both tried at home first.

Symonds Engine[1], must, I believe, be made without a beam (on account of the price) but should certainly be made with common valves. Please therefore to give directions for it to be made in that way & I will endeavour to get all the time allowed I can. —

Our Cornish Debtors are gradually coming to. Their stomachs cannot be brought down all at once to our terms, but to them they must come in a very short time now, or encounter the terrors of the Kings Bench once more.

I am Dr Sir Very truly Yours

J. Watt Junr.

Box 33

[1]A 2 h.p. engine for their Reading brewery.

DOCUMENT 151

To James Watt Junr. Manchester July 16 1800.

Dear Watt

Your last account of Lawson reliev'd my Spirits as I could see clearly it had your's — happy is it for him he was with friends who could treat him with so much Judgement as well as tenderness. As he has experienced but one attack though he has reach'd the Middle of Life I am of Opinion regular Employment and a fix'd object in Life may secure him from Relapse — I wish we could hope he might be successful in his Attachment to Miss R but such a Malady in the Blood forbids any Interest or Interference consistent with Delicacy or Duty even if there was any probability of Success. Your next will I hope bring a favorable account again.

I gave Creighton some Memoranda which I had made about a new Engine but your Letter quite deranged me from transmitting them in a less crude state. The Mill[1] is larger than Marshalls at Shrewsbury where they use or abuse two Engines of Sixty horses power. I am of Opinion much less would do but if I am mistaken, the Remedy would not be easy and the Expence I suppose is by no means proportion'd to the power. Will you have the goodness to transmit me a rough Sketch of a Fifty and of a Sixty horse engine[2] that we may fix the Bearings in the Walls as we proceed, a precaution which an arch'd Building absolutely demands as it would be highly dangerous to derange them.[3]

If the incombustible plan of Building Mills becomes tolerably general and the Machinery is made of Iron which we are likewise adopting, the Consumption of that Metal here must become enormous; As far as I can yet judge the Insurance, which is render'd unnecessary, will be equal to ten p Cent upon the Extra Cost. I hope you have either made some Arrangements for establishing or buying some Iron Works or are still in Pursuit of them as I think it must be an increasing Trade, and who could have advantages equal with you?

We are still in your Debt and if you feel it inconvenient I will remit you the Balance, but we shall be poor because I preferr'd stretching a point by Building this, instead of next, Year when part of the additional Capital is to be advanced.

Write a few lines about Lawson in due time, I am not so unreasonable as to expect a Letter, & remember me to all friends.

Y*rs* truly

Geo Lee.

Pray tell your father there is some Dispute about the promissary note by Bradley & Son claiming upon Bradleys Estate as no such firm existed — I have put it into proper hands & it shall be attended to.

Did Mr Belford recover anything from the Scoundrel at Oxford?

M IV L

---

[1]i.e. Salford Twist Mill.

[2]Two 30 h.p. B & W engines had already been purchased. A 48 in. cyl. engine was constructed in 1801 but in 1805 this company ordered the first 100 h.p. B & W rotative engine.

[3]A reference to the building of the new fireproof mill for the Salford Twist Co.

DOCUMENT 152

Dear Sir,

The bearers Mr. Jones & Tougwell[1] of Bradford Wilts are making a tour thro the north with a view of seeing the various improvements & machinery lately introduced into the cloath manufactory & being particularly anxious to see the steam Engine applied to the turning of Machinery we have taken the liberty to address them to you from the conviction you have both the ability & inclination to gratify in the most satisfactory manner their laudable curiosity. We remain

Dear Sir

your Obed Humb Ser[t]

Boulton & Watt.

Office Letter Book 20, p. 120.

[1]An interesting example of prospective customers being prepared to travel far in search of information, although Boulton and Watt may have mislead them by sending them to a Lancashire cotton mill rather than a Yorkshire woollen mill. They did not order a Boulton and Watt engine and no West of England woollen manufacturer did until the early 19th century.

DOCUMENT 153

To Mr. Burnet, Chippenham. Soho November 20 1799.

Sir,

In reply to your favour of the 19th we have the pleasure to annex estimates of Engines of 4 and ten horses power which we have judged sufficient to give you a general idea as the intermediate sizes will amount to a sum proportioned to their power and as the consumption of fuel will be exactly as the power employed.[1]

We will furnish the Metal Materials including one Iron Boiler, delivered here and Payable in three months after such delivery of a four Horse Engine for £327. Of a 10 Horse for £491. This sum does not include the bolts & straps for the wooden framing and cistern which are most conveniently made on the spot. The whole amount of the woodwork including these bolts etc will be for a four horse about £50. For a 10 Horse £77. The workmen who make the framing will be assisted by us with accurate drawings for which we make no charge.

We shall supply you with an experienced Workman to superintend the erection or fitting together of the Engine. His wages will be paid by you and including them the whole expence of erection may amount to about £40 for a four horse & £60 for the 10H.

When properly loaded the consumption of Coals will amount in the 4H to 50 pound weight of coals in the hour; in the 10H to 110 lbs per hour.

If taken proper care of the Engine will not go out of order and the wear and tear of the parts will be extremely trifling. It will last from 4 to 6 years according to the attention paid to it and to the quality of the water.[2]

There can be no difficulty in finding a man to take charge of the Engine as anyone with common attention may easily learn all that is necessary, and will be instructed in his duty by our workman who puts the Engine together. The wages of your Engine man must of course depend on the common price of labour in your vicinity. No damage can result to the engine from your removing any part of the work, as there is a contrivance by which it is exactly adjusted to the load applied, and the consumption of coal will be proportionably diminished.

Hoping that the information we have had the pleasure to communicate may prove satisfactory

We remain Sir

Your very Obed^r Servants
for Boulton & Watt
Gregory Watt.

Office Letter Book 22, p. 175.

---

[1]No order followed.

[2]In fact, B & W engines lasted a great deal longer.

DOCUMENT 154

To Day, Smith & Alder
Nailsworth, Glos.

Soho Dec 11 1800.

Gentlemen,

If the carding machines employed by you are similar to the scribbling & carding machines used in Yorkshire we apprehend that to drive 30 of them will require an Engine of 30 Horses power.[1]

The Metal Materials of such an Engine including one Iron Boiler, flywheel and rotative shaft delivered here and payable in Three months after such delivery will amount to £750.

This does not comprehend the bolts and screws for the cistern & framing as they are generally found to be most conveniently fabricated upon the spot. Including them the whole expense of cistern & framing may amount to £110 or 115.

The workmen will be assisted by accurate drawings which we furnish without charge. We shall provide an experienced workman to assist and superintend the erection of the Engine. Including his wages which will be charged to you the whole expense of erection may amount to £65. The Engine will consume 210 lbs of tolerable coals per hour.

We remain Gentlemen respectfully,
Your most Obed^{t} Servants
Boulton & Watt & Co.

Office Letter Book 24, P.22.

[1]No order materialised.

DOCUMENT 155

To Boulton & Watt

Sunderland 25th Augt 1794

Gent^{n}

We take the liberty of requesting you will inform us the Conditions on which you furnish your Patent Steam Engines of various Powers as we shall in all probability have in many instances to apply Steam for the working of the Lately invented Patent

Machines, (of which we are the Proprietors)[1] for dressing and spinning of Hemp and making the same into Ropes and Cordage — We have been informed that you calculate by Horse-Power therefore shall be glad to know what you expect a 12 Horse Power[2] will be sufficient to turn, etc, what the probable of Expences of such an Engine will be fitted up on the River Tyne, also sizes of Fly Wheel Cylinder and Building necessary, and in what time you suppose such an Engine could be gotten to work after ordered. Also request to know if Newcastle small Coals will answer, and what Quantity will be required for 12 Horse-Power in 12 Hours — Also the Quantity of water for that time and whether River Water where the Tide flows will answer or not.

I am,

for Grimshaw, Webster, Hills & Self,

Gentlemen,

Your obed^t Hble Servt

W. Scarth

P.S. Please to direct to us under Cover to Rowland Burdon Esqre M.P. Castle Eden, near Durham.

Box 6 G

[1]John Grimshaw did not patent his rope-making machines until 1789 (pat. no. 2335).
[2]A 16 h.p. engine was ordered in 1795.

DOCUMENT 156

To Boulton & Watt — Sunderland 13 March 1795.

Gentlemen,

We were duly favored with Yours of the 18th Ulto since which Mr Barnes has shewn us Mr Brandling's Engine on your construction and he informed us you have shortly to send two more into the River Tyne to be fixed in that Neighbourhood and he thought that you might so contrive that the person you sent down to fix those two might fix one for us at the same time provided ours was ordered in a short time but you will be best able to judge, therefore shall be glad to hear from you on that head — As you are unacquainted with all the parties of our Company we purpose sending this to Rowland Burdon Esqre (now in London) one of the Representatives in Parliament for the County of Durham & a Banker in Newcastle & to request He will forward this to you with a letter of Credit. The Engine (which will be first wanted) will be a 12 Horse power[1] that we understand will carry as much Mill work as 12 Horses would be able to do in 6 Hours. — As its uncertain whether the Water will not be some times rather Salt Mr Barnes advises a Brass Air pump. Shall be glad to know what will be the difference in Cost between such a one and one of Common Cast Metal — We wish the Engine House to be detached from the Mill therefore shall be glad to know the longest length the Fly Wheel Axel may be to lose no power, and whether it will make any difference in preparing your Machinery on which side of the Fly Wheel the Cylinder is placed.

We are Gentlemen,

Your Ob^t^ humble Servants

for Grimshaw Webster Hills & Self

W. Scarth

Box 6 G

[1]Order altered to 16 h.p.

DOCUMENT 157

To J. Southern London 9 June 1796

Sir

I receiv'd a note from Mr Woodward which came from you, the following is the best information I can collect; at Mr. Jones's[1] for Grinding drugs only there is 5 pair of edge stones, & one single one, with 8 sieves, viz.

1 pair	5..4 dia.	14 In thick,	making	5 rev. a minute
2 do	5..3 —	14 In	—	9
1	5..0 —	13 In		9
1	4..6 —	14 In		12

when they grind only bark or some other light drugs, the engine will drive all very well, but when the work is heavier such as grinding linseed powder, almond powder, horse medicines &c 4 mills is sufficient, and sometimes if the engine is not in the best order only 3 pair — at Mr. Rutts[2] Thames Street the same sort of business is carried on, they have 7 pair of stones, their engine the same power will in general turn all, but their stones go much slower, the fastest, not more than 7 or 8 rev a minute; with respect to the grinding of white lead the mills at Brandram's are 2^{ft}..3^{in}dia, 30 revol. a minute; before they had their engine[3] they work'd six horses at a time, attach'd to the horse wheel was 4 pair of stones & 2 mash tubs for mixing the lead and oil together before grinding, and generally mash from 15 to 20 cwt of lead at a time; the 4 mills was sufficient work for the six horses; and when they had to mash, all the mills were taken of, and the strain was then worse than grinding, the tubs about 2^{ft}..9^{in} dia — some not so much; at Mr. Yallop's[4] their Horse Mill had one Mash tub $\frac{F}{2}..\frac{I}{3}$ dia, & 2 pair of Stones each $\frac{F}{2}..\frac{I}{3}$ dia with three horses, and sometimes four, they never could work the Stones and mashing together; so that in general it appears that one pair of flat, white lead or colour stones, is more than one horses work, but not two's, the mashing is at least three horses at Yallop's they say four; this Sir is the best information I can get, which I hope will be satisfactory — please to inform Mr. Watt that I deliver'd his letter for Messrs Cook & Co.[5] I only saw Mr. Sayer, who says that Mr. Cook has been indisposed and they have settled upon the plan; but that he will answer Mr. Watt in a few days — the people at the borough Water Works[6] wish you to give them a drawing for the pump, pipes, &c. and they will get them cast here and put them in themselves, they wish to have an air vessel, and also some conveniency to shut the water off, or on to either engine, as their pipes must both join to the bottom of the same stand pipe, in the tower; Hately says the boiler is not yet order'd you will please to let me know, and also whether the bolts for the condensing cistern are to be sent from Soho or made here; please to inform me what is the size of the chains, that is, from the centre of the pins to the arch head; Messrs Roberts & Co[7] Brentford desire to have another pair of long springs for working gear, which are as above, I remain Sir

Your very humble Servt

Rich. Dayus.[8]

Box 6D

---

[1]T. Jones ordered a 4 h.p. engine for drug grinding in 1790.

[2]J.T. Rutt ordered a 4 h.p. engine for the same purpose in 1787.

[3]They ordered a 20 h.p. engine in 1792.

[4]Yallop, Grace & Jackson ordered an 8 h.p. engine for dye grinding in 1794.

[5]T. Cooke & Co. ordered an engine of unknown power in 1796 but countermanded it.

[6]A 30 in. cyl. 8 ft. stroke engine was ordered in 1795.

[7]Distillers. They purchased a 20 h.p. engine in 1795.

[8]An independent millwright (formerly an employee of B & W's) who acted as their London agent.

DOCUMENT 158

To James Watt

Penrhyn
near Conway
North Wales
July 3[d]
1802.

Sir

Many persons in the neighbourhood of Northwich, are desirous of introducing the Plan of Burning the Smoke — in the Salt Manufactory.

I am desired to ask you — what time & expence is required, to make the fire places, & flues, according to your plan: and, when made, should any Salt proprietor, not be satisfied with the operation of it — whether the fire place, & flues, could be readily altered to the old mode.

And likewise, whether a man could be sent to Northwich to inspect the different works — and explain the new plan; and at what expence.

I am Sir

Y[r] Obed[t] Hum Ser[t]
Penrhyn

I suppose that the new fire place, & plans, can be adapted to the old work.

Box 1 Misc.

DOCUMENT 159

To Thomas Holmes & Co.,
Bristol.

Soho April 21 1800.

Gentlemen,

We have rec[d] your letter declining a 4 Horse Engine as you consider the sum it will cost as more than equivalent to the advantage you could derive from it. If as your Mr. Maltin states you only wish to drive two or three grindstones we are entirely of your opinion that the power of a 4 Horse Engine would be infinitely greater than such an object requires unless you could annex machinery for some other purpose.

We remain Gentlemen

Your very Obed^{r} Servants

for Boulton & Watt

Gregory Watt.

Office letter book 23, p. 56.

DOCUMENT 160

To Boulton & Watt

Keighley Yorkshire.
July 7th 1785

Gent^{n}

Upon our further Enquirey after fire Engines we find your Calculations over rateded so far in the Consumption of Coles by the Common Engine that the Extra Expences attending the Erection of yours with the Yearly Payments of your Premium will not be saved, as we are Situated therefore are about to have a Common one put down — we are much obliged by your Civility & if you make any Demand for the great Trouble we gave you we beg that you will let us know — that we may order some of our friends to call and Pay it — we are your Obed^{nt} Servants

Claytons & Walshman

Box 4C

DOCUMENT 161

To Boulton & Watt

Sutton 30 Jany 1798

Gent^{n}

In answer to your favor, respecting our Steam Engine, it was as you suppose constructed for the purpose of returning the Water, being as appeared by your proposal less expensive than a Rotative one because we had only to pay for the Coals consumed in proportion to the power wanted to make up temporary difficiencies in the Reservoir or Stream. The Cylinder of the Engine we use for this purpose is about 40 Inches consequently the power is considerable, but it is worked only at Intervals which makes a material difference in the Expence of Fire, and we are not certain whether the average power required for regulation is equal to more than that of 8 or 10 Horses. The present times & circumstances, do not warrant much Expense, but by these particulars we suppose you will be enabled to judge whether anything can be done at a small one to be of advantage.[1]

We remain

Gent^n

Your most hble Serv^{ts}

Saml Unwin & Co

Inclosd is a Dft value £24:17:0 which
with 3/- paid for Carriage is in full for the pence —
pray are you making any ½ pieces —

Box 3 V & U.

[1]Unwin did not purchase a B & W engine.

DOCUMENT 162

To Gregory Watt London 7 Decr 1802.

Dear Gregory,

We sometime ago (about two or three months) sent an Estimate of a 40 Horse Engine to a Mr Swann in Goodman's fields London. I am told that he has since ordered one of that size from Murray Fenton & Wood in consequence of their Estimate being £400 less than ours. I shall thank you to let me have a Copy of our letter.

I hope the travellers are safely arrived & beg to be kindly remembered to them. I met Moilliet yesterday in the Street & we purpose coming down together on Saturday or Sunday if nothing occurs to prevent us.

I am Dr Gregory

Yours affectionately

J. Watt Junr.

P.S. Rennie will not be moved by my agreement to dispense with Shakespeares going to Boston in the course of the present Week. As men are scarce there I think it may be as well to comply with his desire of letting him have his brother as an assistant. I should wish enquiry to be made of Mr John Ryland of Birmingham with whom he worked 10 years respecting his character &c — the result transmitted for my Government.

Box 2

DOCUMENT 163

To S. Hornblowers & Co.
Horseley Mill, Glos.

Soho 18 June 1807.

Gentlemen,

We are sorry to be under the necessity of stating that the present state of our orders will oblige us to require from 10 to 12 months for the Execution of a large Engine such as you appear to want.

If this term should suit your convenience we shall thank you to say whether you wish the Engine to be of double power and to what purpose it is to be applied.[1]

Waiting your answers we remain respectfully
Gentlemen
Your Ob^t Hble Serv^ts
Boulton & Watt & Co.

Office Letter Book 24, p. 310.

[1] They did not order an engine. Horseley Mill was a woollen mill.

DOCUMENT 164

To W. Creighton
Soho

London 12 Sepr 1816

Dear Sir

I wrote to you yesterday and now inclose Copy of a letter I have just written to the duke of Bedford's Agent relative to the Engine at Woburn made by Cartwright[1] which will serve for your Government in case an Engine should be ordered. The Duke was not at home, but we understand is very much dissatisfied with the Engine, which is in miserable plight, the boiler leaking at every joint, and the Cylinder, within it, stated to be the only part in tolerable order. This with its metallic segments to the piston kept up by springs, in place of packing, we were anxious to see, but were informed we could not be gratified without waiting till the next day, on Account of the difficulty of taking off the Cover to which all the Valves are fixed. Our curiosity did not extend so far, and we were quite satisfied that their own Miller's statement of its performance was sufficiently correct to render any minute examination unnecessary and we passed sentence upon it accordingly.

Mr Rennie's report of the Millwork, at least of the threshing machine is not much more favourable, only that he thinks it may be greatly improved without going into such a radical reform as we have advised for the Engine.

Should it become an order, Mr Salmon has undertaken to send you the necessary particulars of the present house, which he seems very capable of doing.

Nothing further occurs at present and I remain

Dear Sir
Yours truly
J Watt Junr.

P.S. Mr Howard has been dangerously ill & is not yet sufficiently recovered for me to see him. I understand however that both he & his operator Mr Powell are perfectly satisfied with the performance of the Air pumps & other Machinery made by us & there seems no objection on the part of Mr Hodgson to our going on with any number of them.

Box 46

[1]An earlier enquiry had been made for a B & W engine to which Boulton had made a detailed reply. No order materialised on this second occasion either.

DOCUMENT 165

To James Brown,[1]
London.

Soho May 29th 1818.

Dr Sir

In reply to the enquiry of Mr. Moore[2] the size of millstones or quantity of wheat ground being unknown, it may be assumed that each pair of stones grind 6 winchester bushels pr hour. The metal materials of a Steam Engine to turn 2 pair will amount to very near £1000 delivered here and then payable by draft on London at 2 mo. For 6 pair an Engine would cost about £1780. — Having none suitable for 4 pair in any state of forwardness, it may be observed that to turn 3 pair the Engine would cost near £1230. One boiler of sufficient size to work the Engine by burning coal is supposed ... The exact amt. is in some degree influenced by local circumstances, and a spare boiler may be recommended where the mill works a considerable number of hours daily, to allow for cleaning. Engines of the above sizes could be supplied at a short notice.

For the draining project a single power pumping Engine which may be called 10 horses power, with one boiler and the requisite pump for raising water 24 feet high will cost about £1000, and to give it exactly more particulars of the case are requisite. Such Engine would raise 60,000 gallons of $6^1{}_8$ to the Cubic foot pr hour, & burn 90 or 100 lbs of tolerable coal in that time.

I remain yours &c

W. Creighton for B W & Co.

Box 46

Mr. Brown. May 29.

The Engines proposed for Mr. Moore are 14, 20, & 36 horses power a horses power being reckoned to grind a bushel pr hour. — I suppose it immaterial that Mr. Moore has the sizes in any memorandum given him, if he wants coals consumed a 14 h takes about 150 lbs Staffordshire or 1½ bushel Newcastle. 20 rather more than 2 bushels &c.

The Pumping Engine is supposed 24 inch 5 ft stroke making 15 pr Minute with a 21 inch pump.

There is some doubt as to your enterior circumference of Caledonia[3] Water case, the chimney is given 5..9 and the Case 5..11⅞, now this does not leave ½ inch between the two, and it is considered to be about ⅞ in reality, please examine into this.

W.C.

Price of 10 horse West India
with Duplicates £1040.

Box 46

---

[1]B & W's London office agent, specialising in marine engineering.
[2]No order materialised.
[3]A reference to the Steamship Caledonia which was re-engined by B & W.

DOCUMENT 166

To James Brown,
London.

Soho, June 8th 1819.

D Sir,

I have your letter & Mr. Anderson's of yesterday and will speak to Mr. Watt about it – the Engine required would be a 66 inch 9 or 10 feet stroke, the estimate was made for 9 feet, and hard work then to force the water thro a 24 inch main, only one engine working. Now that the main is longer the load will be greater, as Mr. Anderson gives 5060 yards & Mr. Snee 4500. To have the Engine done in the time you mention appears impracticable. As the size is so very great it becomes a matter of consideration whether double engines and pumps could be applied. Mr. A. says The Butterley estimate is 2000 under B & W's, (6200).[1]

Being much hurried yesterday I omitted to inform Messrs. Davidsons Barkly & Co. that the fire places such as go with west india engines are considered proper either for coal or wood – if not of the best construction for the latter no information has been received of what change may improve them. I should think the people in Jamaica quite competent to set boilers properly.

Please tell Mr. Murdock that a letter is just come from his nephew in Jamaica who is well & wants answers to several questions respecting a scheme for supplying Kingston with water from a spring 4½ miles distant.

I remain your &c

W. Creighton.

Congo Boilers[2] are come and will be transmogrified

Box 46

[1]This possibly concerns Glasgow Waterworks for which a 54 in. cyl. 8 ft. stroke engine was ordered in 1819.

[2]A reference to the B & W engine for a steamship ordered for exploring the Congo River. The ship did not answer and the engine was removed and used elsewhere as a land engine.

DOCUMENT 167

To Boulton Watt & Co.
London.

Hastings Jany 28th 1824

Gent^n

We have wrote to your House in Birmingham respecting a Steam engine & they refer us to you to make the Present enquiry — we will thank you therefore to inform us by return of post what the expence of a Person would be to come to this Place to survey our bath rooms &c & to give us your idea on the Subject. We wish a Small Steam engine fixd for the purpose of Pumping the Water &c also the same Boiler to supply the Bath with Steam of sufficient quantity to make a Bath of any heat required. We think an engine of Half Horse power will be sufficient the price of which you will also send that we may buy it before the Proprietors.[1]

Your Ob^t Ser^t
for T & J Marrington
J Bartlett

P.S. Any references you may require of us be beg to refer you to Messrs Daniel & Payne Frig Iron Wharf

[1]No order materialised.

The overseas market

Documents 168-194

DOCUMENT 168

To Ambrose Weston
London

Birmingham May 26 1792.

Dear Sir,

We are obliged to you for the Caution contained in yours of ye 24th. We did not so much as know of the act 23 G 2. We know indeed of an act 25 G III but understand the prohibition only to extend to sending artificers out of the Kingdom to carry on any of our manufactures, a thing which independent of the prohibition, we never should have done. We considered the sending a man to put together an Engine (which we can only send in pieces), and which was entirely made here in so far as concerned the manufacturing thereof, as a very different affair. However now the affair is started it may be proper to have an opinion thereon, say Mr. Anstruthers or Mr. Plomers.

Case — There is no law known against the exportation of Steam Engines or any of their parts.

The machines being very bulky & necessarily fix'd in a house which serves as a support to & is part of the machine, they must be exported in separate pieces, some of them packed up in boxes, and all of them requiring to be adjusted to one another at the spot where the Engine is to stand.

They cannot be put together in an effectual manner without the aid of a skilful workman.

Therefore if workmen cannot be sent abroad to erect them then the exportation must either, virtually, be considered as prohibited, or exporters must train up forreigners in the art, which in the end will be a loss to the manufacturers of Engines in this country.

NB at present steam Engines are very well made at more than one manufacture in France[1] where they be under no restrictions but can send the Engines and workmen to any part of Europe.

Query – if all the parts of an Engine are manufactured in Great Britain, and adjusted to one another as far as can be done here, whether sending a workman to put them together in a forreign country can be considered as a breach of the Statutes 23 G II & 25 G III. Such workmen being employed abroad in no other art or manufacture & returning as soon as he has compleated his business.

Query 2nd whether there is any law against the Employing forreigners so instructed abroad would be a breach of the Statute. We shall be obliged to you to take one of these Gentlemens opinion on the foregoing case & report to us as soon as you conveniently can. It appears to us Thus if the opinions are in the negative, that a workman cannot be sent abroad to put up a Steeple Clock, an astronomical instrument or any other bulky & complicated piece of machinery nor perhaps even to sollicit orders for our manufacturers.

As to R. Cameron we are not over fond of having any thing to do with him & therefore we shall take some other means of relieving him.[2]

During the time of the American & French Wars we sent two very large Engines to Paris by virtue of passports obtained from the Privy Council. We sent no workman, being war time, & by that omission we were obliged to send such instructions for the erection of them as enabled Perier to set up his manufacture in Paris where he now makes almost as many Engines as we do.

In short we believe that many workmen have been sent abroad by the various Iron Founders for this very business of putting Engines together & it has never been considered as contrary to Law.[3]

We remain Dear Sir

Your Obed^{t} hble Servants

Boulton & Watt

(J. Watt's hand)

Office Letter Book 15 p. 162.

---

[1]A reference to Perier who was making Watt-type engines only shortly after importing a Boulton and Watt one.

[2]See [100].

[3]Weston advised to look at each case individually. The laws relating to the export of artisans and machinery did not prohibit the export of steam engines and where a customs official was suspicious the partners were known to resort, on occasion, to a little gentle bribery.

DOCUMENT 169

To Boulton & Watt

Balbriggen 2[d] Jan[y] 1788
Ireland

Gent[n]

I humbly request you will be pleas'd to acquaint Me what the Expence of a Steam Engine will be that can work a Hammer of 3 Cw[t] for a forge, likewise one of 6 Cw[t] with other necessary Works; — Any Information you can gie Me will be thankfully rec[d] & as possibly I may be from Home please direct to M[r] Jo[s] Smith Balbriggen.[1]

I am Gent[n] Your most obt Ser[t]

T King

Box 4 K

[1]No order materialised.

DOCUMENT 170

To J. Staples Dublin

Birmingham, Sep[t] 9 1784.

Our friend Mr. William Chapman having advised us that you have determined to erect on your colliery at Drum Glass, a steam Engine of our construction, with a cylinder 36 inches dia[r] and 8 feet stroke, & having desired us to write to you on the terms & to give orders for the materials necessary to be sent from this country, we send you inclosed drawings for the Engine house and shall as soon as we can, send the other drawings & order for materials. Mr. Chapman, we suppose, will give such particular directions as may be necessary in building and situating the house in the most convenient place, so that we conceive that anything more particular from us on that head is unnecessary.

As our Mr. Boulton is now in Cornwall and J. Watt does not know what may have passed betwixt him & you concerning our terms, we proceed on the supposition of nothing having past. Our common way of agreeing is, in cases where coals used by the Engine are valued at or above 4/- pr ton, and the Engine is required to work at the rate of 6 strokes or upwards in every minute of the 24 hours or to work 12 hours at the rate of 12 strokes pr minute. In such cases we agree to take ⅓ of the value of the fuel saved by the use of our Engine when compared with a common one of the same power and doing the same work. In the beginning of our business many comparative trials were made between our engines and those of the common construction, in every one of which cases, it appeared that our Engines used less than ⅓ of the fuel used by the common ones in doing the same work, in many cases less than ¼, but as such trials were troublesome & expensive & opened a way for disputes, we now agree that our Engines shall be rated as being only 3 times better than common ones or that they burn ⅓ of the fuel in doing the same work. A common engine of the same power & length of stroke with the one which you desire for Drum Glass, would burn when going 12 strokes pr minute, at least 705 pounds weight of good coals pr hour, rating it at 3 times the consumption of our Engine. Our Engine will require of good coals 235 pounds pr hour at the same rate of going, the total saving will then be 470 lbs pr hour, of which our third will amount to 156 lb pr hour.

In order to ascertain the work done or the number of hours during which the Engine works it is usual for us in this country (where the engines are not too far distant from us) to fix to the working beam of the Engine a machine which keeps an account of the number of strokes, which we examine twice a year or oftener to settle the premiums due to us, but where Engines are distant, or other circumstances render that mode of agreement improper we agree for a lump annual sum — some way proportioned to the price of coals and to the use which is purposed to be made of the Engine. Now we conceive your Engine to be in such circumstances on account of its being distant from us, and of the expense which may attend sending to take account of the number of strokes made; therefore we would propose, if agreeable to you, that the mode of an annual sum should be adopted, but we cannot at this distance venture to propose the sum, as though we are informed that the value of the coals which it will use is about 4/- British Sterling pr ton of 20 cwt, yet we are ignorant of the work it may be required to do.

In order therefore to shorten the business as we cannot have the pleasure of a personal conference, we write to our common friend Mr Chapman to confer with you on the subject, and expect he will be able to make proposals or to settle it in some way which may prove agreeable to both parties, it being very much our wish to render it so to you and on receiving his opinion we shall immediately write to you in answer.

The remainder of our terms are 1st that we furnish gratis drawings & written or printed directions for erecting & working the Engine, and also provide at the expense of our Employers such materials as may not suit them to provide themselves, and we also send at their expense an experienced workman to put the Engine together (which workman need not come untill the materials are all on the spot.)

2[ndly] that the Engine shall not be sold or removed without our consent which we oblige ourselves to grant, on terms proportioned to the price of coals & the work to be done by it in its new situation.

3[rdly] the annual premium to be paid us half yearly during the term of our patent in Ireland, or during the term the Engine is used at the Colliery if shorter than the term of the patent — with other common provisos.

We shall be obliged to you to appoint a meeting with Mr Chapman as soon as suits your convenience and remain

With due Respect
Sir
Your Obliged
humble servants
Boulton & Watt
(in hand of James Watt).

Office Letter Book 9, p. 35.

DOCUMENT 171

Draft letter J. Watt to Lord Dartmouth — n.d.

I take the liberty of begging your lordships advice on the following subject — Since our Steam Engines have become famous we have had repeated demands for them from foreign Countries particularly from france but have constantly refused unless we had exclusive privileges granted us in those Countries. This the Court of france have complied with after being thouragh(ly) apprized of the difficulties they must have of obtaining them otherwise — In consequence we have received some orders for fire engines from that Country, particularly to supply the city of paris with water[1] and a smaller one for a private person in Brittany[2] — In the present warlike state of the two nations we cannot transport them hither without proper passport. — Our friends there offer to procure a passport from their coast for an english vessel to carry them directly to france provided a similar one can be obtained from the proper office in England — But as I am ignorant of the proper mode of application for such a favour & also of the propriety of that application therefore entreat your Lordship's friendly advice upon the subject.

I should be sorry if your Lord^{p} should look upon what I have now mentioned, as an attempt to transplant a manufacture into an enemy country — The things we propose to send are completely finished here and consequently must be looked upon as a part of the manufacture of England — and did we not supply them with these articles the natural consequence would be that they would establish works of their own for the purpose of making them. We have very good reason to believe that we have already quashed an attempt of that kind which but for us had certainly been carried into execution, — but how long we shall have it in our power to retain the manufacture here it is difficult to say without we are permitted to supply them with the goods. They have several Engines of the Common Cons^{n} in France all of which were made in Britain, and unless they meet with particular obstruction here it is not probable they will change their marks.

We have lately had some german Barons here who have used even the meanest arts to rob us of the invention and the Country of the Manufacture, we believe they are still visiting the several manufacturing towns. We have guarded ourselves as well as we could and cautioned others.

Box 1 Misc.

[1]Perier for Paris Waterworks: two 63 in. cyl. 9 ft. stroke engines (actually one complete engine and parts of the second).
[2]Jary of the Mine du Nord, a 44 in. cyl. 8 ft. stroke engine.

DOCUMENT 172

To His Excellency Count Cavalli
La Rochelle.

Birmingham Oct. 16 1780.

Sir,

Your Excellency's obliging letter to my partner Mr. Boulton, who is at present in Cornwall, was this day put into my hands to answer.

We have never made any steam engine (machine a vapeur)

that cost less than £400, and we have for some time declined making any so small as they did not turn out sufficiently profitable. We have indeed proposed to ourselves as soon as our more Capital engagements would permitt us, to make a number of small portable Engines which we think may be sold for about £150 each and by making a great number at one time we might indemnify ourselves for the smallness of the profit on each. Happily for us we are not yett permitted to have any leisure for a manufacture of that sort, having just contracted for the erection of 6 very large Engines in Cornwall which will take up our whole time and attention for a year to come and will merit such attention as they are to pay us above £2,000 pr annum premium.

The small Engines I have mentioned are proposed to have cylinders of 12 Inches dia^r and 2 feet long; they would be capable of raising a weight of 1120 pounds 24 feet high in one minute or would work a pump of 4 inches in dia^r and 70 yards {feet (210)} high at the rate of 12 strokes of 2 feet long in every minute and at that rate of going would consume about 20 pounds of coals in an hour. Our large Engines will raise about 500,000 cubic feet of water to one one foot high by 112 pounds of coals or proportionally less to greater height. This is a much greater performance than I have stated above but all small engines burn more coals in proportion to the work they do than large ones.

I am very sorry that at present we are not in a situation to supply Your Excellency with a small Engine. I cannot say how far it might be agreeable to Mr. Boulton to permitt such an Engine to go out of the Kingdom but I am certain that there is no person he has a more earnest wish to oblige.

We return our most Grateful acknowledgment for the honour you have done us in recommending the copying machine, and I beg leave to remain with the utmost respect

Sir

Your Excellency's
Most Obliged
& most obedient
humble serv^t
James Watt.

Office Letter Book 5. pp. 95-96.

DOCUMENT 173

To Sir John Dalrymple Bart. Soho 29 March 1788.

Dear Sir,

Since our last we have received the Honor of your Letters of the 25th & 26th Inst.

You may depend on our not taking any part in the Projects you communicate to us, that has any tendency to run counter to your Interest, but we cannot help thinking you are wrong in making so many applications before you have actually executed, & completed the Invention & proved its utility by satisfactory experiments.[1]

If the thing prove equal to your sanguine wishes there will be little need of the many introductions you mention as the most effective one will be the utility of the Machine itself.

We are both acquainted with Monsieur de Callonne[2] & entertain the highest Respect for him, but we cannot say that we are upon such a footing with him as to take the liberty of giving an Introduction to him, but we have no doubt you will find many Persons in London in another station that can do it more effectively.

In respect to the intended French canal you mention we must observe that as we know not the particular Ground it is to pass through nor the general situation of it, nor the number of Locks that will be required, & we have no doubt but it will require many, as the Seine is very rapid, & hath more than 100 feet fall below Paris.

We say as we are Strangers to these essential Data, we cannot pretend to give the least Guess at the Expense, particularly in a Foreign Country but we know there are very able Engineers in France in these matters.

You will please to observe that Locks are made as narrow as possible in order to save Expense & to save Water, which in general is not in sufficient Quantities for inland Navigation, particularly in dry summers.

We conceive that such double Boats as could pass any of the English Canals would not contain a sufficient burthen to answer any good purpose.

We do not apprehend that Water Wheels can be applied to single Boats so as to pass Locks.

We have laid it down as an Axiom, that a Fire Engine ought never to be applied in any Cases where the work can be done with one or two Horses even where coals are as Cheap as in this Country.

That many things can be done in this Country where coals are from 1/6 to 6/- pr ton which cannot answer in Countries where coals are from 1/- to 2/6 pr cwt & to work them with wood would require near double the weight that it would do of coals.[3]

We shall be very happy to see Mr. Miller if he makes Birmingham in his Road, & hope by the time he arrives our Mr. Watt may be in better health, for at present he is realy incapable of attending to any business. We are with great Regard

Dear Sir

Your Faithfull
& Obed hble servts
Boulton & Watt.

Office Letter Book 6 p. 146.

---

[1]In fact he took out no patents.
[2]C.A. de Calonne (1734-1802), became Controller-general of finance in 1783 Attempted to revive French trade by a programme of extensive public works. Left France in 1787 and from 1790 till the fall of the French monarchy acted as chief adviser to *émigré* princes.
[3]Sir John Dalrymple (4th baronet of Cranstoun) experimented in many of the useful arts, discovering, amongst other things, how to make soap from herrings and a method of distilling tar.

DOCUMENT 174

The following Questions are proposed to Boulton & Watt by Messrs Signoretti, Rodella Assistant Astronomer at Padua & Dr Scandella, on the part of the Chevalier Pisani of Venice. —

Viz: It is proposed to raise 810,000,000 paduan Cubic feet of Water to the height of Fifteen paduan feet, in Ninety Days.
Qn 1st How many Engines will be necessary to produce that effect in that time & what will be the size of the pumps?
Qn 2nd What Quantity of Coals will it require?
Qn 3d What will be the probable Expence or cost of such Engines?
NB The paduan foot in length is to the English as 1580 is to $1351\frac{2}{3}$.

1st In answer to the first question, it will require 5 Engines, to work five pumps, and each pump to be 57½ paduan inches in Diameter and 6 ft .. in paduan measure long in the stroke.

2nd It will require 600 Bushels of the best Newcastle or Swansey

Coals p^r day, or it will require 600 Quintals of Staffordshire Coals which is equal to 600 × $\overset{lb}{112}$ or 67,200 pounds weight English, or paduan weight — 62,000 pounds.

3^d It is impossible to give an accurate calculation in so short a space of time of the Cost of such Engines, but to the best of our Judgment, The 5 Engines with all their Boilers, their Beams and other timber work for the Engines, including their 5 pumps and $\overset{£}{500}$ for their Erection (but exclusive of the buildings for the Engine houses which we cannot guess at as we know not the expence of building in Italy,) we say to the best of our Judgment they will cost 20,000 pounds Sterling.[1]

Soho, 18 July 1794.

The weight of one engine, with its boiler & wood work, together with its Pump is estimated to be betwixt 130 and 150 tons.

Box 5 XIII

[1]B & W added in a covering letter 'We must say that though this drainage by the means indicated appears perfectly practicable yet it seems not to be advisable.'
B & W to Pisani, Nov. 1793.
The engines were not ordered.

DOCUMENT 175

To Fermin de Tastet & Co.

Birmingham
Feb 19 1793.

Gentlemen,

Our J.W. has received a letter from his son at Naples dated Jany 15 of which the following is an extract. "I have seen your Engine at Capua[1] which is extremely well put up and every thing very neat, but very little used — The King & every body else are wonderfully pleased with the Engine, which has puzzled all the Engineers here not one of whom comprehended it. Malcolm Logan is now putting up a mill for the King at Carditelli about 5 miles nearer Naples than the Engine, it is nearly finished & he hopes to have it at work in a fortnight, after which he has the Kings leave to go to Spain, but under promise of returning when he has finished the Engine there. He drinks much less than he used to do and has recovered his health." When this letter was written J.W. Junr could not have received his fathers letter of the 8 Jany to which a part answer will probably be given. Logans address is at Messrs Cutler & Hagan's Naples.

The Engine no. 1 Gardoqui[2] is all packed up upon account of the embargo, and sent from hence, as wharfage at Hull would turn out very expensive if it lay any time, we therefore wait your orders.

The remainder of the goods for that Engine say a Boiler, five grates and some few other articles which are at Chester we have stopt shipping for London on account of the rise of insurance & if you approve shall try to get them shipt at Liverpool for Spain on being informed of the port, otherwise shall send them to London when convoy is appointed.

We remain very respectfully
Gentlemen
Your faithful hum^{ble} Serv^{ts}
Boulton & Watt
(J. Watt's Hand)

If we delay too long there is a danger of the Trent becoming dry betwixt here & Hull.

Office Letter Book 15 p. 227.

---

[1]A pumping engine owned by the King of Naples with a horse power equivalent of 28 h.p., ordered in 1787.
[2]One of a pair of 55 h.p. blowing engines for a Spanish Ironworks.

DOCUMENT 176

Mr Dillon's Translation of a Spanish paper delivered to Mr Pasqual Menza Merch^{t} in Cadiz now in London N° 45 Basinghall Street — London 30th Jan 1789.[1]

1	1
What will be the cost of a fire Engine similar to the one Erected at the Albion mills? —	About £2400 or £2500 including all profits & prem^{ms} —
2	2
How much time is required for the constructing of such an Engine? —	About 6 or 7 months from the time of being ordered till shipped. —

3

What time is necessary for erecting the whole, on the other side? —

4

About 4 or 6 months with one Englishman & 5 or 6 common carpenters & smiths of the country. —

4

What will be the expence of such an Erection? —

4

This is answered by article 3, as there only remains to calculate the pay of the workmen according to the time of their labour, the Englishman is to have 25^{s}/- p^{r} day.[2] —

5

What will be the monthly expenditure of this Engine?

5

To this Mr Boulton answers that, The shadwell Engine burns 1 Bush[1] of coals p^{r} hour
"makes 1000 strokes p^{r} hour
"raises the water 90 f^{t} high
"& raises about 6 cub f^{t} p^{r} Stroke

$\overset{\text{strokes}}{1000} \times 90 = 90{,}000 \overset{\text{cub ft}}{\times 6} =$ 540,000 p^{r} bush Coal but Boulton & Watt will not guarantee more than 500,000.

6

What may be its daily result with respect to grinding corn? —

6

It works 8 pair of Stones & each grinds 8 bushells p^{r} hour or 1800 or 2000 quar^{rs} p^{r} week or more properly 5 bush^{ls} p^{r} hour taking the week round on a general scale. —

NB. 120 revolutions p^{r} minute for stones of 4½ f^{t}. diameter

7

How many people are necessary to superintend the Engine?

7

Only 1 man at a time.

J. T. D.

General Observations

A Calculation is still wanted of the aggregate rotative work both in Iron & wood which must be annexed to the Engine & its primitive relative parts, this addition belongs to the owner of the Engine. — The expence thereof will be proportionate to the operations to which the Engine is applied. This is no part of Mr Boulton's concern; however he has promised to obtain for us, the calculation of the millwright's work for this part. —

Mr Boulton has set up above 300 fire Engines in these Kingdoms & is still going on with the greatest success & the general approbation & thanks of the concerned exclusive of what he has done in France, Flanders, Holland & Italy. — he obliges himself to render his work perfectly complete. —

His terms are, to be paid one half in London at the time of the delivery of the Engine on board. — the other half in 6 months payable in London without interest — he is further willing to accommodate with time on being paid the national interest & to do everything reasonable that may be expected towards promoting this business in a mercantile way. —

London 28th Jany 1789

J. T. D.

Box 36

[1]These enquiries resulted in an order for a 55 h.p. engine for a Spanish flour mill, the design being based on Albion Mill. John Rennie designed and made the millwork and milling machinery. [177]

[2]This must be a mistake, 25/- per week being a usual wage for an engine erector.

DOCUMENT 177

Mr Rennies list of Men for the Mill & price of Mill work for Mr Mensa Mar. 1789

List of Men necessary to work ye Mill daily

1	Clerk	at 40/- p^r week
1	Englishman to superintend ye Steam Engine	at 25/- p^r week

2	Men to help him 1 the other	at 21/- p^r week 14/- d^o
1	Englishman to superintend the Mill	at 25/- p^r week
2	Men to attend the fire	12/- p^r week
2	Master Millers	at 25/- p^r week each
5	Millers	at 18/- p^r d^o
4	Dressers	at 16/- p^r D^o
3	Men to carry the sacks	9/- a 10^s/- p^r week
1	Sack Mender	at 9/- p^r week
22	Men employed daily	

This List is made according to Mr Rennie's advice.

Box 36

Estimate of the Mill Work of a Mill to be worked by one of Messrs Boulton & Watt's Steam Engines, consisting of Ten pair of Mill Stones, Dressing Machinery &c (1789)

	Ton,	Cwt	q^r
86 Iron wheels	17	16	3
40 Shafts & Spindles	8	16	2
324 Plummer Blocks	5	8	3
270 Brasses		16	
wrought Iron	4	11	2

About 1800 Cubic feet of Oak Scantlin for frames &c

Fourteen dressing Mills

About 2950 feet Oak Plank

About 680 Elm d^o

Ten Pair Mill stones

£3350. 3. 11.

This is a Copy of that wch Mr Rennie delivered to Mr Mensa in London

Box 36

Plate [10]

✠

REAL CEDULA DE S. M.

Y SEÑORES DEL CONSEJO,

EN QUE SE CONCEDE privilegio exclusivo por término de veinte años á Don Simon Plá y Mensa, y Compañía, para la introduccion en el Reyno de las Bombas de fuego, llamadas de doble inyeccion, baxo las condiciones que se expresan.

AÑO 1790.

EN MADRID:

EN LA OFICINA DE LA VIUDA DE MARIN.

DOCUMENT 178

To B. Bewicke Esq. Birmingham

Oct 8 1789.

Dear Sir,

I arrived here in good time yesterday but much indisposed from my journey.

In a very short conversation with Mr. B last night I learnt that Mr. Torres wants the full power, expressed in pounds of a double 72.[1] We shall start every objection, but if we begin to treat & he should come to our terms we should be impowered to act. Now Mr. Mensa confines us to the sizes in the tariff[2] & this is 1000 degrees beyond it. It is exceedingly probable when Torres hears of the coals & other expenses he will stop short, but if he should not & what he desires should so far seem practicable as to merit a calculation it will enable us to add many stages to the tariff;[3] though I conceive to little real advantage, as few will chuse to enter on such enormous enterprizes. I know not what Mr. B has written to you but thought it best to know Mr. Ms sentiments.

With Comp^s to him & to all your good family, I remain Dear Sir

respectfully
yr Obed^t hmble Servt
James Watt.

Office Letter Book 14 p. 142.

---

[1]a 60 in. cyl. 9 ft. stroke of 90 h.p. was ordered for this Spanish saw mill.

[2]Mensa had the sole agency for B & W engines in Spain having been instrumental in obtaining the Spanish patent. The tariff referred to was the agreed price list.

[3]B & W by-passed Mensa with Torres' order as they by-passed Jary, their sole agent in France, on receipt of the Perier order.

DOCUMENT 179

To Malcolm Logan
Cadiz.

Soho 10 Aug 1795.

Dear Sir,

Our last to you was of the 30th March since which we have received yours of the 8th Feb, & 11th & 29th April, from the latter of which we are glad to learn that there is a prospect of your procuring more orders at Seville, as soon as Mr. Wetherall's Engine[1] is put up. It is now compleatly ready and would have been sent to Hull to go with the convoy which sails on the 30th instant, but that we have not yet finished the millwork which is to go along with it. We did not receive the order to prepare the latter until the end of June, and it will be October before it can be compleated, it being our intention to make it as perfect as possible. We have made some slight alterations in your plan of the mill, which render it less expensive and more convenient. These we will point out when we sent the drawings of the Engine, which shall be forwarded to Messrs Wetherall by the first opportunity. You may then be going on with the Engine house immediately that there may be no stoppage when the Engine arrives. As we were afraid that new wood would not stand in the heat of a Spanish climate, we have purchased the framing of an Engine which had stood a couple of years and is perfectly sound in every respect. We have had it and the Engine put together here, and shall mark the parts so that you will have very little difficulty in putting them together again. We shall send two boilers & duplicates of the furniture of the nozzles, of all the brasses, and of the Rotative wheels, which are the only parts liable to wear or accident.

Inclosed we send you a list of the prices of our Engines of from 8 to 40 horses. We do not intend to send abroad any of less than eight horses and we do not recommend using any larger than 40 horses without a double rotative motion, which very much increases the price. We think that if any of a greater power are wanted it will be preferable to employ more than one, for we think it will be difficult to teach the Spaniards how to manage such large Machines.

You will please to take notice that the prices here mentioned are for the metal materials (including the boiler and necessary duplicates) only. We do not comprehend in that estimate either the wood or ironwork of the framing, which we think may as well be made in Spain, after one or two have been erected. However we shall not object to undertaking the framing if particularly desired, as in this instance of Mr. Wetheralls, but we have not experience enough of the cost to be able to determine the price for different sized Engines. We have engaged to furnish Mr. Wetheralls Engine with framing compleat for £630 computing that the framing would amount to about £80, but in future we would not undertake to furnish it at that price for Engines of the same power.

You will also observe that the sums we have specified are to be paid upon delivery of the Materials at Hull, and that we undertake no foreign orders without having a Guarantee in this country, being Engineers and not Merchants.

You will please not to give out any copies of this estimate which is intended for your own use only.

We shall write again with the Drawings and in the meantime remain very sincerely yours

for B & W
Jas Watt Junr.

Prices of Engines of various sizes, delivered at Hull, including all the Metal Materials, the Boiler (of Iron), the Fly Wheel, Rotative Shaft and Sun and Planet Wheels, with Duplicates of the Furniture of the Nozzles, the Brasses and Rotative Wheels — August 1795.

An Engine of the	power of	8	Horses as above will cost	£550
Do.	Do.	10	Do.	£607
Do.	Do.	12	Do.	£660
Do.	Do.	14	Do.	£824
Do.	Do.	16	Do.	£885
Do.	Do.	18	Do.	£935
Do.	Do.	20	Do.	£984
Do.	Do.	24	Do.	£1205
Do.	Do.	28	Do.	£1300
Do.	Do.	32	Do.	£1606
Do.	Do.	36	Do.	£1728
Do.	Do.	40	Do.	£1840

Nb The Engines of the power from 8 to 12 horses have a four foot stroke in the Cylinder; those from 14 to 20 horses a five foot stroke; those from 24 to 28 a six foot stroke; and those from 32 to 40 horse a seven foot stroke. You will observe that the price is considerably increased by lengthening the stroke.

Office Letter Book 17 pp. 251-252.

---

[1]A second-hand 8 h.p. engine for a leather and cornmill.

DOCUMENT 180

To Boulton & Watt | Munich April 25th 1791

Gentlemen:

Your obliging Favor dated the 31th March has reached me in due time, it has been immediately laid before his Highness The Elector, who is perfectly satisfied with your proposals & quite determined to have such an Engine at Manheim.[1] — There being a sufficient number of Springs within the walls, the town being Situated between two rivers on a flat Sandy ground, we Shall certainly adopt the plan of raising the Water by Steam. It is, therefore, fixed upon that a 4 Horse Engine (the Size which you propose) Should be erected as soon as possible. — The Size of the reservoir we are not yet able to determine, of course nothing can be fixed about the proportions of the pumps, or the time of drawing water: but it is agreed upon that the Engine Should be worked continually, &, besides the raising of water, give motion alternately to a Corn mill & to a Fulling mill. The Machinery for both Should be entirely executed in England after the most improved principles; even the Grinding Stones should be made there from the best Quality. — The Fulling-mill should be of the Construction comonly used in England for fulling ordinary & midling Cloth[2]. Besides it is the wish of the Elector to have a Small jet d'eau playing at times about 50 or 60 feet high above the ground, which, I presume, might be obtained by two forcing pumps with an air-vessel, to which the power of the Engine could be applied occasionally. — This, however, is but a Secondary object, & of no great Importance, if it cannot be conveniently connected with the rest of the plan.

The chief points of the order being thus previously Settled, we beg that you may begin without delay to execute the main parts of the Machine, & to send us, as soon as possible, a compleat Set of plans & Elevations for the Engine-house & Mill-houses, in order to enable us to finish the necessary buildings in the mean time. — As I Shall only Stay a Fortnight longer in this place, I beg to direct the Same immediately to Sir Benjamin Thompson Adjutant General to His Electoral Highness, à Munic en Baviere,[3] who will order immediate payment through his Agent in London.

The Engine being intended to work day & night, I think it will be adviseable to have two Boilers, in order to prevent Stoppage in case of any accident happening to one of them.

I hope to have the pleasure of paying You my visit at

Birmingham in a few weeks, & Shall bring with me a young Smith from this Country whom the Elector intends to Send over to You in order that you may instruct him to put up & keep the Engine. In the mean time I remain with respect

Gentlemen,

Your most obed^t^ humble Servt

Joseph Baader.

P.S. In case you should think the parts of the Fulling mill too heavy & expensive for the Transport, you are requested to Send compleat Drawings of the Same after which they may be executed by our Millwrights at Manheim. But the Cog-wheels belonging to the Mill must be made in England, because they cannot cast them So well in this Country.

Dr Baader returns respectful Compliments to Mr Boulton, has the honnor to acquaint him that his Highness the Elector will be very much pleased with a Specimen of his improved Coin & the paper relative to his Several Improvements in the art of Coining. Mr Boulton is, therefore, requested to Send the Same to Sir Benjamin Thompson under the above direction, who will present it to the Elector in M^r^ B^s^ name. — The parcel will be conveyed with perfect Safety if comitted to the Care of the Electors Envoy Count Haslang, Golden square, London. —

His Highness being a great patron to arts & Sciences, & very desirous of adopting any new Invention or Improvement that may be of Service to his Country, Mr Boulton may at the same time propose, if he pleases, his Conditions under which he would chuse to communicate to the Elector the Secret of his Inventions in that line.

Dr Baader will enquire after Such an Engraver, as Mr Boulton is in Search of — he knows the Man at Manheim, to whom Mr B. alludes, but fears that he will not be allowed to leave the Country.

MB-B1/4

---

[1]an 8 h.p. engine was ordered.

[2]Benjamin Gott of Leeds advised about the power requirements of fulling machinery.

[3]Count Rumford.

DOCUMENT 181

To M.R. Boulton
Soho. London St 27 June 1803.

Dear Sir,

I saw the Chevalier Edelcrantz yesterday, who wants the Engine for the purpose of grinding Corn; also for bruising Malt and raising Water for a distillery. I supposed about 8 horses would be sufficient and recollecting a later foreign Estimate, stated verbally the price at £630 delivered at Hull, compleat with Iron Beam &c and payable upon such delivery. Drawings & Instructions to be furnished, but no man, the Chevalier being a philosopher & a Mechanician, undertakes the erection himself provided we give him Instructions upon a working Engine before he leaves the Country. Time 6 to 7 Mos. — I suppose he will consider it at leisure, and compare this price with that of others. He talks of calling at Soho in about a Month.[1]

The following Memorandum was brought here by Mr Lockey from South Carolina on the part of a Mr Lucas formerly of Cumberland, but now settled there; who he says is, or was, a millwright, and acquainted with my father.

"The Power of a Steam Engine required for the manufacturing of Rice — one that would be equal to the driving of 3 pair of Stones for grinding, 5 ft 6 Inches Diameter; or the power of a Water Wheel 18 feet Diamr 6 feet 6 Inches wide of buckett, with 10 feet head and fall included; either of those powers would answer. The Expence of an Engine of the above power, and the quantity of Coal it will burn in the 24 Hours; also the quantity of fresh Water necessary for the same — or if Salt Water would not answer by making some of the most delicate parts of the Engine of Copper or brass."

I tried to dispence with giving this Estimate by stating the difficulty about Engine Erections &c, but in vain, as he seemed to think Mr Lucas competent & at all events that too much time would be lost in transmitting that answer, and he assured me the application was not one of mere curiosity, but was perfectly serious. I however expect nothing from it;[2] but as I have promised him an answer shall thank you to furnish me with Estimate &c.

Dayus has not yet made his appearance; and Jackson & Dampierre[3] object to Pooles leaving them; indeed though his part of the work seems finished, the presence of one of our men may be necessary for some days. The Woodwork is far from being compleat and the Engine has only hitherto worked on trial. I am quite at a loss what to do about Symonds, as I have not been able to find out where Newman is, or how employed. I shall however institute an enquiry at Bateman's and Brandrams in the course of the day upon the supposition that he may be putting up the new Nozzles at one of those places. I am to call at Jackson & Dampierre again in the course of the day and shall use my rhetoric to get Poole released.

I am sorry to learn there was a mistake in the Parallel bars & radius bars of this Engine which had to be shortened 4 Inches. I must also observe that the Piston and Case being left black whilst all the rest of the work is polished, has an indifferent appearance.

I have seen Mr Smith of the Gravesend Tunnel[4] who has paid £200, and promises the remainder in a couple of Months. The concern is compleatly abandoned

I am Dr Sir
Yours truly
Watt Junr.

Box 2.

---

[1]No order materialised.
[2]Watt was correct, no order resulted. See [194]
[3]Oil and Mustard millers, purchasers of a 14 h.p. engine.
[4]A 52 in. cyl. 8 ft. stroke engine was ordered in 1800.

DOCUMENT 182

To Boulton & Watt Copenhagen 4 Novr 1794

Gent^{n}

It being our Intentions to Erect a Corn Mill here to go with Steam for Working 2 pairs Stones[1] — you will therefore be pleased to Inform us if you wou'd Undertake to Erect such a Mill for us & what the Expence wou'd be = you providing Every thing (Except the Buildings) and sending a proper person over here to Erect the Same — wee wish to have french stones & the last New Invented Sifts — if you do not find the last 2 Articles — wee wou'd willingly procure them Ourselves tho' at the same time you will oblige us by Informing us what a pair of french stones Costs & also the Sifts, as soon as wee receive your Ans^{r} wee will address you to some respectable Houses in London for our Sufficiency & remain truly

Gent

Your very Ob. Ser^{ts}

Nelthropp & Harris

Box 4 N

[1]No order materialised.

DOCUMENT 183

To Boulton & Watt Constantinople 10th Octr 1794

Gent^{n}

You will infintely Oblige me to Furnish Me with the Cost and charge of erecting one of your Patent Steam Engines for the purpose of raising Water — as well as Cost of Iron pipes For the conveyance of the water.[1] The Engine must possess the power of raising a large Column, you will please therefore specify the quantity of Water to be discharged per minute, in Fact you will be so good as to enter into every necessary particular.

At the same time you will Furnish Me with the cost and charge of the Mechanical powers to be attached to your Steam

Engine for the purpose of grinding Corn[2]

You will please excuse the trouble I give you
and remain
Y^{r} M^{t} Ob^{t} Hb Ser^{t}
F. Daniell

Direction
M^{r} F Daniell
Pera, Constantinople.

Box 4 D

[1]No order materialised.

[2]Further evidence of the impact of Albion Mill as an advertisement of steam corn milling.

DOCUMENT 184

To Boulton & Watt Vienna August 27th 1796

Gent^{n}

I have the pleasure to inform you on the receipt of your much esteemed favour of 24th March but which only came to hand 28th May by reason that some time after my arrival in Vienna I was oblidged to return to Prague for I intended to make a very usefull speculation in Wheat for England, which for to let you know all the particulars of it might be to tiresome for you to read, but should you like to know something of it Mr Henry Hadley can give you informations, as I mentioned it to him during my last stay in Birmingham, and wrote him since my arrival in Vienna, after which I only got your esteemed letter. According to your advise we found it necessary to find out the place first where the Engine was to be errected before we could give you a true information wheighter the Boiler was to be placed in the Engine house or distinct. It was not in our power to say which plan we would adopt, as we made a double proposition to the Court to sell us one of two Buildings of which both are of no use to them, it is but two days that we got the resolution, for which we have been waiting every day this two Months. We are now resolved to construct it with a distinct Building for the Boiler, as you recommend it yourself, and therefore do not find it necessary to make any alteration in the general disposition, and the Plan may be followed entirely as it is drawn in your Sketch[1]

As for the plummer bloks and brasses belonging to the rotative Shaft, we shall have them made here, as soon as you will be pleased to send us the general drawings of the Engine by which I doubt not we shall be enabled to give the necessary information, and instructions to our foundrys.

We are now in Vienna in a very Critical Situation, as we are almost surrounded by the french, and I have not the Courrage to say that there is no danger to see therfor here. Our Army's have suffered most terribly by a continual march, and warr of 13 years, since the year 83 they are in continual Marching & fighting. This is the epoch, as you know they went to the Low Countrys from there to the frontiers of Turkey, and then to france, every body who knows what warr is may easily judge that an Army who fought 13 years can not be blamed if tired.

I can assure you that I dont know how to get the Engine in so critical a moment, as even the Post course is interrupted, and we have no letters this five Weeks from Hambro. I therefore take the liberty to communicate you my opinion and wishes on this Subject.

When I had the pleasure to be introduced to your acquaintance, no body in the World was able to foresee the present situation, and I desired you would be kind enough to let me have one of your famous Engines; the present State of affairs is so much Changed, that it is almost an impossibility to get it even with a considerable risk of loosing the whole, and therefore would be oblidged to let it lie in Hull for many Months, and perhaps to the end of the Warr. I therefore take the liberty to desire as a particular favour that you may be pleased not to compleat nor to deliver the Engine till Circumstances changes, and am willing to bonify on delivery the loss which perhaps might arise by retard.[2] But in case that it would not agree with your dispositions, or that you should not like to fullfill my desire, I wish to hear your oppinion as soon as possible that in case you dont aprove my disposition, I may give the necessary information to Messrs Battier Zornlin & Co in London, to accept your Draft at two Months after the delivery of the Engine in Hull agreeable to our agreement.

His Majesty has granted us all the privill: we desired and as soon as our public plan will be ready I shall do myself the honour to communicate it to you.

I have only to add the sincere wish that this may find you Gent^n^ and your dear family in good health. M^r^ (Har)dinger and my other friends join in best respects & Compliments with Gent^n^

Your Sincere and

Ob^t^ humble Servt

B Tschoffen

SCHEDULE

Referred to in the annexed ~~Agreement~~, Letter

BEING

A LIST OF THE METAL MATERIALS,

To be furnished by

BOULTON & WATT,

To B: Tschoffen Esqr

---

1. CYLINDER; its Top and Plate; its Bottom, Gland, Brasses, &c. compleat.
2. Bolts, with their Nuts, Washers, Plates and Dogs, for holding down the Cylinder.
3. Piston; its Cover, Bottom-Plate, and Spanners compleat.
4. Nozzles compleat, with a spare Set of Valves, Racks, and Sectors.
5. Working-Gear compleat.
6. Perpendicular Steam-Pipe and Eduction-Pipe.
7. Condenser-Vessel, with Blow-Pipe and Blowing Valve.
8. Injection-Cock, Rod, Handle, and Index.
9. Air-Pump; its Bucket, and top and bottom Valves compleat.
10. Bolts and Dogs for holding down the Air-Pump.
11. Air-Pump-Bucket-Rod, with Cap and Bracket.
12. Piston-Rod, its Cap and Cutters.
13. Parallel Motion, with Saddle-Plates, Plummer Blocks, Brasses, &c. compleat.
14. Bolts for fastening ditto to the Working-Beam and Spring-Beams.
15. Main Gudgeon, its Plummer Blocks, Brasses, Bolts, Beam-Straps, and Gland.
16. Saddle-Plate, Gland, Brasses, and Bolts for outer End of Beam.
17. Gudgeon, Hoops, and Bolts for Top of Connecting Rod.
18. Connecting Rod, Bolts, and turned Pin for it.
19. One Sun-Wheel, and one Planet-Wheel.
20. Fastnings for Connecting Rod and Planet-Wheel.
21. Connecting-Link, with Gibs, Cutters, and Brasses, compleat.
22. Rotative Shaft, Plummer Blocks, Brasses, and Bolts.
23. Fly Wheel, with Pins, Nuts, and Washers.
24. Working Barrel, Bucket, Clack, and Gearing for Cold Water Pump.
25. Hot Water Pump, Bucket, Clack, Suction Pipe, and Gearing compleat.
26. Boiler-Steam-Pipes; Throttle Pipe and Valve; and Governor.

~~27. Iron Boiler~~, with Man-hole Pipe and Cover.

28. Safety Valve and Pipe, and Reverse Valve.

28. Gage Pipes and Cocks, Feed Pipes and Feed Apparatus.

29. Feeding Mouth, Dead Plate and Frame, and Fire Door.
30. One Set of Bearer and Grate Bars, and Damper-Plate and Frame.
31. Barometer-Pipe, Cock and Scale, and Steam-Gage.
32. Valve for Cold Water Cistern.
33. Screws and Nuts for all the Joints.
34. Cement, and a Box of Pomatum.

N B. It is mutually understood that the iron boiler in the articles N° 27 is not to be furnished by Boulton & Watt to Mr Tschoffen and is not included in the estimate

Bernard Tschoffen

Plate [12]

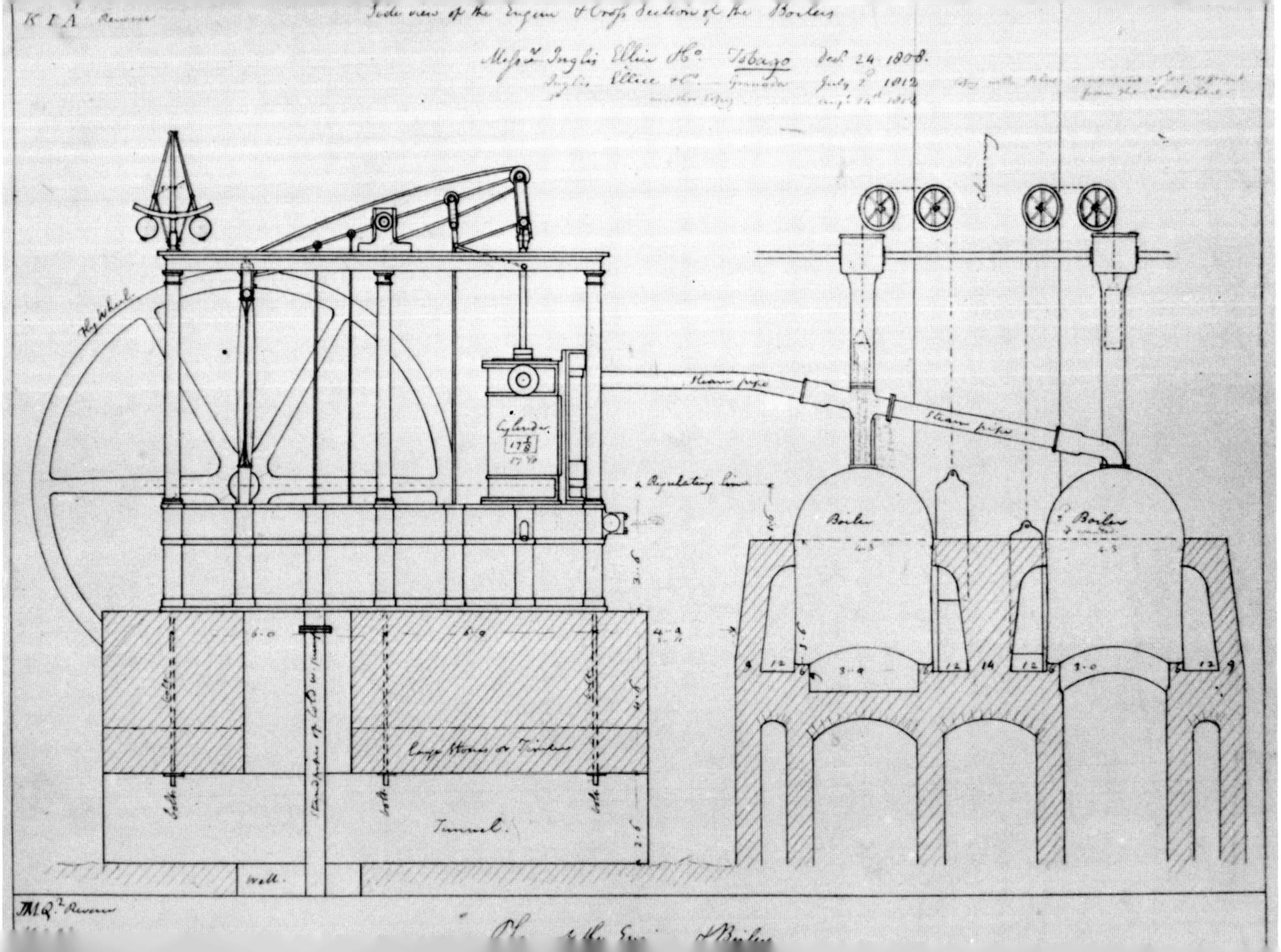
Side view of the Engine & Cross Sections of the Boilers
Messrs. Inglis Ellice & Co. Tobago Dec. 24 1808.
Fly Wheel
Cylinder
Steam pipe
Steam pipe
Regulating line
Boiler
Boiler
Chimney
Large Stones or Timber
Tunnel
Well.
bolt
bolt
bolt

P.S.
I am desired in a letter wrote in London, that you may be pleased to let me know the Amount of the beutifull Coins you have been kind enough to juse but have not been favourd with an answer.

Box 5T

[1]This was a 10 h.p. engine.
[2]The engine was not sent.

DOCUMENT 185

Manchester Nov 19 1815.
Sunday Evening.

To J. Watt Junr.

Dear Watt

Though we have been so long without Communication you have not been the more from my thoughts; I have had so much to retrieve after four Months Absence that all my disposiable time has been engaged in it except a little Wear & Tear in the Shape of violent Rheumatism in the face which has confin'd me to the Cottage for a fortnight — but by the usual Medecines finishing with a Blister upon each Ear I expect to be releas'd from my Confinement tomorrow — meantime I send you these few lines as the first fruits of Recovery. I shall be extremely glad to hear that you have found some Means of lessening your Load of Business though even in Prospect — the Demand here for such Persons as you want is so great that I hardly expect to meet with any which would correspond with your wants & wishes.

A young Man in our Counting House has a younger Brother (a Youth) whom he wishes to place with us, but though the one who is with us is quick & steady and has the rare Gift of Combination and Sagacity yet I decline taking the Youth only because I do not like to have two Brothers in one Counting House — if you chuse to venture upon so young a Subject (about 17) I will propose it to him.

Notwithstanding your Regulation of Exclusion of Visitors to y[r] Manufactory I have ventur'd to promise to two, your Exception & Indulgence; and though Brothers to an Emperor you will find the Arch-Dukes John[1] & Louis of Austria anxious to acquire Information and the former very much enlightened in political Economy, for <u>such</u> a Personage. At present they are visiting

Liverpool & Glasgow, intending to make an Excursion to the Highlands to cool themselves, or their Curiosity, and probably will not return your way before Christmas — the Etiquette is that they will signify their Arrival to you & to expect you then to wait upon them — they had no Spies under the Name of Attendants and I do not think they will themselves carry away any of y^{r} Mysteries.

I have been very much surpriz'd and shock'd at hearing thro' the Newspaper the Disaster at the Sugar House in London and am still at a Loss to imagine how they could be so negligent as to confine the Steam to such a prodigious Elasticity as to produce such a tremendous Explosion. I never understood that it was intended to use more than the ordinary Pressure as so little additional temperature is gained by any greater Degree which is safe. It will operate for a time against the Plan, but will be forgotten in its Utility — It is however desirable to know the Cause of such a terrible Event.

It was with much pleasure I heard from Creighton that Murdock was going on so well and I hope he has now the full use of his Limbs & will keep it.

Kind Regards from my Sister & myself to all friends at Heathfield & Soho

Yrs mo truly
G.A. Lee.

M IV L

[1]A 6 h.p. engine was ordered in 1817.

DOCUMENT 186

To Boulton Watt & Co St Petersburg 4th April 1805

Gent^{n},

I have to acknowledge your favour of 31st Jany last rec^{d} about a fortnight ago, (& the only one rec^{d} from Birm^{m} since October) which seems to have been retarded by the uncertain Winter-passage of the Gulph of Bothnia; but I deferred replying for a few Posts wishing previously to attain some information essentially connected with the appearances which your projected Engine Establishments in Russia assume at present.[1] It gives me much pleasure to find that as far as that Subject has hitherto been touched upon your Ideas seem to coincide generally with mine, since it leaves us no steps to retrace, & as my opinion contained in the slight development originally given you on the principles & mode of what to aim at or what to avoid in Engine business here, are confirmed in my own mind by subsequent reflexion & enquiry.

So very long a period having elapsed previous to learning your Sentiments, it must naturally be expected that precisely the same Offers then made me no longer exist, & that the Parties have either dropt their projects, or framed other arrangements. Count Soblekost for example, (who has a sort of Inspectorship of the Imperial Manufacteries) has procured the Engine which he wanted for the Cotton-works of the Dowager Empress at Alexanderofsky from England thro' the medium of a Baron Bodé, a person employed until lately by Sir Charles Gascoigne[2] — from what I can learn it was only a 6 Horse-Engine of yours which had been sold to a Mr Cooper of Leith, & by him to Bodé who had an intention of employing it to saw Planks at Narva, but finding it too small resold it to Gascoigne who set his "Scavens" to dissect it & note all the wonderful; then disposed of it to the above Works, of which he is appointed Manager in chief — I suppose it was bought a cheap pennyworth by Bodé as the success of his purchase has induced him to take an expensive Journey again to England thro' Sweden to try if he cannot pick up a few more at the same rate for joint-acc^t of himself & Gascoigne, as well as any other thing or person in the mechanical way that comes within their Grasp for Gov^t employment. They are not over nice, almost all are Fish that come to the Net. — A small Model of this Engine is making by Gascoigne to be presented to the Dowager Empress for her amusement.

Now that you have enabled me to give a decisive answer to any Russian demands for Engines which may be made, I am of course not only prepared to accept the Commissions on proper terms, but shall put myself on the inquisitive, & keep a look out as far as my limited situation admits wherever there appears a probability of doing anything eligible.

The information afforded me in your last Letters, together with the original Documents received from Mr M.R. Boulton, will, I presume, furnish sufficient Data from whence to give satisfactory answers to all ordinary Queries; tho', were a Continental Trade so far advanced as to be productive of Orders for a great variety of Works, some further details ought perhaps become requisite to reply with facility & certitude to the multiplicity of mechanical Questions which might be suggested by the different nature of situations & Establishments. — Ex^{le} such as

1st — How to measure & estimate the forces of Overshot or Under-shot-Wheels, or, of Water-Mills in general — since all the Labour performed by Machines in this Country, whether in Saw-

Mills, Corn-grinding, the working of Blast-furnaces & Forges, the rolling & slitting of Iron, &c is chiefly done by Water, & frequently on Rivers & Falls very inconveniently & expensively situated for conducting the Business.

2nd — Having the means of ascertaining those forces — How to turn such calculation into Horses Power — or, rather say, — What is the Ratio or principle used in estimating a Horses power? I believe it is made to correspond with a given weight raised up to a given space in a given time. — But, — What are these Data?

3rd — The number of Horses power being thus ascertained — How to estimate the Diameter of a Steam-Engine-Cylr capable of doing the same work? as I perceive people not only like to know the price of an Engine of a certain number of Horses power, but, also, it's size. —

From these Heads, joined to Mr M R Boulton's Doct it appears to me, that, without possessing mechanical knowledge, I might be enabled in any part of the World to state to an enquirer the exact price & size of a Steam Engine adequate to the performance of such a quantity of work as may be demanded; — or, capable of replacing any Water-Mill actually employed for the work; & consequently prevent the necessity, trouble, & loss of time occasioned by either their or my writing to Soho for information previous to giving an Order.

What I could most wish to have the satisfaction of accomplishing 'ere I quit this Country is, — the fixing of a permanent & invariable general Basis on which for you to act in future, independent of any accidental Commissns that may fall in my way here; & which, whilst it supersedes the necessity of dispatching an Agent for the purpose of forming Connexions, will also leave you a safe & unerring Road open by which at all times to proceed in the acquisition of such Comns as are worth acceptance, without incurring superfluous expences in their attainment.— Probably the interchange of 2 or 3 Letters betwixt us might enable me to effect the desired Object; & for this end I shall take the liberty of submitting to your better Judgement some additional Observations to those contained in my first Letter, explanatory of the chief points of view in which I have been led to regard the Subject.

I infer from your silence that you are satisfied with the validity of the reasons already assigned against forming an Establishment of your own in this Country. Those reasons not only retain their full force in my mind, but, it may be proper to add that — tho' your Agent must keep his general Residence here as the Centre of Russian Commerce, yet, the places where he is most likely to receive his Orders from will generally be found to lie either in the Corn-Districts South-East of Mosco' & 1000 Miles from hence, or, in the Mining, smelting, & forging Districts of Siberia above 2000 Miles off, which will always leave the Workmen too far removed from his immediate Controul & consequently more independent of him than could be wished in a Country where their value is comparatively so much more important than at Home, that it becomes scarcely possible to prevent their being immediately spoiled. — These circumstances might also oblige him to take frequent long & expensive Journies which would augment the cost of this uncertain Establishment to an amount that ought not in prudence to be set down at one farthing under £1000 Sterl^{g} p^{r} annum; & it would require a person not easy to be spared from Soho, say of mechanical Talents equal to those of Mr Southern or Mr Lawson joined to a knowledge & practice in the Routine of Business seldom possessed by persons confined to that profession.

But it ought not to be inferred from hence that waiting tranquilly at Soho the arrival of chance Orders from Russia, would prove the most proper means of introducing the Engines on this part of the Continent. — Besides the usual disadvantages of distance, some others exist of a local nature, the chief of which arise from the circumstance that, either the Crown or principal Nobilities are directly or indirectly Proprietors of nearly all Establishments of magnitude, You will easily admit these two Classes are too far removed in most Countries from the Mass of Society to be the best adapted for a proper choice & accomplishment of useful Works: in this Empire I make no scruple of saying they are the very worst; & yet its political & commercial situation is such that all valuable improvements will long stand in need of one or the other of those supports to carry them to effect. — Generally speaking, the undertakings of the Crown will be managed by those persons, who, thro Intrigue or personal Interest can obtain a sanction for their Plans, & an Order on the Treasury for the means of execution, & consequently will too often become totally indifferent to the success of the Object itself provided, they can divide the funds destin'd for its support amongst their friends & followers; & this will more easily be arranged & held read amongst the Parties by connexion with adventurers & Charletans, than with Mechanical Men of probity & talents. — They of course will keep as clear of you & your Machines as possible.

Let me however remark respecting any thing that might be attempted on your part for the Crown — that, the principal person who can be disposed to throw Obstacles in the way is Mr Boulton's old friend "the Baron"[3]; but, advancing Age, & misfortune of becoming at length rather too well known, have diminished the capacity, (tho' not the intention) for mischief too far to leave little apprehension of much future danger from that Quarter.

On the other hand, — I feel every reason to presume that Mr Novapittzoff, who is the close confidant, one might almost say Factotum of H.I.M., remains, as he has invariably professed himself, frankly & sincerely the friend of Mr Boulton, & disposed to extend that friendship to his connexions, especially where advantages to the progress of useful Arts throughout the Russian Empire could be combin'd with it. Consequently, as Mr Boulton's Mint-undertaking has proved long-winded, expensive, & unprofitable, may at least be well that the Engine C^o^ (for no other branch can) should try to reap some advantage from the favourable disposition of the Russian Gov^t^ towards Mr Boulton for his able & honourable conduct in the execution of this Mint, which is what they are so little accustomed to meet with at the hands of those who make public contracts with them, that they look to finding themselves cheated as an affair of course.

The Nobility will seldom be found either so safe or easy to deal with as the Crown — The antiquated part of them who remain chiefly immured in Mosco' or their native Provinces entertain no very high opinion or knowledge of the progress of national improvement since the good old days of Peter the Great. — The modern portion (who in general have made "Grand Tours" of various denominations thro' Europe) whilst abroad, must be allowed indefatigable in the chase after Novelty useful or ornamental, — listen apparently with the most eager attention to every hint that may tend to benefit their Country, — &, one would imagine, were hurrying home to put them into immediate effect; but, no sooner arrived, than the remnant of what they recollect is doom'd to evaporate by way of News in giving a transient lustre to the "convezationi" of Courtly Circle; & tho' they may often sit down to project Castles in the Air, yet they seldom reach further in more solid Plans than some needy patronised French pretender to Science, or Englishman who has taken french leave of his Country, may be able to lead them: — are never clear in what they want, or aim at, — & would feel very sensibly gratified by succeeding in engaging you in a tiresome & endless correspondence without specific Object. — In the few instances where any thing effectual can be done with them, one should be personally present to strike whilst the Iron is hot, & even then it would prove a complex & difficult task to bring their Ideas & yours to accord about the main point, Money, — as ordering & paying are by no means with them either synonymous terms or natural cause & consequence. — There are doubtless in this vast Empire several honourable & distinguished exceptions to this general Rule, but, their whole number are comparatively few, & of course the chances are against those individuals being found amongst your Customers. — Whenever such a one does occur, I shall be very ready to accept his Comm^n for you.

For Manufacturers, or Merchants resembling those who in England combine Capital with enterprise, talents, & Industry, it would be vain to look amongst the Russians, — few indeed would be discovered; for the Departments of Gov^t & a small portion of the Nobility comprise nearly all that are employd in promoting useful Arts: and, amongst these, we must not expect to find the short & simple principles (of) English Trade prevalent, or, that they will order an Engine & send the money for it without a great deal (of) circumlocution & previous management in talk or writing, the first of which is impracticable, & the later too troublesome & contrary to the value you place on the employment of time to be long complied with. — Even were a person from your own House sent out & established here, he could not be directed to meet with early success, as the character, manners, & peculiar tact which the Russians have in their mode of conducting affairs of personal Interest would require a close & long study before he could hope to acquire their confidence, & from thence be enabled to make his arrangements with certitude (to) the best advantage.

Now, in the more effectually to overcome these material difficulties, without which nothing like a Trade wth cultivating can be well expected, (& as we are agreed that a solitary Order now & then is not likely to repay the extra trouble it will occasion) I am decidedly convinced on the matured reflexion, that the shortest & useful method would be for you to fix upon some mechanical Englishman resident here, of good general experience in Business, & a particular acquaintance with the language & people, & make an agreement to constitute him your Agent, on terms of reciprocal advantage. — It has never yet been your custom within my knowledge to form connexions of this nature either at home or abroad, & therefore suppose that you intentionally avoid acting on such a principle from motives which it is not my business to enquire into. — I only consider the present Object in its relationship with the nature & situation of this Country in particular, & am led in consequence to propose this System as the least expensive, most feasible, & competent to the end in view. — Tho' the point is a little delicate, — yet, such being my fix'd opinion, & feeling anxious, if possible to serve you as the Instrument by which to settle the Basis of a good Engine-Trade in future to this Country, I conceived these circumstances would warrant my looking round for a fit person to fill such an Office. —

It would be a useless waste of Paper to explain why the English mercantile people are altogether unqualified for the Task. It is amongst those who were originally mechanical Adventurers & Speculators that we must enquire, many of whom having come out with the true Corinthian polish on their foreheads, have in consequence advanced rapidly on the road to favour & fortune. — The greatest part of this worthy Community need only to be known to be avoided by you — And in all the List of those worth notice, I can hear of no one so well adapted either by Character, knowledge, or Property, to promote your Interests & Views in Russia as Mr Charles Baird. —

I don't know whether you are acquainted with any branch of his family, but, he has an older Brother in Edinburgh, & a younger one here whom probably Mr James Watt may recollect having seen at Soho about 3 years ago. — He came to this Country originally for account of Sir Charles Gascoigne (as most of the Projectors have done) & in company with Mr Roebuck of Birmingham, both of them having been employed in the Carron-Iron Works; & about 14 years ago had acquired sufficient property & knowledge of the Country under Mr Gascoigne to venture setting up for himself as founder, forger, Steam-Engine Maker &c, but, has chiefly been connected with the Board of Admiralty. He also undertakes all the Iron work for the Mint-Department, & has done a good deal for us there, (at least as much as they have paid above £1000 Sterl^g^ for, tho' I understood on arrival scarcely any thing in that way remained to be made) — Mr Baird is supposed a Man of very solid property at present, tho' from his expensive mode of living necessarily adopted by all the English in business here, it is not easy to attain a precise knowledge of those facts without experiment, for mere appearances would indicate many to be rich who are overwhelmed with Debts. He is about 40 years of age — certainly very active & intelligent, — knows the Russian language well, which rest assured is a very essential point & a work of time: — is pretty intimately acquainted with the Mechanical Professors in general, both Russian & Foreigners — also with many of the principal Nobility; & with the proper mode of applying the Key to the private Doors of the Chief-Officers in most of the Gov^t^ Departments; & as you will have observed, comes from the North side of the Tweed which is the best recommendation a man can bring to this City, the Caledonian Phalanx being the strongest & most numerous, & moving always in the closest Union. — If not received as a friend, he is in my estimation, the person who by his experience & address might in the event of your engaging extensively in this Country, become the most troublesome Opponent. —

I have been careful to avoid committing either You or myself in regard to Mr Baird, tho' in order to put the matter into your train, I have held a conversation with him, the main purpose of which was to learn whether or not in case of turning your attention towards the North of Europe, & appointing a confidential Agent, he would feel disposed to accept such a Charge on reasonable

Conditions? Or, — rather, — if he would undertake to procure Orders for your Engines on such Terms as he might be enabled to bargain for with the persons wanting them, but to order them from you in his own name & for his own Account & Risk, paying at the rate & in the mode stipulated by your Letter to me, & not to be deprived of any Commissions thro' applications by yourselves direct.

He seemed to enter with more facility into this latter Plan, which also appears to me upon the whole, considering Distance &c, best suited to your Interests; because, if, as I apprehend, & with which Mr Baird coincide, the commercial & manufacturing state of this Country is not sufficiently ripe as yet to admit a uniform & regular business of that kind, you will by the mode above described, be at least screened from trouble, expence or loss, or seeking or taking Orders; &, on the other hand, — if he, who being a fixed resident & incurring in obtaining them, no matter to you from whom, or on what terms, your Object is fulfilled by the certainty of payment before the Engines leave England, & time & trouble saved by the Correspondence being limited to one intelligent person; whilst his profit will be calculated to arise chiefly from furnishing those parts, Machinery & Iron work which are the general concomitants wherever Steam Engines are employed.

But he would deem it necessary in the commencement to be provided with a good Workman from Soho, who might in time form as many others amongst his own people as requisite. You may easily suit him with one without much loss; for (entre-nous) Hailey will never be so useful for any other place: he is become too much a Gentleman, but here will do well enough as they are all as proud & lazy as himself. The Mint & his Engines must certainly be finished, & he at liberty in the course of next Winter.

I have informed Mr Baird that I would state to you the substance of our Conversation, & immediately on receiving your answer, should hope to find myself authorized to offer either the Propositions already mentioned, or such modifications of them as might be more consonant to your Judgement for his final acceptance or rejection; & if we came to an Agreement, a direct Correspondence might then be opened betwixt you which would embrace all the minutae of the Subject in a fuller & clearer manner than lies within my power. — I have therefore to solicit you will favour me with a reply to the above, which will enable me to understand myself distinctly & definitely with Mr Baird, & also in case of your rejecting all connexion with him that I may lose no time in receiving & acting upon such other Instructions as you may deem fit to transmit. Tho' I doubt not but, should the Noise of War betwixt France & England be once drowned in the Acclamations for a solid Peace, & that energy now employed in inventing & preparing new modes of devastation turned to the cultivation of Commerce &

the useful Arts; that there will soon be displayed in the middle of Europe a much ampler & richer field for the employment of your Steam-Engines than what this corner of it can vie with for half a Century to come.

I have previously alluded to Novelty as the grand source of actions amongst the Russian "Scavans" who in that, as in too many other qualifications copy closely after their great Models the French, & as a proof of its force I cannot refrain from mentioning a new Phenomenon in the mechanical World. — A Mr Major, originally from Birm^{m}, & whose History I believe will easily recur to Mr Boulton's recollection, after long roving about like a wandering Arab thro' Austria, Poland, &c has pitched his Tent for these 2 or 3 Years past in Siberia, in which philosophic retreat he has had leisure to contemplate uninterruptedly the profoundest depths of Mechanics, & has at last brought forth a new Steam-Engine which for simplicity & cheapness in proportion to its power is to set all that has hitherto been done or thought of at nought. A Model of the Machine has been introduced at Court along with it's Inventor & made to move in presence of their I.M. & the most renowned Nobles & Sages of the Land, to the astonishment & admiration of the beholders; but, the best motion made on the occasion was that this modern Archimedes should be rewarded with a place which, with it's appurtenances will nett about £2000 Sterl^{g} p^{r} annum in Siberia, together with an exclusive priviledge of erecting the Machine.

An Engine intended for real work has been put up by him in Siberia, but in that frigid Climate does not seem to have succeeded so well as the Model did in the warm sun-shine of the Court. It would keep moving for about Half an Hour at a time, then invariably turned restive. — It differs essentially from the usual Objects Inventors have in view of discovering some new principle or improving a known one, as it's tendency would be to carry the Arts backwards, working by the expansive force of Steam alone without employing the improvement of a Condenser to produce a Vacuum: now so far I understand was in contemplation even in the time of the celebrated Marquis of Worcester, & subsequently executed by Captain Savory. — His Cyl^{r} is placed in the Boiler that it may always remain in an equal & comfortable degree of heat, — an humane Idea, suggested we may presume by the experience of a Siberian Winter. — In short, — the whole, I am pretty well informed, is a piece of Mechanical Charlatanism well calculated for the purpose it has answered of surprizing & amusing Lords & fine Ladies in a Saloon, & receiving to the fortunate Projector a certain & liberal annual emolument in the shape of Pension; & may serve as a measure by which to regulate your Judgement when the progressive state of Arts & Sciences in this City is the theme in question.

Before I close this tiresome Epistle, it appears still proper to mention that soon after the receipt of your Letter I waited on Mr Novapittzoff to solicit his patronage if perchance the future views of the Russian Gov^t should be directed at any time to Objects requiring the employment of Steam-Engines for their execution, & easily obtained from him the promise of a cordial exertion of his best influence in your favour in all cases where they could be usefully adopted. Amonst other things he stated to me a project he had much at Heart of establishing by Gov^t something like an Albion-Mill for the occasional supply of this City as well as for part of the Army & Navy. — I did not scruple to promise that not only the Steam-Engine for the concern would be readily furnished by you, but also that tho' forming the general Plans for such a Building & Machinery did not fall within the precise limits of your occupations; yet, if they would make an Establishment large enough to employ 3 or 4 Engines of 36 Inch Cyl^r each, I did not doubt of your willingness to procure him a proper person to undertake the Mill-work, (having Mr Rennie in mind) if you could not with propriety take more upon yourselves, than the setting up of the Engines. — He observed with expressions of regret that his attention was forcibly turned from it's primary Objects (viz, the encouragement of Arts & Sciences & improvement of public-works, throughout the interior of the Empire) to others purely of a Diplomatic nature, & of such importance as to preclude for the moment all those considerations most congenial to his former studies & habits; but, that he certainly would not lose sight of the Subject in particular, & probably might be able before we have done with the Mint-Machine to arrange matters so as to come to some specific proposition for the undertaking.

Now, — as you can easily refer to the original Drawings of the Mill, — to the costs of the Engines as well as of the Building & Mill-work, — & both to the calculations of what it was ultimately intended & expected to do, as well as of what was actually done in a given time by the Engines erected; — probably, if some of your young folks were desired to make out a Mem^m of those points, I might be able to use it advantageously with Mr Novapittzoff; for it is absolutely requisite that he should be furnished with more Data than I could otherwise give him to admit of his forming any probable estimate of the Expences in Building & Machinery necessary for a Mill grinding a certain quantity of Grain. — As this would cost you little more trouble tha setting some person to look over a few old Drawings & Calculations, — I take the liberty at the same time to suggest the propriety of stating to me what any of your

Engines have done applied to Forges & Founderies: Ex^{le} — that at the King & Queen foundry at Redriff — Also what has been the quantity of Water drawn up by some of the principal Cornish-Engines, — say the large one on North Downs; for it strikes me that some general out-line of the knowledge of what has been actually performed by the Engines in their various applications would both give a Minister like Mr Novapitzoff a more accurate, & impress him with a much higher notion of what might be effected by their general introduction in this Country than he would ever conceive by any other means. — The above being information of that general species, which, as far as I can judge, there could be no imprudence in divulging, more especially to him.

In the interim — should either a good Comm^{n} fall in my way, or any new & important information on this Subject, I shall take care to transmit you immediate intelligence. — Excuse the tediousness of this long desultory Letter, & impute it solely to a wish to give you the best Ideas I can of the mechanical state of this Country in such matters as bear any reference to the further introduction of Steam-Engines.[4]

I beg to be remembered with Sentiments of the most cordial respect & esteem to every individual branch of your Firm.

Gent^{n}, Your most Obed^{t} Hble. Servant,

Z Walker

PS Yours is the first Letter received from Birm^{m} since I was favoured with Mr Boulton's of 8 Octr last.

box 20

[1]The writer was superintending the erection of the two mint engines (50 and 20 h.p.) at St. Petersburg.

[2]Formerly of Carron Ironworks.

[3]Charles Gascoigne who had emigrated to Russia from Carron Ironworks. For 20 years he was principal adviser on technology to the Russian court.

[4]No Russian agency was established.

DOCUMENT 187

To J. Watt London 1st May 1776

Sir

I understand that some attempt has been made to apply the Fire Engine to the purpose of turning Mills for squeesing the Sugar Cane in the West Indies, & that it is supposed that the same fire which is used to boil the sugar, will be sufficient to produced Steam to work the Mill — If this should be practicable, it would be a matter of infinite consequence to the Islands & to the Public & would occasion a very great demand for your Invention — one Steward has already got a patent for such a Machine & one Clark has got another patent but neither of them have yet been able to make it succeed — Stewart has published a pamphlet describing his machine which I will send you if you think this affair is worth your attention, & I can hardly conceive any thing which is more likely to bring you a great return[1]

I am sir

Your obed^t Servt
William Pulteney

Box 2 P.

[1]Although considerable interest was expressed in steam power by West Indian Planters in the late 18th century, no order was placed with B & W until 1801. Pulteney was correct in his prediction of the profits to be made by selling engines to the West Indies. Both B & W and Fawcett & Littledale of Liverpool gained in this market.

DOCUMENT 188

To Boulton & Watt London May 15th 1787

Gentlemen

Mr Wyatt having informed me that you intend to erect a Steam Engine on the Island of Jamaica, the latter end of this year, I take this liberty of informing you that I purpose leaving England for Jamaica about September next as a Mechanic: (if therefore it should seem proper for you to employ me (& should you not be already engaged with a person to fix the Engine at Jamaica) I shall beg leave to offer you my Service for that purpose, and flatter myself that, having already acquired some practical as well a theoretical knowledge of your Engines, my service if accepted will ultimately lead to your Satisfaction & my advantage there.

I have been bred a Mill wright, was employed at the Albion Mills from nearly the beginning until July last since which time I have applyed Myself to Mathematics &c.

I hope you will pardon me for thus addressing you, it being previously assented to by Mr Wyatt[1], who as I have already stated, was so good as to give me the information, and at the same time advised me to make immediate Application.

A Line directed for me at N° 150 Fleet Street — London, informing me whether my Services are accepted, and what Terms you propose, or appointing an interview, in London or near it, will be most gratefully acknowledged by

Gentlemen

Your most Obedient humble Servant

Geo Glenny Junr

Box 4 G

[1]Samuel Wyatt architect of Albion Mill and a partner in the concern.

DOCUMENT 189

To Boulton & Watt — Gustavia in St Bartholomew 1st May 1788

Sirs

Your M^r^ B will recollect that in the month of February 1778 he paid me a Visit, at my House in St Martins Lane — Cannon Street — London, when he gave me some particulars of the performance of your Steam Engines — I have since that time quitted England and made my Residence in the West Indies — I have visited Several of these Islands and taken notice of there being many ponds & Stagnating Waters, which the Inhabitants would spare neither pains nor Cost in removing — The which brought to my mind the Conversation I had with your s^d^ M^r^ B — But as the particulars he gave me & which I have by me in writing, are not sufficient to Convey a proper Idea of your s^d^ Engines — I must have recourse to you & beg that you'll be pleased to give me as soon as possible every particular or details relating to the same That I may be enabled to Converse pertinently on the Subject & Convince People of the Superior advantages attending them.

Something truly Capital, may be done, for they are not wanting in one particular place but, in many, and, as it is owing to the Conversation I had with your M^r^ B that I know more about them, than I did before; — I think, I must in return aply to you & to none Else.

I presume your Steam Engine is upon the principle of our Chelsea Water Works, about London; and that also used occasionally, at the London Bridge Water Works when the River is too low, or, perhaps an Improvem^t^ upon both — The Quantity of Water a Determinate quantity of Coals will raise, I am acquainted with But, not the Space of time in which it is done — a Circumstance which must be known — a Plan and a very accurate one of the Machinery should be Sent, and what would answer far better, a Modell, and an Estimate, not only of the Cost of the Engine itself, but also of the fixing of it & the probable Expences attending the keeping of it in repair and in use — which Implies the Quantity of Coals requisite to work it & the price & duties at Evaporation.

Since the above, I have at the request of the Governor of a Capital Island waited on him & Council when after a pretty long Conversation I had with them on the Subject — They were pleased to assure me that I could depend upon the Greatest encouragement — But, that they must be better informed — which I make no doubt you'll enable me to do — and as soon as possible.

I am

Gentlemen

Your most hble Servt

Anthony Villion

PS you'll be pleased to Direct your Letters for me Post Office St Christopher and all pr packett as the safest & Speediest way —

Box 4 V

DOCUMENT 190

To Boulton & Watt London May 13th 1789

Gent^{n}

The Subject of the Slave Trade was extraordinary well discuss'd in the House last night by Mr Wilberforce who mov'd it — Mr Burke Mr Pitt & Mr Fox and unanimously for a total abolition, no division but to proceed to devise means and begin on Monday next —

The Same Idea that I mention'd to you when my Engine was first put up about their use in the West Indies Mr Wilberforce has inquired of me about, and I think you told me some West India Planter had consulted you upon them and that you have some reasons why you thought they would not answer there, but must beg the favor of you to give me a sufficient explanation herein which I will lay before him —

They think (as we know) they would be very usefull if the Climate will do.

I am Gent^{n}

Your most Hble Ser^{t}
S^{l} Whitbread

Pray do you know anything
about the Copper Coin

Beg your immediate Answer

Box 4 W

DOCUMENT 191

To Boulton & Watt London, 3 October, 1794.
Birmingham

Gentlemen,

We have had an application from the Conde de Casa Montalvo & Don Francesco de Arango, of the Havana, for a letter of recommendation for you, which we intend to give them, as they really are two gentlemen of rank & influence.

On this occasion they communicated to us their intention to have two Steam Engines of you[1], & requested to know our opinion respecting the mode of payment which they might propose to you. We told them that this proposal was rather to come from you, & that your general plan was to be paid on delivery, when the Engines were ordered of you, by or thro' an established house in England, & that no London house would undertake to give orders for account of Gentlemen in America, unless they had the counter guarantee of some very solid house at Madrid or Cadiz. They appeared desirous to have time on account of the present ruinous exchange. We repeated to them that they could not well ask it unless the order was given thro' a known London house. The Conde thought that nothing was easier than to obtain a counter guarantee at Madrid or Cadiz; but Mr Arango was of opinion that the shortest way would be to deposit in your hands £400 until this point should be settled between you.

We have thought it proper to acquaint you with all these particulars for your government; & we shall, on all occasions, be happy to shew you how sincerely we are,

Gentlemen,

Your most obedient
& humble servants,
Fermin de Tastet & Co

You need not take any notice to these Gentlemen of our having had any correspondence with you about their affair.

Box 36 Spanish Engines.

[1]No order materialised.

DOCUMENT 192

To Adam Lymburner. Birmingham 12 March 1789.

Sir,

Having never erected an engine for working a mill precisely of the power your friend at Quebec requires, and very seldom in this country having occasion to provide the whole materials of an Engine we are not able without a long & laborious calculation to say exactly at what sum we could deliver such an Engine. But as near as we can at present inform you The Hammered Iron, cast Iron, Brass

& copper materials, including an Iron boiler, the fly wheels & the 2 horizontal shafts or axles in Cast Iron, for an Engine capable of working effectually 6 pair of millstones of 4 feet 2 inches dia^r would cost put on board at Liverpool about £2200. The above sum includes our premium for the use of our invention & for our trouble, but includes no woodwork nor any part of the mill except the two primary horizontal shafts. In such great undertakings in this country we give the preference to the latter, because though it is more expensive than wood it is much more durable & gives less interruption by repairs which is an object of much consequence; at Quebec it may be otherwise as it might happen that in case of breaking any of the cast Iron wheels they might not be able to get them repaired. The carpenter work & millwork lye out of our common line of business and are what we never interfere in, except to give directions as to dimensions etc. We have reason to say that the carpenter work of such an engine with the principal parts executed in good English oak would cost in London above £200 but in Quebec perhaps might not cost half the money. The millwork, exclusive of the building, but including the woodwork of the machinery, with Cast Iron wheels & shafts & 8 pair of millstones, with dressing mills for wheat & flour, proportioned, would cost in London £2500 to £3000. The value with wooden wheels we are intirely ignorant of nor would we wish to have anything further to do with the millwork than to give our advice as to the best disposition or construction of it.

An Engine working 6 pairs of stones would grind & dress from 36 to 40 Winchester bushels of wheat pr hour according to the order the machine is kept in, the quality of the wheat & fineness of the flour.

The consumption in Newcastle coals would be about 4 bushels an hour, and of our coals in this country about 480 pounds weight.

It will be necessary to send a good workman from hence to erect the Engine, but it will be extremely difficult to procure a proper man & his wages would not be less than 30/- pr week, his freight & other charges paid for him.

In regard to payment we should ask to be paid in London by bills accepted by some good house there, payable in 6 months after the delivery of the bill of Lading.[1] Admittance to the Albion Mill cannot be obtained without an order from the Committee which we expect will be granted on your application to S. Wyatt Esq. the Manager.

We shall be much obliged to you to inform us of your friends answer when you receive it, and remain

Sir

Your most Obed^{t}
humble Serv^{ts}
Boulton & Watt.
(J. Watt's hand)

Office Letter Book 14 p. 96.

[1]In spite of considerable interest in B & W engines in Canada, no order was placed until 1811.

DOCUMENT 193

To Richard Conquest,
Birmingham

Orleans 5th November, 1788

My dear Sir

Long have I promised myself the pleasure of writing to you, nay more I had flatterd myself with the hopes of paying my respects to you & your Lady, in person before this — I have however been once only in England since I was <u>fully</u> established in this country & then set out, with a full determination, to have the happiness to see you at Birmingham but was obliged to return sooner than I at first purposed. I was even so straightend in time that I could not wait the return of my Brother Captain Murray, who happen'd then to be at Scarborough, so that I came back without seeing him — You, my dear Sir, I know, used to take trips upon the Continent, let me engage you & Mrs Conquest in your next jaunt to make M^{rs} F & myself so happy as to pay us a little visit — We have been, now some time, comfortably situated, in a very good town, with a charming Country in the neighbourhood — & with such a prospect, in respect to business, as will I am very certain give my most esteemed friend Mr Conquest great pleasure — I have the honor to have Partners with me, in a very large undertaking for the spinning of Cotton, the Duke of Orleans himself, three others also of the nobility, the Marquis De Gorig D'Arcy, the Marquis Duirest, the Baron De Boisbaudron & two private Gentlemen — I have already between two & three thousand spindles, with the accessary machines for spinning Cotton, mounted, & shall have upwards of three thousand in three Months from this time. It will give me prodigious pleasure to shew them to you — We will work in a very superior manner & yet have not a single machine in our Manufactory like any that exist in England — We mean to put the whole in work by the means of a Fire Engine in the course of this next year & upon this head I know you will have the goodness to excuse my venturing to give you a little trouble — One of the Gentlemen mentioned above who has a small share in our concern is the Person who has made several fire Engines in this Country — but it is uncertain, from some circumstances, whether he will be employed to make our's — I then beg the favour of you (if you can do it without giving yourself much trouble) to enquire of Messieurs Bolton & Watts — the expence of their Engines with a rotation (I think they call it) or fly wheel, of different diameters of the Cylinder[1] — particularly of one of 30 Inches, of 36 or 40 — with the degree of power they attribute to each — the consumption of Coal (by weight of the Coal, or I shall be at a loss respecting the Measure) — I could wish to know nearly the expence, as well of their charge for each size, compleat, as of the carriage to some English Port where it might be shipped for Nantz — we shall of course have two pairs as it is our plan to work in general, night & day — I saw some fire Engines the last time I was in England with rotation Wheels, particularly one at Sheffield for the turning of Stones for grinding knives &c — but I presume those of Bolton & Watts are infinitely superior, in every respect, as well as in the economy of the matter of Coal — Will you be so kind to ask in your information, in what time the work for an Engine of 30, 36 or 40 Inches diameter may be got ready in England —

I could wish that, at the present moment, it might not be mentioned to Bolton & Watts for what situation this Engine is intended, only that it is for a Cotton work — as I know they are acquainted with the Person who has the small concern above mentioned with us — & who at this moment is involved in a law suit respecting the "Eaux de Paris"[2] — with one of the other Principal Partners of our undertaking — & who seems determined that he shall not make the Engine for us — I ought to make many apologies to you, my dear Sir, for thus troubling you with these details — but I know the goodness of your heart & make free with it —

My Wife is just delighted with the acquisition of an excellent Instrument, a grand Forte Piano — She earnestly joins me in the request that you will make us so happy to join in a little concert with us very soon & with the company of your good & amiable Lady — to both of you, we join in every heart felt good wish & the utmost esteem & regard

I am, my dear Sir,

Your most faithful

affectionate & obliged h'ble Servt

T Foxlow

Please to direct to me
A'Monsr Foxlow
a Orleans
France

post paid to London or I shall not get it but this you know as well as me.

Box 4 F

[1]No order was placed.
[2]Presumably J.C. Perier of Chaillot.

DOCUMENT 194

To James Watt Charleston 3 July 1792

Sir

When in Briton I showed you a piece of Clay Which you Said was good pipe Clay and desired that Mr Bowman would let you know what price it Could be delivered at Charleston or in London By the Ton. I am now out of Mr Bowmans Employ and Can asure you thare is any quantiy of that Clay. Thare is nobody heare at presant that knows what it is. The Black lead I told you of is Too distant From this place as it Would not Bear Cartage, the other is on the Banks of the River and Can very Easey (be) got as it formes the Bank of the River. I(f) you will Be so good as To infer me the price you Could afford To give For that Clay pr Ton delivered on Board in Charleston If I find it worth While I will then Set about it and will Secure a large quantity of it as I Can very Easey do it at an easey Rate. If thare is any other Articles hear that you may Want, perhaps live oak For cogs may Be Wanting, if so please Send your Dimentions and I on the Receipt of your letter I will Instantly Ship To you any pieces you may Want. The price of Ship timber in Charleston is from 10^{d} To 12 & 14^{d} pr foot Cube measurement Which I do not doubt Will Surprise y this knowing that this is a timber country but labour hear is verry high and the quantity Sold but very little, only What Ship carpenters use and thare very little done at presant in Ship Building. I am now Imploid in making Rice mills By Water To pound and Shill the husks of it. It is don By millstones To tak of the Roug husk and then put into mortars and pounded White. The Morters is Wood and hould 3½ Bushels at a time there are 16 pestoles and mortars in Some mills of 200 lb Each

Striking 40 to 47 Strokes pr. minute lifting 3 = 3 Inches high. I made 5 of them Last year and Shal make as Maney this year. The Water Wheel is 7 feet Broad and 18 feet high. The head of Water is from 5 To 9 feet But Water fales is not plenty heare among the Rice plantations. We are obliged To Bank acros Butwen hills and Stop Kane Water and flo(od) Sometimes 2 or 3 hundred acres of land. When Wonce fill Will Clean out a crop of Rice of 700 or 1000 Barels. As Such a mill Will Clean out 40 Barels pr day Steem Engines Would answer very Well provide that thay Would not Cum Too high & that Wood Would answer For Fuel as pit Coal cannot be had hear at Less than 30 Shillings pr Ton and verry Commonly 40 Shillings pr ton. We have plenty of fare Wood I therefore thank you To let me know What Such a Steam Engine Will Cost to Work With Wood and To acomplish as much Work as the water mill I have described above, allso the price of one that is Equal To the poure of one Hors as I want one of them For my one use To grind Corn and tirn Wood or grinstones. I am sure thay Will answer Better then any Water mill In Such Countrys as this For our dams is Caried away By hard Raines in Spite of What Can be done. I am thiraly aquented With your Engines I therefore Will be at no los To put them up. Thare Was a man By the name of Rumsy that applyd For a pattent For Steem Engines[1] I had it Stoped as I found that it was of your Construction. Rumsey Was or is in England under a pretence of giting a patent For a Steem Engine —

I am Sir your humble Servant

Jonathan Lucas

please Direct To me To the Care of
Mr John Nrabowski Merchant in Broad Stree
Charleston
or To Mr Bowman

Box 4 L

[1]See document no. 35.

Boat business

Documents 195-206

DOCUMENT 195

To Wm Creighton — Paris 10th Octr 1816

Dear Sir

I have received at Bruxelles & here your several letters of the 23rd 24th & 30th Sepr and Mr Wm Murdocks of the 26th D^{o}, in consequence of which I have written the annexed letter to Mr Banks, which Mr Rennie has enclosed in one from himself and most probably will become an Order. In that case I should wish every possible attention to be given to the workmanship & appearance of the materials, as I believe it is intended for a kind of State Barge for the Lord Mayor & his friends.

I am surprized you have not yet had any Account of Mr Humphreys first Boat[1], and am very anxious to hear about it.

We made the whole of our intended tour through the Netherlands and called at Essen, where we saw the Engine at work and Mr Scheibler[2] highly satisfied. Johnson had conducted himself extremely well & was to leave in a few days & to proceed to Rotterdam. We were there little more than an hour & indeed the whole of our Journey has been made with great dispatch, not having tarried in any place longer than to take a general view of it & of the principal sights. We passed through St Quentin but I did not call on Mr Joly. I shall see Mr Oberkamp.

We shall probably remain here 10 or 14 days

I am Dear Sir

Yours truly

J Watt Junr.

Box 33

[1]A reference to a marine engine for Germany.

[2]A 10 h.p. engine for cotton spinning ordered in 1816.

DOCUMENT 196

To J. Banks London | Copy | Paris 8 Act^{r} 1816

Dear Sir

Mr Murdock informs me that he has had some conversation with you respecting an Engine for the new Steam Boat upon the Thames and that you were so obliging as to offer us your Interest to procure the Order if our prices were not materially different from those of others. As far as I recollect, our price for the compleat Metal Materials of a 14 Horse Engine & of the Paddle Wheels & Machinery for raising & lowering them is £1250, but as I am extremely desirous of having a good Engine introduced upon the Thames, I should, if this is objected to, be willing to undertake the delivery of the materials at Soho for £50 less; or if it is insisted upon that we should ourselves undertake the erection (which has always appeared to us an objectionable plan for both parties) we will in this instance undertake it and our charge for the whole including materials, carriage & erection will be £1300, pay^{ble} in cash upon completion of the erection. It is to be understood however that no part of the woodwork is included in this Sum, and that every thing is to be prepared by the Boat builder previous to the arrival of the materials, that the erection may be proceeded in without delay.

I am confident that the Engine of that power we are makeing at Soho, will be found not only superior in every point of construction but I conceive also lighter than any the performance of which can at all be relied upon. I think the weight including Water in the Boiler is about 16 Tons.

If the order should be given us, it will be very desireable that your Boat builder or yourself, or some other intelligent member of your Committee should go to Soho to confer with Mr Murdock & Mr Creighton about the construction & dimensions of your Boat, as I am sure that as much remains to be done in improving the shape of the Vessels as in the Steam Engine.

Mr Rennie & myself have had a pleasant tour, the greatest part of which has been rendered more so by the company of your partner Mr Joliffe

I am Dr Sir
your ob^{t} Ser^{t}
J Watt J^{r}

Box 33

DOCUMENT 197

To James Watt Junr. Heathfield Jany 8th 1818

Dear James

I have received your letter of the 30th ult^{o}. My last to you was of the 1st ins^{t} — I think Capt Wagers memorial[1] is very properly drawn up but I doubt much whether the prayer will be granted. The most desireable part for you is the exclusive right of supplying the engines, as to that of the navigation without you can get some able and active associates in the Country I should fear for its success, and without you can find some persons to take that burthen off your Shoulders I should feel very unhappy. I hope if the patent is to be clogged with too many things to be performed on your part & restrictions, which may prove vexatious, that you will decline it altogether. As to the rest The ferry you mention seems a very proper situation for one or two boats. From Rotterdam to Antwerp for an equal number and the Rhine as far as Dusseldorf or Cologne for several, but I doubt much whether they will every pay in the more rapid parts of that river, on account of the slowness of their progress & the great expence of fuel. I am not acquainted with the inland trade of Holland but should suppose those boats would be very useful in navigating the channels which separate the various Islands between the Meus and the Scheld.

The price of coals seems to form a great obstacle to these navigations, but it is said in the newspapers that it is in agitation to lower the export duty on English coals, that however would I fear only tempt the dutch to increase their duty, to encourage their own collieries.

As I can give you no further advice that can be of use to you in the present circumstances, I must now leave the matter to your own discretion hopeing you will soon be able to bring it to a conclusion one way or other; but I hope that you will not in any case again venture across the channel in the Caledonia at this season however flattering appearances of the weather you may have. In my opinion & that of more of your friends you have ventured much already.

Mr Boulton was here yesterday & read over Capt Wagers memorial on which he had no material observations. He is today settling quarter day accounts. He says there are no orders of much magnitude come in & that a boat Engine of 60 Horse power is finished or nearly so ...

Dear James
Yours affect^{ly}
James Watt

Box MIV Watt 27.

[1]This refers to a 'Memorial to the King of the Netherlands' drawn up in Dec. 1817 in which an exclusive privilege for 14 years for steam navigation in all waters of the Netherlands was sought by Wager and B & W. MIV Misc(1). The privilege does not seem to have been granted.

DOCUMENT 198

To Jas. Brown,
London

Soho, 9th Oct. 1818.

Dear Sir,

You will have learnt from my letter of yesterday, to Mr. Mosley, that Count Conzaloneri has been here and has determined upon a 14 Horse Engine to be placed in the same Barge with the Goods to be carried. He brought with him particulars, which seemed accurate, of the Current and Depth of the Po: the former, he makes to be 3 to 4 Italian Miles an hour from the Mouth of the River to Piacenza (150 miles) where for 2 miles there is a rapid of 5 miles per hour, but from thence to Pavia (30 miles) the current again becomes 3 to 4 miles per hour. The total distance is 182 Italian miles (60 to the Degree) = 210 English, which they take 30 days to perform with the present method of navigating. Eight horses are applied to tow 3 Barges, which are loaded sometimes with 200 Tons, say between 60 & 70 in each, and draw when loaded 3½ feet Water. But as there is in times of drought only 3 feet water in many parts of the River, the Load is then distributed into more Barges. The feet above mentioned are French feet.

28 French feet, or nearly 30 English, may be allowed for the extreme breadth of the Barge and Paddle Wheels: of which we propose to take 16 for the width of the Barges, which being usually built of Larch, we conceive may be built not to draw more than 1 foot when empty; and by giving it the length of 120 to 130 feet, we calculate that it may carry 105 Tons (viz 18 Engine, Paddle Wheels & water, 20 Coals for the Voyage up & down, which is more than requisite and 67 Tons of Goods), without drawing more than 3 feet of Water. At this draft we calculate that a 14 Horse Engine, with improved paddle wheels, will move it at the rate of 5½ Italian (6½ English) Miles through still water; consequently 1½ to 2½ Italian Miles per hour against the general Current, which should enable them to perform the voyage in about 90 hours.

By the same rule of calculating, we made it out that if the Engine were placed in a towing boat 9ft wide & 100ft long, drawing

2ft 6 of water with its coals & machinery on board, it would only move a barge containing 67 Tons at the rate of 4¾ Italian Miles per hour; which, if there was any adverse wind, would hardly have stemmed the ordinary Current. The towing plan was therefore wisely abandoned.

The only doubt that remained in the Count's mind, was whether the circumstances of the River would admit of a boat of the length of 120 to 130 feet; but for this there is no remedy, the width & depth remaining as before stated. The paddle wheels are to be made to rise one foot & fall one foot, to adapt themselves to a lighter cargo in going up, or a heavier one in descending. The Barge to have a cabin with beds for 6 persons (including 2 Engine men) and to have two light masts. The goods may be stowed part fore & part aft to balance the vessel.

We are to send drawings for them to build the Barge by; and as the Count presses to have them as soon as possible, and all hands are occupied in the Drawing office, we must call for your assistance in preparing them. I think the best way will be for you to come here and either remain to compleat them, or return to London after the plan is settled, but the former will be most preferable, and I am not aware of anything urgent in town at present, as I suppose that W. Wight & his assistants will be able to compleat the taking out & sending off the paddle wheels of the Caledonia.

You will therefore please to see Mr. Richards with the inclosed letter and if there is no difficulty on his part as to issuing the order, you will proceed hither as early as you conveniently can, and will find a bed at my house, as long as I remain in it, which will be at least long enough to enable you to look out for another.

I observe what you say in yours of the 6th about the Richmond Company. The Engines are going on and if they do not speedily decide upon what they will do, another Company must be formed, which can readily be done. But I hope Mr. Tyrell will have influence enough to get them to adopt the resolution for one or more new Boats and leave the execution to him & me. I fear however that the Boat will be bungled, if it is to be built above bridge, as he proposed; and if you have any conversation with him you must try to get him into the notion of having it done by Mr. Brent.

If Mr. Ralph Walker, whom you will find at the Jamaica Coffee house, has finished his drawings of the Boat, I should wish you to bring them with you, and remain

Dear Sir

Yours truly

J. Watt Jn.

Box 46

DOCUMENT 199

To Mr. Pauly Soho 21 Decr. 1818.

Sir,

In reply to your favour of the 19th the observations made upon the operation of your boat[1] have been communicated, but it would be much more satisfactory both to yourself and others if a comparative experiment was made by applying the common propelling wheels to your boat using the same power of a spring to put them in motion; this was mentioned to Mr. Brown to be proposed to you, but it does not appear from your letter that any experiment has been made.

It must be observed that the opinion entertained here of your method of Driving Boats is not in favour of its proving advantageous either in effect or economy; and it has always been a rule to make trials of any new project that has secured here, by models and experiment before incurring any expense on a large scale where such could be avoided. In the absence of Mr. Watt nothing further can be said and I have the honour to remain Sir

your most obt. pub. sert.

signed — W.C.[2] for B.W. & Co.

Box 46.

[1]Pauly had made suggestions for various improvements in steamboats. See [200].
[2]William Creighton.

DOCUMENT 200

To Jas. Brown,
London. Soho. Decr. 21st 1818.

Dr Sir

Mr. Pauly has wrote to say that Capts Brown, Wager & Mosley have witnessed an experiment with his model of a boat and seemed perfectly satisfied at its performance — he therefore wishes for some determination respecting it. I can only answer by the inclosed letter which you will please forward him[1].

It would save expense to adapt paddle wheels to the present model which would not be difficult to do and work them by the same spring, but perhaps another boat the same size and an equal spring would be more satisfactory and give a fairer race between the two vessels — besides his spring might expend more power in one case than the other for a given distance, and as it is possibly not more than equal to 5 rats, some dispute in the race might arise, which two boats could avoid — however, one may afford a satisfactory experiment (this is rather an ambiguous term and may mean good or bad).

Suppose wheels 3 inches wide and 6 or 8 diamr. made of turned iron, worked by a belt or wheelwork from the spring so that speed could be varied from 3 to 6 miles pr hour at the circumference?

I send back Mr. Pauly's specimen of metal and a similar one of Mr. Murdock consisting of 4 iron and 1 nickel, he tried this and numerous others many years ago just as matter of curiosity; and has heard that some of the makers of cheap iron articles hereabouts add nickel to their iron to assist in getting a good polish.

You will find on examination that the pipe for returning condensed water along the sides of central hexagon of the penitentiary has been sent, except pieces at the corners and some apparatus to join the boilers.

I remain &c

W. Creighton
for B.W. & Co.

Box 46

Mr. Murdock enquires if you have not his copy of D. Robison's account of Mr. Watt's steam Engine. It is considered that all the pipes ordered for Woolich have been sent. The 2 wanting are making.

---

[1]Document 199.

DOCUMENT 201

Memo from Mr. Watt's letter to J. Rennie
respecting scheme for Propelling Frygates — dated 7 July 1819.

If the Paddle wheels are only (as proposed) 13 ft diameter making 9 Revo. p.m: this would give for the Circumference of the Paddles a velocity of little more than 4 miles per hour, but it may be doubtful whether half this velocity would be gained for the Vessel — more velocity would be obtained from the use of the Paddle wheels of larger diameter — and possibly it might be expedient to go to the length of 20 feet retaining the proposed breadth of 6 feet, which we suppose is all that could be conveniently applied, as no outer bearing for the shafts is intended. It is stated that 180 men could be employed at the 3 Capstans, but when we consider the full load under which steam Engines work on board of ships, we scarsely think the effect produced by them would be equal to that of a 20 horse engine — and if an Engine of that power were applied to drive the frigate with paddle wheels of 20 ft Dr. we do not think she would obtain a velocity of more than 2 miles per hour.[1]

Box 46.

[1]The early Navy orders for marine engines were for Tugs.

DOCUMENT 202

To James Brown, Soho, 2 March 1819.
London.

Dear Sir,

I am in the receipt of yours & Mr. Mosleys letters of the 26th ult^{o} and of the 1st inst^{t}. As Messrs. Bigg, Heighington, Knight &c are so desirous of having two ten Horse Engines in preference to one 14 or 15, and are willing to grant time for getting them ready, we shall comply with their request, though this will add considerably to the weight of the Machinery and perhaps give a greater velocity than may be required. My original proposal was certainly to have two Engines of this power, but upon conference with Mr. Tyrrel and a further acquaintance with the navigation of the River, it seemed more than was wanted. However, if our friends are desirous of adhering to the original power, I am not disposed to object, the only difficulty I feel being with regard to the draft of Water, which I am afraid can hardly be brought so low as two feet, and it appears to me very desirable it should not exceed it by many Inches.

I do not see any objections to the dimensions you have proposed, except that I think the height of the Cabin may be less than 8 feet. Most men, with their hats on, can stand in 7 feet; W. Creighton is not at hand to say if the Engines can work in that height, but if not, it would in this case perhaps be desirable to have the Engine house higher than the rest of the vessel.

The Paddle Wheels I should still propose to be 10 feet, or 10 feet 6 diameter, as we must endeavour to make up by diameter for the small width to which we are confined.

I am sorry that Mr. Brent[1] declines giving in an Estimate, as he will find it difficult to deal with a large Company, like that about to be formed, upon any other footing. The experience he has had may make his Estimate higher, but should not I conceive preclude him from giving in one, but if he apprehends that advantage will be taken of his drawings &c, he had better for the first give in only a general Estimate upon the dimensions & data you have communicated.

I agree with Mr. Mosley that the next best persons to employ will be Mess[rs] Wallis & Shore.

Mr. Ralph Walker having written to say he had prepared a design for a Richmond Boat, it would be as well for you & Mr. Murdock to see him with the inclosed letter, as if he has anything new we may as well reap the advantage of it. I perfectly agree with him in one thing, that the boat should carry out her width as near to the bow and stern as possible.

I am
Dear Sir
Yours truly
J. Watt Jn.

Box 46

[1]A London boat builder who had executed hulls for several boats engined by B & W.

DOCUMENT 203

To James Brown, London.

Soho, 26 March 1819.

Dear Sir,

By this day's coach we return Mr. Brent's drawing and specification of the Swiftsure.[1] The former seems to give nearly all the buoyancy that can be obtained, which in this case is perhaps the first object to be aimed at, as we believe much of the misfortunes that have befallen the other Richmond Boats have originated in their taking the ground. We have made some pencil lines upon the drawing to shew a still fuller bow, which though it may detract somewhat from her appearance will serve to keep her head out of the Water, but these are only meant as a suggestion which Mr. Brent and the Committee may adopt, or not, as they see proper. You have not said any thing upon the subject of outriggers, and we should wish to have Mr. Brent's opinion upon them. If the boat only draws 2 feet of water, which we hope may be effected, there will (according to the reduced height of the Cabins) be 6 feet 2 or 3 Inches out of water, which we presume is more in proportion to the depth in the water than any of the other vessels, and must render her more crank than they are, except in so far as that defect is remedied by keeping the weight low down. If the outriggers are not adopted, would it not be proper to keep the bottom of the gangways very near the water, so that when the vessel heels, they may come into contact with the water and stop the progress of her side motion. For this purpose they must of course be planked under, as the Caledonia and most of the other vessels have been, and we do not think that the lowness of the Gangways would be liable to objection where they are little exposed to the action of the waves. With respect to the dimensions of the timber, Mr. Brent is the best judge except in what relates to the Engine house, where we think it would be adviseable to have the depth of the Keelsons 9 instead of 7 Inches.

The amount of the Estimate undoubtedly strikes us as considerable, being we believe double the Sum which the London cost, but we are aware that by using the best materials and workmanship, the cost must be greatly enhanced. The Committee will undoubtedly give this matter the full consideration to which Mr. Brent's character and the manner in which he has fulfilled his former Contract entitle him. We had certainly laid our account that the sum of £2000 would have compleated the vessel including Joiners

Work & painting, and in any case it will necessary to know what those additional articles will amount to.

There can be no doubt that a good vessel in the Gravesend trade would pay, and we think no vessel can be better adapted for it than the Caledonia. It is most likely that nothing will come of the negociation with Messrs. Goldschmidt and we are very anxious to have the Experiments tried with her, which may decide her future employment.

We remain
Dear Sir
Yours truly
Boulton & Watt & Co.
(in J Watt jnr's hand)

P.S. Please to present Mr. Gilbert with
my father's publication & explain the cause
of his not having received it sooner.

Box 46

[1]This was driven by a pair of 15 h.p. engines.

DOCUMENT 204

To James Brown,
London.

Soho, 2nd May 1819.

Dear Sir,

Having executed the documents for the Sale and Delivery of the Caledonia, I have to thank Mr. Mosley, Cap[n] Wager & yourself, for the attention you have throughout bestowed upon her, and particularly in the late Experiments and final disposal of her. I part with her with regret, and am particularly sorry that I have not been able to witness her performance with the new paddle wheels and to take a final leave of her. As Commodore Bille proposes taking his departure on so early a day, it will be quite impossible to prepare any drawing of the Machinery, however slight, as the sketches we have, require to be adapted to its present state. If Commodore Bille judges it necessary, it must be sent after him, but as it can be of little use, it had perhaps better be dispensed with.

The account you give of the performance of the Eclipse[1] in yours of 29th & 30th ult[o], is exceedingly mortifying, and as far as we can conjecture here, must be occasioned by her Draft of Water & build. How does it happen that after we had so fully explained to Mr. Brent the advantages that would attend her load line being kept near to 4 feet, and in no event exceeding 4ft 6 with the fullest cargo, she should draw empty, & I presume even without all her stores, 4ft 7 mean Draft, but with an actual draft forward of considerably more? How is it, that after having from former experience been fully aware of the necessity of her being light by the Bow, after having in all our conferences with Mr. Brent endeavoured to impress the necessity of this, after having given him the weight and the Centre of Gravity of the Machinery accurately, the vessel is built to draw Six Inches more Water fore than aft? I have referred back to my Mem^{ms} & letters home, written at the time of our Conferences, and I find my recollections of the full manner in which these matters were stated, confirmed. Mr. Brent at the time expressed no doubt of keeping much within the limits of Draft that had been fixed, and the vessel was lengthened & the situation of the Engine altered, in order to meet his ideas of the Centre of Gravity of the vessel!! We have indeed learnt since from Cap^n Wager's & your letters, that the Bow was made somewhat fine, & the timbers of the vessel of full dimensions, but have constantly been given to expect that she would not exceed the proposed Draft.

By an account in one of your letters, her Draft when launched was

	ft.
Bow	1. 0¼
Stern	2.11½

	2. 6 mean
Cap^n Wages informed me that the Engines & Boiler sunk her	9¼
in addition to what the Beams projections & Paddle cases had before sunk viz	4½
Suppose water 12 to 14 tons	2
Coals &c	1¼
Gives for total Draft	3.11

I am aware of nothing, except the soakages, that can have added to this, and request that after having conferred with Cap^n Wager & Mr. Brent, you will inform us whence the excess of Draft proceeds, and why the Centre of Gravity is so much misplaced. These Circumstances united do not however fully account for her deficient performance, and there must be something either in her mode of entering, or delivering the water, which is adverse to her speed.

In the present state of our information here, the strong leaning of my opinion is, that it would be the interest of the

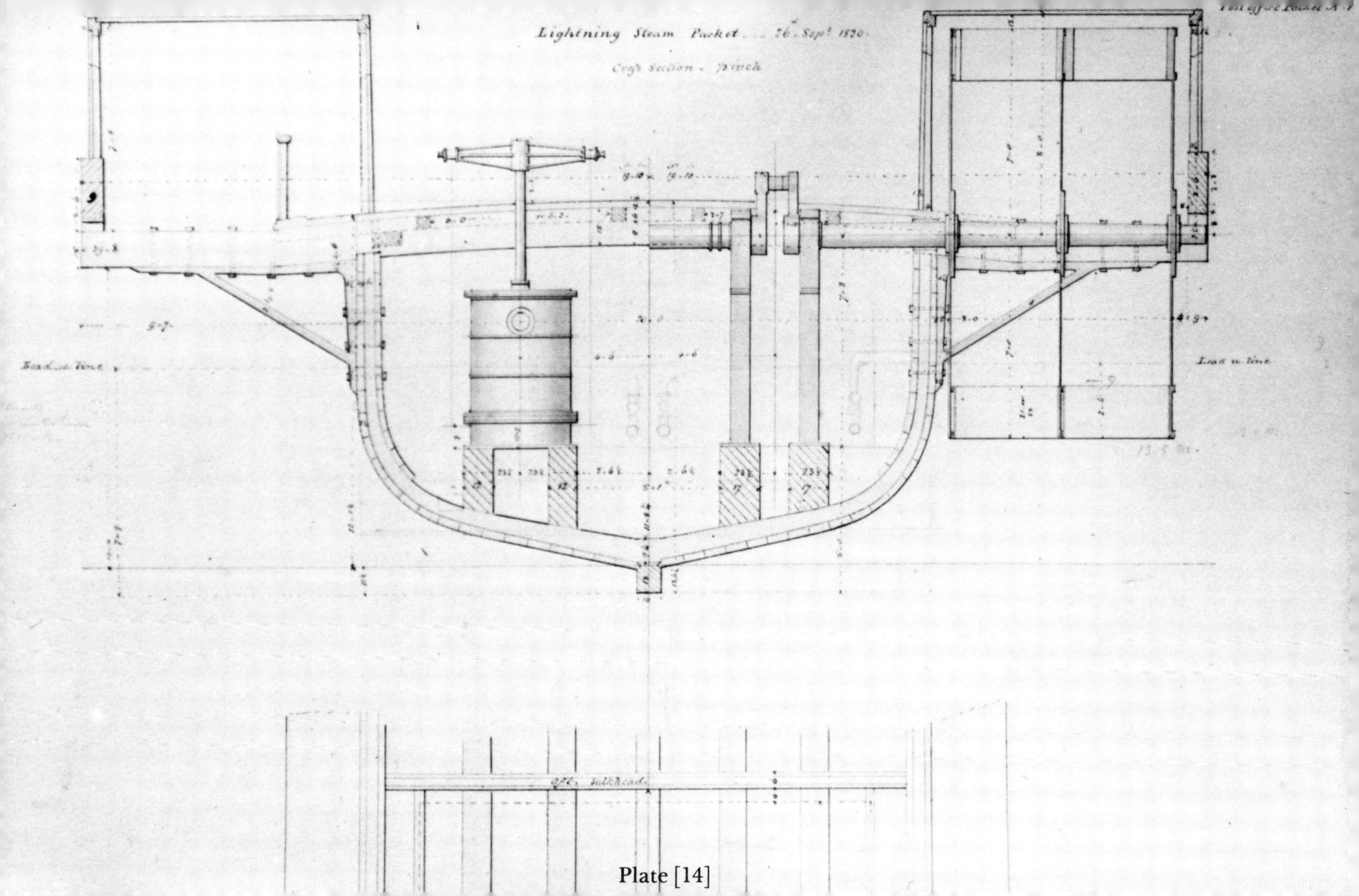
Lightning Steam Packet. 26th Sept 1820.
Cross Section
Load w. line
Load w. line
Aft. bulkhead

Plate [14]

An asses head would be preferable to a horses, as in point of carrying live lumber is probably accustomed to convey more at once either on back or in panniers.

But as Eclipse will be Heclipse by the Cockneys I propose a Hieroglyphic of a man, say a Taylor, or Gardener, with an enormous pair of sheers, for the figure head, in attitude of cutting the bowsprit. — it will then be said "He clips" — the boats name at once.

Plate [15]

Proprietors to put her into Dock again & widen her Bow so as to get rid at least of one apparent fault of construction. The Committee should allow her to be put into your & Mr. Murdock's hands for the purpose of trying what Experiments appear necessary, and among these the first that occurs, would be that of bringing her to an even keel by weighting her aft.

I remain

Dear Sir

Yours truly

J. Watt Jn.

Box 46

[1]Driven by two 30 h.p. engines.

DOCUMENT 205

To Jas. Brown,
London.

Soho, 15th May 1819.

Dear Sir,

Inclosed is a letter to the Committee of the Eclipse Steam Boat, which I wish to be presented, as no answer has been returned to my former one. I think they should call upon Mr. Brent to account for his having so much exceeded the Draft of Water he had specified, and you may repeat to them the contents of my letter to you of the 5th Inst^{r}., as I feel it due to ourselves that the matter should be investigated.

The new Articles preparing, will be ready on Tuesday, and as the Canal is likely to continue stopped, they will be sent by land, in which case they may reach you by the end of the week, and it will be advisable to have as many hands in readiness as can work at them, in order to compleat the alteration in a short time.

I am Dear Sir

Yours truly,

J. Watt Jn.

Box 46

DOCUMENT 206

To James Brown 1821.

I did intend to have prepared a scheme for the distribution of Engines & boilers for Leith Packet[1] or Smack but have been so bothered have neither spirits nor resolution to do any one thing & such as are attempted are sure to be full of blunders. Could not the boilers be crammed into the hold & have cabins over them? Firing place to be the full depth of boat by way of being more airy.

Engineers have got a method of puffing that is new to me — got themselves prosecuted for carrying 99 passengers!!! I have not "Critic" by me to examine the nomenclature therein given — tis hardly the puff oblique much less direct — perhaps "puff by implication" is the thing.

Scandal says a thief was lately caught in a chimney at Aston hall[2] — found by leaving his legs too low down — best part of story is his lugging a piece of ordnance out of his inexpressibles in order to frighten the women but Mrs. Perman knew men & things too well to mistake the article for a horse pistol — man then pretended to be mad but was speedily lodged in quod.

W. Creighton

Box 46

[1]The Leith Steam Boat Co. ordered 3 pairs of B & W engines before 1825, the first, a pair of 40 h.p., was for The City of Edinburgh in 1821.
[2]Home of J. Watt jnr.

After-Sales Service

Documents 207-213

DOCUMENT 207

To John Southern Newcastle 26 June 1798

Dear Sir

I wish to be informed what size Engine you would recommend to work an 11½ Inch Pump 37 fathoms, 7½ feet stroke, with 25 fathom of dry Rods & to raise its own Injection Water 25 fathoms & what will be the price of the metal materials supposing us to furnish the same as for the 60 Inch ordered by Mr Johnson.[1] It is for Flats Colliery, a new undertaking about to be opened by Mr Johnson for the Ex'rs of the late Mr Lambton.[2]

I have been this day to Burn moor with G. Taylor, to set out the Engine house for the 50 Inch.[3] They have a very fine stream of Water running within 50 Yds distance from the pit, which turns a Corn mill about half a mile below & will not have to be raised more than a few feet to the Condensing Cistern. — The Agents say it is probable they will have to apply a 9 Inch pump to the Engine in addition to the one specified in the data given to us by Mr Johnson, (19⅓ D^r^ 40 fathoms) to raise the Water from a Seam of Coal lying 22 fathoms deeper, or say in all 62 fathoms; please to say us in this case how much you would diminish the size of the 19⅓ Inch pump? And whether in case our Engine should be applied solely to raise the water 22 fathoms in a 9 Inch Pump & it is then raised today by an atmospheric Engine already upon the Colliery, you would in that case think the 19⅓ Inch pumps too large.

In making out the Agreement, you will please to notice that the Engine is to be erected at Burn moor in the Country of Durham and you will insert the name of Mr Featherstonehaugh "for & on behalf of the Trustees of the late William Lambton Esqr. —

I have now seen most of the Engines. Walls end[4] is by far the worst and I fear is incurably so, for Buddle either not being able, or not willing, to comprehend the Engine, is at great pains to magnify its defects & to cry it down, if I am rightly informed; but I shall probe that matter to the bottom. All our other Engines give very great satisfaction, although some of them are (god knows) bad enough. The owners are all tractable and give me leave to propose any alterations or improvements I think proper. I would therefore wish you to consider what additional quantity of Fly will be proper & in what manner it should be applied. The Breaks are all in the Fly wheels, which are faced with red beech on the outer Rim, secured by wooden segments on the sides, & if you change these wooden segments into Iron ones, it may be proper to provide some easy means of fastening the beech rim whenever it requires renewing, which it appears likely to need from time to time. I find that in all the Engines they use the breaks far too much & the throttle Valve too little.

Campbell & Gibson[5] have had a sore job of their well in sinking through Quicksands. It is at length compleated this day & their builder has contracted to have the Engine house, (which it was not prudent to begin upon sooner) compleated in a fortnight. —

Mr Marley of Sheildrow expressed the utmost satisfaction with his Engine.[6] He says it went remarkably well until one Lawson from Soho called to see it & ordered some alterations, which being executed, it went worse & worse until it would not go at all. Upon which emergency G. Taylor being sent for, said that all Lawson had ordered was wrong, replaced matters in their former situation & since then she has again given compleat satisfaction; but the Engineman was highly indignant at having been taken in (as he calls it) by Lawson, wept bitterly at his own credulity & swore that if ever he caught him there he would inflict his vengeance on him. This is Mr Marleys information. I shall see that Engine & Pontop pike[7] tomorrow. — I hope Mrs Southern is well. Barnes lady is in a very thriving condition. Your other friends are well as is also your humble Servt

J Watt Junr.

Box 33

[1]Gateshead Park Colliery, 1794.
[2]No engine was ordered.
[3]A 60 in. cyl. 9 ft. stroke engine was ordered in 1799.
[4]A 30 h.p. rotative engine ordered in 1796.
[5]Bill Quay — they ordered an 8 h.p. engine.
[6]A 49 in. cyl. 8 ft. stroke engine ordered in 1796.
[7]A 24 in. cyl. winding engine ordered in 1795.

DOCUMENT 208

To J. Watt Junr. Glasgow Ocr 5th 1800

Dear James

I have yours of 30th & am glad to find you safe returned as I began to think your absence from the home business rather too long considering G^s[1] inexperience — It also gives me pleasure to hear of L^s[2] amendment. I fancy his relieving Creighton here will be useful, though I hear of no new orders Trade continuing still very dull, I thank you for the account of your tour, but since my last can send you no counterpart having only been a few days at Strathblane, mostly confined to the house by squally rainy weather, when we were recalled hither by the sudden death of our friend James Campbell, whose funeral I attended on Friday. He is much lamented by all who knew him as a worthy benevolent man & an honest Tradesman. The loss to his family is very great, as independent of their affection for him which was of the most tender kind, I fear he has not been able to provide for them as he could have wished. Very few now remain of my old acquaintances & these are gradually leaving the Stage, of my schoofellows at Greenock I could collect only 3 but heard of 2 more being still alive! Those left were however very kind & I should have spent a longer time with them, had it not rained so much that I could scarce get out of the house.

I am very sorry to hear of T. Walkers[3] circumstances but fear that his usual imprudence will continue to keep him poor. I should hope however that his spirit will impel him to do something for himself & family before it is too late.

Creighton has got Mr Pattisons Engine[4] going some time but the machinery is not yet ready. Bartholemews Engine[5] is also nearly finished but the machinery in the same state of backwardness. C. is today gone to Edinburgh & the vicinity & proposes if he can find a conveyance to visit Dundee before he returns. I understand there are several of the minor poets here, who get <u>some</u> business etc. as they pretend to work at lower prices. To preserve our preference the utmost attention ought to be paid to the goodness of our materials & workmanship for there lies our forte. I go on Sunday to visit Mr Love at paisley, who is one of our greatest panegyrists & I expect will be able to surmount the obstacles to the establishment of a Steam corn mill there.[6] This dry summer & the increasing population of this neighbourhood seem to convince every body of the necessity of Mills of that kind, as the want of them has contributed very much to the dearth.

I have just parted from Mr Miller & his children who from their stay at Gourock where I visited them are at present in good health & Margarets eyes quite well, by all accounts she was in a miserable state when she went from hence, her eyes so sore that she could not look up & very much emaciated, she remains still very thin but seems otherwise well & her temper seems much mended I fear however the influence of the winter upon her health.

Gregorys letter to his mother came to hand today. I think the journey he proposes very proper & desire he will remember me kindly to all enquiring friends, but at present recollect no commissions for him, except that if when at Clifton he can hear of any estate thereabouts which would suit me to let me know — In regard to the new Machine proposed by the Gentleman, (Mr Mathews I think) at Bristol, he may be wrote to, that from my time of life I must decline engaging in any business which does not assimilate with the Engine business, but when I return I shall be ready to give him my advice upon the subject if he chuses it. Thank Mr Wedgwood for the venison, and as the receipt has been acknowledged shall defer writing him until I return —

I am very glad that Mr R.B. has settled with wheel Unity[7] & very well satisfied with the terms, I wish he may be able to make similar ones with the rest of the delinquents, please remember me cordially to him when you write to him.

M^rs W, joins me in sympathy for Miss Boulton & in prayers that she may soon recover from the disease in her knee — Please remember me kindly to Mr B. & family & to all other friends — Mrs W. joins me in love to you & Gregory & I remain

Dr James

Yours affectionately
James Watt

MIV Watt 9

[1]Gregory Watt.
[2]James Lawson.
[3]Of Richard & Thomas Walker, Manchester merchants and manufacturers.
[4]A 32 h.p. engine.
[5]A 10 h.p. engine.
[6]A 20 h.p. engine was ordered in 1808.
[7]A reference to M.R. Boulton collecting premiums still owing in Cornwall.

DOCUMENT 209

To Gregory Watt. Soho 10th Sepr 1803.

Dear Gregory,

Yours of the 6th Inst^{t} is just received with the proceeds of the Dft on Eadon and advice of your Settlement with Sword,[1] who seems in every respect a worthy competitor of McGeorge[2] or of any other of these northern swindlers. We shall without hesitation adopt the resolution you recommended with respect to both of them & it may be proper to instruct Creighton & our men to the same effect. I could not advise you from Manch^{r} respecting the time of executing this Engine and I conceived it too late upon my return here. It is certainly so now, I shall however briefly state the facts. The Engine was ordered provisionally by Lawson the 4th May when he observes "they are anxious to get the Engine as soon as possible, I have stated 9 to 10 Months, but could wish you would say less"; in our reply, we stated 9 Mo^{s} & the order was confirmed through him upon the 25 May. It became due on the 25 Feb^{y} & was then and even before then, ready, but retarded by the frost. The order for the boiler was not confirmed by Sword & Co until the 22 Feby and the order for the Boiler Apparatus was only sent on the 25th of that Month — yet the whole of the Articles were dispatched upon the breaking up of the frost the 8th & 17th March! —

Considering how complicated the transaction has been and the difficulty and expence that might have attended the production of legal proof, I am satisfied that you have done perfectly right in following Mr Hamilton's advice & that we should have been in a worse situation by postponing the settlement. We shall take care in future to make our bargains with Gentlemen on that side the Tweed so direct & specific as to make our proof easy should we called upon to produce it.

From the number of Engineers who are starting up there and the present badness of their trade, it may be questionable whether it is worth while to continue our establishment at Glasgow. I should think Creighton and his adjutants might leave as soon as they have finished Wilsons & Stirlings[3], but any resolution upon that head may be deferred until we see you.

The Engine for Sir Jno Hope, Midfield Colliery near Inveresk, of 8 Horses power was sent on the 2nd July and is probably now arrived. It may be proper to call there and appoint one of the hands for the erection. Vivian we suppose is disengaged, if he is thought capable of it.

We expect to send off Houstoun's Engine in the course of the present or early in the following month.[4] We have a larger one for Gilbert Meason & Partners at the Leadhills (say a 60 Inch 8 ft St.) in hand, but it is not wanted before March. It might be as well if you were to call to inspect the progress made in the Engine house and see that it is done according to the drawings and to Mr Southern's instructions whose correspondence with Mr Laing the resident partner, it may be right to refer to. As they wished to have the boiler as soon as it was ready, it has been already sent.

These are the whole of the Scotch orders which I at present recollect. We have one for a 24 Horse with two 28 Horse boilers for Mr Richard Jackson of Carlisle, where Lawson has been recently to set out the house &c, so that there is no necessity for making that your road, but I believe there is still less to do upon the other. I shall look over our Northern Acc^ts^ and send you in a post or two, directed as this, such as are due.

I remain

Dr Gregory

Yours aff^tely^

Watt Junr.

P.S. Is Vivian properly informed that the charge paid for his board by the Perth C^o^ goes to his debit? We generally leave the adjustment of this article to the men themselves, as they sometimes pretend it was promised to be allowed them by the parties, and we have in a recent instance been obliged to sustain the loss ourselves.

Box 2

---

[1]Sword, Heywood & Co. of Johnstown, purchased a 20 h.p. engine.

[2]Purchased a 4 h.p. engine.

[3]Probably Wilson's 48 in. cyl. 8 ft. stroke blowing engine and Stirling's 20 h.p. engine.

[4]Either the 33 in. cyl. 7 ft. stroke or the 20 h.p. engine.

DOCUMENT 210

To M.R. Boulton — Edinburgh 8 Novr 1808

Dear Sir

Your letter of last Month reached me at Glasgow, but I have deferred writing you, chiefly from having nothing very particular to communicate, and presuming you would learn enough of my motions from my father. —

During my residence at Glasgow, I visited all the Engines that had been erected since I was last there and most of the old ones. The greater part of the latter, though in tolerable working order, have relapsed into a state of filth, and many of the new ones are hastening to the same. A few remarks upon each may not be useless, and shall be stated in the order in which they occur.

1. Glasgow Water Works.[1] These Engines are exceedingly well put up, and the house both elegant & commodious; it has not given way farther than it had done when Taylor left it, and I should hope is now out of danger. The working gear makes rather more noise than it should do, but in other respects I found nothing to reprehend. The job indeed altogether does G. Taylor great credit.

2. H. Houldsworth.[2] The irregular Motion of the pump had occasioned the fracture of one of the Connecting Wheels a few days before I called, and although it was got mended and to work again, the house was rather in confusion. The Engine appears however to have been well put up, and to be neatly kept Mr H. complains of the narrowness of the house rendering access to the subterraneous parts inconvenient, and indeed this is increased by the Governor being placed in that situation. He has very little load upon the Engine as yet, besides the pump of the Cranstone Water Works which is pretty constantly at work and is a most wretched performance. Both Houldsworth & his Engineer Buchanan assert that they did not suppose we would have undertaken it, or should have applied to us![3]

3. Girdwood & Co[4] This Engine is tolerably erected; But as the wooden Spring Beams have shrunk a little, they regret much that they had not iron ones and even complain of their not having been suggested to them. I believe it will be best in future to represent them upon our drawings, or to make a Memorandum to state they may be substituted for wood. Girdwood consulted me about the proposed increase of the Velocity of his Engine, without mentioning that he had wrote the house, and of course received the same advice which you have given. I presume he will gain the increased speed by an alteration of his Millwork. The Hot Water pump was not working well, and the Engine but indifferently kept.

4. Mr Dunn (6H) This Engine is put at the dark end of a large workshop and partitioned off, so that it cannot be inspected at midday without the aid of a candle. It is not neatly kept, but was working well and gives satisfaction.

5. Mr Robertson (14H) His Mill has not yet got to work and the Engine is only occasionally exercised. I did not see it at work. It seems a firm job, but very awkwardly contrived. The Governor placed below the floor where access cannot readily be had to it when at work, and no communication with the Beam floor but through the Mill. The basis of the Nozzle pillars sunk below the floor, which gives them a very awkward appearance. The Mill very

neatly finished and filled with good Machinery. Mr Robertson was absent.
6. Ja^{s} Finlay & Co Balfron (14H) Not at work, having plenty of Water. It is pretty well put up & gives satisfaction, but there are symptoms of incipient dirt, which Mr Finlay has promised to correct. The throttle valve is not very good.
7. Gourock Ropework C^{o} (32H) Very excellently put up in the first instance (by W. Wright) but very slovenly kept, and the Governor laid aside, as they said it would not act. Upon enquiry of Mr Murdock who had been there lately, he alledges the throttle Valve to be lobb sided and never to have acted properly. The Engine is very lightly loaded, and they seem insensible to its defective state.
7. Jno Laird & Co (4H) Not at work, for want of hemp; but very neat, and reported to give great satisfaction. —
8. Wm Stirling & Son (14) a very firm job & working well, but dirty. Their 20H still worse, and out of order.
9-11. Jno Pattison, Dunlop & Co & Houldsworth & Hussey were in very indifferent order[5].
12. Mr Houstoun's (6H) doing its work excellently & kept tolerably for a Colliery Engine. His Pumping Engine not in such good order as I have seen it. He is much pleased with his Winding Engine, particularly as he brought it to bear himself, after he had discharged our man for idleness & obstinacy. —
13. The Engine house & Mill for the Corporation of Bakers at Paisley[6] are roofed in, and they propose beginning the erection of the Engine as soon as Mr Murdock can come, which I suppose may be a week or fortnight yet.
14. Hugh Vallance & Co (14H) The building about level with the ground, but propose going on with it through the Winter, and will be ready for the Engine as soon as it can be had. Mr Houston complains much of Hutton for having put the Engine house too low, there being no drain from the foundations. They are likely also to be short of Injection water on the surface, and I have advised a Well to be sunk. They determine to employ Iron Spring Beams & wish to have a drawing sent them, as they can get them made on the spot.
Please send Hutton a drawing of Iron Spring beams for the 20H Corp. of Bakers, Paisley.
15. On my way here, I called at Stein & Dewars[7] where I found the erection of the Engine nearly compleated, and the operation of plaistering & finishing off, going on. It is by much the best job I have seen of Mr Murdock's, but the basis of the Nozzle Pillars are sunk here also, and indeed they can not come in upon a level with the bottom of the main pillar, unless the latter is lengthened. The Bell cranks &c of the Governor gearing in the new way were carried up to the floor over the beam, & there is not, in the 4 feet strokes,

head room to allow of their being placed under the Spring beams. — This had also necessitated some other alterations, and the construction had better be reconsidered, as far as respects this length of stroke. The Millwork would not be compleated for 3 weeks or a Month.

16. Jn° Stein's (20H) was not at work. Was erected by Pool, and seems a tolerable job; but has too much communication with the Mill and some wheels within it, which I daresay are rather discordant. He will want a Boiler shortly in lieu of an old one which is applied to the Engine, to be of 2 (to) 4 Horses more power than the Engine. Will either order it himself or instruct Hutton.

He is likely to want a new Engine soon[8], to do the work of two Atmospheric ones; (a 16 Inch, 5 feet Stroke, 18 per M. & a 16 Inch 4 ft 25 p^r M.) Will determine soon & set Hutton to make a plan of the situation & machinery to be driven by it. It must either be a 6H or a 10H on the new construction. —

17. Geddes & C° (16H) at Leith is in most wretched order, and I have desired Wm Murdock to come over for a few days to repair it, before he goes to Paisley. It is absolutely unsafe to be near it. I never saw a case of more complicated disorder. —

18. I expect you will soon receive an Order from M^r Jn° Wyld of Glasgow for a 24 or 32H for a Cornmill[9], the expectation of which detained me there a week longer than I intended. — The Company is not yet compleated & until then, the matter is not to be spoken of. Mr Wyld is a wholesale Grocer & was lately a Bankrupt. It will therefore be necessary to make some enquiry about the C° when formed, upon which head I have instructed Hutton. Mr Wyld was introduced to me by young Mr Laird, who is a very prudent, businesslike man. I understand too he has a brother[10] who is wealthy & in good ...[10] and may be probably become connected with him. —

I received at Glasgow £32..17 from Mr Laird & £18 from Mr John Campbell in payment of their respective Accounts, which please to pass to my debit.

The Accounts against the Paisley Corp. of Bakers & against Messrs Stein & Dewar may be sent to Mr Hutton.

I have received a letter from Mr Hirst requesting me to call at Northallerton; which I shall do, having previously visited the Lead Hills. — I propose setting out for Arbroath tomorrow, to inspect our Engines in that quarter; but I shall not go further North, as I understand that a visit to Aberdeen would answer little purpose.

I have not heard from Soho for some time & shall be glad to receive a good acc^t of yours & your fathers health.

The inclosed from Hutton reached me yesterday. I find his friends here (Mr Robinson & Prof^r Playfair) know nothing of the affair, of which I had heard some undistinct rumours at Glasgow. I have informed him in reply that there is no probability of my being there by the time he mentions. I am Dr Sir

Yours truly
J Watt Junr.

Box 33

[1]A pair of 36 in. cyl. 7 ft. stroke engines ordered in 1806.
[2]H. Houldsworth had a 45 h.p. engine (1805).
[3]Cranston Hill Waterworks Co applied to B & W and ordered a 36 in. cyl. 7 ft. stroke pumping engine in 1810.
[4]A 32 h.p. engine for a saw mill.
[5]Pattison ordered a 32 h.p. engine in 1800, Dunlop a 24 h.p. engine in 1802 and Houldsworth & Hussey a 14 h.p. engine in 1804. All three were cotton spinners.
[6]A 20 h.p. engine.
[7]A 14 h.p. engine.
[8]He did not order one.
[9]No order materialised.
[10]Letter damaged.

DOCUMENT 211

To M.R. Boulton — Edinburgh 16 Novr 1808

Dear Sir

I returned on Monday Evening from my northern excursion. Mr Renny's Engine[1] at Arbroath was working well, and is very neat & well erected; I think the most so in point of convenience and appearance of any I have seen on this side the Tweed, and gives great satisfaction. Mr Louson's[2] is pretty well erected, except the mistake of placing the Cistern 2 to 3 Inches farther off the end of the flyshaft than ought to have been done. The old gentleman, (who has a will of his own), is not to be dissuaded from having a new double crank strengthened and the toe lengthened, and proposes transmitting his resolve to the house himself. The end of the present one, where the brass takes hold of it, is a good deal worn; and from what I can understand, the Engine has not always been kept in as good order as I found it; he has now got a new Engineman who seems intelligent and I dare say will take care of the essential points, but cleanliness is not the order of the day.—

Mr P. Hutton's (6 Horse)[3] at Dundee is very indifferently kept; and though he seems sensible that the defect lies in his management, he will (with) difficulty be brought to make the exertions requisite for a reform. Mr Wilkie's (20H)[4] is in the same state of filth in which it has always been. It was standing for want of flax, which is the case with several mills at Dundee. It is reported to be in good working order, and to burn less coal in proportion to the work it does, than any Engine in that neighbourhood: A new one of Murray's close to it, (which, I was informed, is very neat and handsome), is reported to burn double its proportion of coals. Mr Wilkie's beam is a good deal twisted & he has some thoughts of having one of Iron with a parallel Motion, blacked & ground, and Iron Connecting Rod. Please to send him an Estimate, and also an Estimate of an Iron entablature plate & pillar; the foundations of his house have given way on one side, which has cracked the lever wall and it is said to partake of the motion of the Engine when at work. — New bottoms will also be wanted for his boilers, which seem in bad condition, and I advised their being taken out and examined. His millwright or foreman will report upon them & order what plates are necessary from Soho.

Mr Wilkie is highly satisfied with his Engine, which he considers by much the best we ever sent across the Tweed. A new condensing cistern must be put in, the present one being compleatly rotten. I think it will be right to include arm Brackets for the Working Gear & an iron plug & guide, in the Estimate of the Beam.

Mr Wilkie introduced me to a Mr Sharp[5] who has lately bought a flax Mill with an Atmospheric Engine; which he means to replace by one of ours (20 Horse power,) and if the flax trade does not mend, to convert his Mill into a corn mill. He told me he had already written to a friend of his in Birm^m^, a Mr Wood, to make enquiry as to the cost of one of our Engines of the above power. He has a new oblong boiler, similar to ours in shape & appearance, 13f^t^ 7 long, 5f^t^ 3 width, & 5f^t^ 8 extreme depth, which I suppose will do again. I have little doubt of this becoming an order, and have taken a rough sketch of the present Engine house. — Mr Sharp is a sadler by trade, and sufficiently responsible; he intends this concern for one of his sons.

As Mr Sandeman's[6] & Col. Mylne's[7] Engine lay rather out of the road, I did not visit them.

At Perth, I saw the foundry Engine[8] working well and properly taken care of; no person allowed to enter the house but the Engine man. None of the Proprietors were at home.

I reached St Andrews on Saturday Evening just in time to see Mr Dempsters 6 Horse[9] at work. It is a good job and well kept: he has had all the wrought iron parts polished, and is determined it shall have justice done it. Mr Dempster has taken out a patent for a new mode of making Sail cloth, which consists in his doubling & twisting the threads both for the Warp & Woof: the canvas produced is of a very even texture and stated by him to be of double the strength of the common sort. He has not been able to get any Contracts from Government, owning to difficulties thrown in his way by Mr Turner, the Inspecter general; at least, he suspects this to be the case. But it has already acquired a great character in the Merchant Service, and he has orders for more than he can execute. In consequence of which, he is proposing to form a company and set about the manufacture upon a large scale. He has already purchased a plot of land of 2 to 3 Acres for its erection, and when his plan is arranged will probably give us an order for a 36[10] Horse Engine — He is trying the weaving by machinery but has made little progress yet: He has only one loom and that very ill constructed; much inferior to P. Marslands. — There is no other manufactory but his at St Andrews: and I should think the situation a good one, as hands are plentiful and provisions cheap.

At Kinghorn I found Messrs Walter Fergus & C^{os} Engine[11] at work, spinning flax. It is but indifferently kept, and was consuming more coal than it ought; 1½ Cwt per hour, turning only 450 spindles with a Water wheel to help it. The millwork is wretched; wooden shafts & wheels never cleaned from the dirt and tow, which hangs from them in flakes & is clotted in with the grease upon the gudgeons &c. The last boiler that we sent them, has been rivetted together & put up by themselves, and is very ill done, leaking much at the rivets and the damper apparatus so hung as scarcely to act. They are however highly satisfied with the Engine, which has never given them any trouble, and propose whenever times mend, that is when flax becomes cheap, to order another from us to replace an atmospheric Engine (30 Inch D^{r} 4½ feet St., 22p^{r}M) now at work in another of their mills at Kinghorn.

Mr Young's Engine[12] at Burnt Island was grinding wheat, the distillery being at a stand. It is well erected and well kept; and both M^{r} Y and his son seem to take a pride in it. —

I went out to Sir Jno Hope's collieries[13] on Tuesday. The Bart was not at home, and his agent Mr Jno Grieve was likewise out of the way. I found Mr George Grieve occupied with the erection of the new 64 Inch at Buccleugh Colliery. It is to replace a 72 Inch atmospheric, which is still at work. The new cast iron cistern & Air pump & condensor are in their places, and an extension of the Cylinder platform going on. The materials are all there with the exception of the Cylinder & the broken parts of the Steam case.

Cairnie 60 Inch was at work; the main plummer blocks not firmly bolted down, and the beam waggling sideways at the end of the stroke; the working gear noisy, and the house showing symptoms of Slovenliness. In fact, I suspect that but for the arrival of G. Grieve it would soon have been brought to a very different state from that in which G. Taylor left it.

The 60 Inch on Midfield Colliery & the 8 Horse and one of the 6 Horses have reached a maximum of filth and disorganisation, which reflects much upon Jn° Grieves management. His brother is of opinion that it will be necessary to have new working gear with iron posts & plug and brackets for the 60 Inch, upon which subject he will probably write to Soho. He seems to look forward with some dread to the difficulties he will have to encounter to get the Engine into tolerable order and their keepers under the regulation. He complains of finding things dearer and worse than at Birmingham, and his better half does not like the country and would have given vent to her regret at quitting the foundry had I encouraged the subject.

They are building a new Engine house at Cairnie, which upon enquiry from Grieve, I learnt was to receive a pumping Engine 42 Inch D^r^ from Butterley! so much for Cosenage. Grieve said it was ordered before his arrival, and that his brother already thought he had acted rashly, but was tempted by price.

Sir Jn° has taken a 21 years lease from the Duke of Buccleugh of his colliery, for which he is to pay £2500 per Annum. —

Mr Geddes Engine[14] has been repaired by Wm Murdock, who is gone again to Alloa. The millwright burst his new wheel in putting in one of the last cogs, and another has to be cast and cogged.

Mr Rennie is here, on his way to Arbroath. I understand that E.S. Hutton & his lady are also among the arrivals. My present intention is to leave this early in the ensuing week, — and to proceed to the Lead hills. — I am Dr Sir

Yours truly

J Watt J^r^.

P S. In the narration in my last, I omitted to mention having seen Halbeath 63 Inch, which was working extremely well & kept very clean. Anderson of Leith who is son in law to Mr Young the Manager, has been putting up some atmospheric Engines for winding Coal there. —

I observe I was under an error with respect to the cause of the nozzle columns being sunk so much in the ground in the 14 Horse Engines lately erected here. I see now, it has depended solely upon the erector of the Engines.

Box 33

---

[1]A 14 h.p. used for flax spinning.

[2]A 6 h.p. used for flax spinning.

[3]Used for flax spinning.

[4]Used for flax spinning.

[5]W. Sharp ordered a 32 h.p. engine for flour milling in 1809.

[6]A 14 h.p. engine used in bleaching.
[7]A 6 h.p. winding engine was ordered for Mylne Colliery, Co. Angus, in 1804.
[8]A 6 h.p. engine ordered in 1802.
[9]Ordered by R. Miller in 1804.
[10]No order materialised.
[11]A 12 h.p. engine ordered in 1800 for a cotton mill.
[12]A 14 h.p. engine ordered in 1805.
[13]Six B & W engines had been erected for Sir John Hope:

	reciprocating		rotative
1799	60in.	9ft.	
1803			8h.p.
1805			6h.p.
1806			6h.p. (Cairnie)
1806	60	8	
1808	64	8	

[14]A. Geddes' 16 h.p. engine was ordered in 1800 for a glassworks.

DOCUMENT 212

To John Southern — London 5 Sepr 1810

Dear Sir,

I have yours of the 3rd — It was my intention to have accompanied Mr Rennie to Guiting on Thursday and to have proceeded with him from thence to Soho; but as I have since learnt that Mr Miller is upon his way to town, it is probable that I shall prolong my stay here another week & shall thank you to forward my letters as late as Monday's box.

Mr Rennie will expect to meet Mr Murdock at Guiting on Friday Morning and will accompany him to Soho, where he will finally settle with you the situation of the Engine for the York Building Water Works.[1] That C^o have not been able to purchase the adjoining premises at any moderate rate, and having how obtained permission from the City to erect a Dolphin in the river at the end of their Tunnel, have determined to build the new Engine upon their present wharf, as I believe was before proposed. All the particulars of it remain as settled with Mr Rennie when he was last at Soho, and the Estimate may remain over until my return. Mr Lumley is now anxious to have the drawings of the house, which he proposes to have built during the present season. — Mr Rennie will wish to take back with him the plans he left at Soho of these premises, it may therefore be as well to be makeing Copies of these in the interim.

I have seen Mr Millington more than once about the West Middlesex levels. He says he has himself taken the level of the Kensington bason forwards & back again, & makes it 105 feet above high water mark at Hammersmith. Mr White has taken it separately & makes it the same. The latter, or his partner, has also taken the difference of level between the Kensington bason & the surface of the Grand Junction Canal at Paddington, which he makes to be 22ft 9 Inches. You will see that if these measurements are correct, the high water of the Thames at Hammersmith must be 6 feet 1 in above the level of high water at the London Docks; say 88 ft 4 + 22 ft 9 = 111 ft 1 Inches for height of the Kensington bason above high water at the London Docks, and only 105 feet for height of said bason above high water at Hammersmith. This Mr Millington is to clear up this & to level from the same high water mark at Hammersmith down to the pond in St James's Park, or to one of the Bridges. The point is of little importance as concerns us, but in the mean time the determination of the Committee is suspended. Mr Lumley is to consult with them and finally to settle with Mr Rennie. Our plan for the three Engine houses is approved of, the Chimney to be 80 ft high. As there have been so many mistakes about these Waterworks, I have requested that when they have settled their plans, an official letter may be written to us stating the Data upon which we are proceed.

Mr Simpson is getting on very rapidly at Chelsea;[2] the walls are already ten feet high above the piles and the work very solid & good. He had written to Soho about some trifling difficulties he had met with in considering the drawings, which of course you will answer. His cast iron wells were arrived, but I did not see them.

He desires we will enter an order for an Engine for the Lambeth Water Work[3] which is to be one third more powerful than their present one, a 24 Inch 6 feet St. double, which has a fly wheel & works a pump on each side of the Center of the beam: these draw their water from a Suction pipe 12 Inch Dr which extends about 100 yards into the River. He says they cannot sink a well and lay a pipe to supply it from the River, on account of the Land springs, without going to a very great expense. But we are to have all the data laid before us & then to decide upon this point, and also whether the Engine is to be single, or double; a fly wheel he will have at all events.

Mr Lawson has promised to send the particulars of Lewis & Hughes[4] Engine house today. I hope my letters with those of the houses for the Grand Junction Water Works [5] & Gardner Howard & Co[6] have been received.

I hardly think we shall be able to execute the Engines for the Birmingham Canal, if large ones, in less than 9 to 12 Months, from the period we received intimation of their intended orders, although I should be very desirous of doing every thing in our power for such old customers.[7]

I will learn whether the Covent Garden gentlemen have got the Apparatus[8] brought to the spot and are quite ready for a man, but in the meantime I think that either Varley, or Robert M^c^Murdo should hold themselves in readiness to proceed hither at short notice. Whoever comes will be wanted afterwards to erect Leader & C^os^ Engine[9], for which I conceive Smith unfit, and Wright & Shakespear cannot well be either of them spared out of town at this season.

I am Dear Sir

Yours truly
J Watt Junr.

Box 33

---

[1]A 54 in. cyl. engine was ordered.
[2]For a 50 in. cyl. 8 ft. stroke engine.
[3]A 36 h.p. rotative engine was ordered.
[4]A 10 h.p. engine for a dyeworks.
[5]A 54 in. cyl. pumping engine.
[6]Of the King and Queen Foundry. A 14 h.p. engine had been ordered. (Their third B & W engine.)
[7]Three engines were ordered: one 36 in. cyl. 7 ft. stroke and two 54 in. cyl. 8 ft. stroke.
[8]Heating apparatus (convection principle).
[9]A 40 h.p. engine, their second B & W one.

DOCUMENT 213

To J. Southern London 8th Feby 1811

Dear Sir,

I have yours of the 4th & 6th and Mr Hutton's of the 2nd. I have not had leisure to consider the contents of the latter fully, but certainly a more particular statement of the Property of Monteith Bogle & Co[1] as well as of the principal debts is necessary for our government, and these I shall request Mr Hutton to procure.

The present circumstances of many of the Commercial houses at Glasgow affords an additional motive for the detention of Messrs Garden & Sons 2nd Boiler at the Foundry. I scarcely think they can from thence raise any question about the time of payment, as it must be considered to have been delivered when they were advised of its being ready.

I was at the Kent Water Works[2] on Wednesday with Mr Rennie. The Engine house is built but the roof not compleat. The Boilers, Air Vessel, Flywheel, Entablature & Spring Beams &

Columns are in their places & the Cylinder was about to be fixed. The house is well built & the walls very strong: so far indeed every thing seems well done and Smith is anxious to have the finishing of it committed to himself, but I think it will be right for either Varley or Wright to be there a day or two before it starts. The Beam has not arrived. I suppose Nozzles & working gear must now be ready, if so please let the carriers have directions to deliver them without delay. They were laying the 12 Inch Main along the road to Greenwich, but I believe the greater part of their Mains are yet to lay, and the reservoirs at the Engine are not yet compleatly excavated. — Mr Wise was from home, but I must seek some other opportunity of seeing him as he wants Estimates of several Engines for Canals & other works in which he is concerned, or for which he is employed as Engineer.

I wish Mr Pearson to send me a statement of the account of the Woolwich Water Works,[3] chargeing Interest upon the Amount since it became due. I find there is little chance of getting immediate payment, but think they cannot refuse to allow the Interest. —

Mr Pearson will likewise note that Mr Rennie has given me a check on Child & Co for £490 on account of the Com^{rs} of Howth harbour which I have paid to the London Ac^t House.

At York buildings they have now secured the foundations of the old buildings by Sheet piling and Mr Rennie is about to order the Suction or Tunnel pipe, which he proposes to make 3ft. 6

diam^r^ and to extend 200 Yards into the River: it will enter our pump well at the bottom, which purports to be 4 feet under low water. Do you see any objection to this? I think it is conformable to what was talked of when he was at Soho.

The Strand Bridge Engine[4] is arrived and they are going to commence their proceedings on the South side immediately. —

Wm Brown says that when he took off the Slidevalve case at Mr Hoddinots,[5] he found all the Cement out at the bottom (I suppose the same as at Guinnesses) and they had been consuming much coal. The Spindle of the throttle Valve was easily made to fit; but in taking off the Slide Valve Case he broke 4 of the six Screws which fasten it to the Cylinder & was obliged to make a tap before he could proceed. He says the pins were so corroded by the Cement that they could not be turned. I am afraid this must happen in most similar cases & it will be well to reconsider the construction of this part of the Engine, with a view both to preventing the leaks & to facilitate the taking it off. Brown says he was detained 4 or 5 days longer than he should have been, by the breaking of the pins. — The Spindle of the throttle Valve has been broken by the Engine man striking it with a hammer in order to open the Valve when he started the Engine. It seems that when the Steam is allowed to enter with the valve shut it cannot be opened afterwards without violence, but why the valve should not be previously opened I do not perceive. Mr Hoddinot desires a box of Cement may be sent to the Bell Inn Thomas St^t^ Bristol, with directions to be forwarded to him at Milton per Boyce's Waggon. Please give the enclosed to Mr Pearson.

Yours truly

J Watt Junr.

I think it may be a week yet before I shall be able to leave town.

Box 33

---

[1]Cotton spinners of Lanarkshire. They had purchased two engines: a 30 h.p. in 1802 and a 53 h.p. in 1809.

[2]A 36 in. cyl. 8 ft. stroke engine had been ordered.

[3]A 21 in. cyl. 5 ft. stroke engine was ordered in 1809.

[4]A 6 h.p. engine.

[5]A 6 h.p. engine employed at a silk throwing mill at Bruton, Somerset.

Select bibliography on steam power and its application

(1) *Books 18th and 19th Centuries*

M.A. Alderson, *An Essay on the Nature and Application of Steam,* London, 1834.
G. Birkbeck and H. & J. Adcock, *The Steam Engine Theoretically and Practically Displayed,* London, 1827.
J. Bourne, *A Treatise on the Steam Engine,* London, 1846.
J. Bourne, *A Catechism of the Steam Engine,* London, 1847.
J. Bourne, *A Popular Description of the Steam Engine,* London, 1856.
J. Bourne, *Recent Improvements in the Steam Engine,* London, 1856.
C.D. Colden, *The Life of Robert Fulton,* New York, 1817.
J. Curr, *The Coal Viewer and Engine Builder's Practical Companion,* Sheffield, 1797 and London, 1970.
O. Evans, *The Abortion of the Young Steam Engineer's Guide,* Philadelphia, 1805.
J. Farey, *A Treatise on the Steam Engine,* Vol. 1, London, 1827 and Newton Abbot, 1971.
J. Farey, *Vol. 2,* Newton Abbot, 1971.
E. Galloway, *History and Progress of the Steam Engine,* London, 1830.
R.L. Galloway, *The Steam Engine and its Inventors,* London, 1881.
J.N.P. Hachette, *Histoire des Machines à Vapeurs,* Paris, 1830.

J. Hann, *A Short Treatise on the Steam Engine,* London, 1847.

P.R. Hodge, *Analytical Principles of the Expansive Steam Engine,* London, 1849.

J.M. Hoëne-Wronski, *Machines à Vapeur: Apercu leur état actuel,* Paris, 1829.

D. Lardner, *The Steam Engine Familiarly Explained and Illustrated,* London, 1836.

D. Lardner, *A Rudimentary Treatise on the Steam Engine,* London, 1848.

T. & J. Lean, *Historical Statement of the Improvements made in the Duty Performed by Steam Engines in Cornwall,* London, 1839.

M. Marestier, *Mémoire sur les Bateaux à vapeur des États-Unis d'Amérique,* Paris, 1824.

(Meikleham) R. Stuart, *A Descriptive History of the Steam Engine,* London, 1824.

(Meikleham) R. Stuart, *Historical and Descriptive Anecdotes of Steam Engines,* London, 1829.

F.N. Mellet, *Traité des Machines à Vapeur et leur application de la Navigation, aux Mines, aux Manufactures, aux Chemins de Fer . . .* 2nd Ed. Paris, 1838.

J.P. Muirhead, *Origin and Progress of the Mechanical Inventions of James Watt,* 3 vols., 1854.

J.P. Muirhead, *Life of James Watt with Selections from his Correspondence,* 1859.

M. Mulhall, *Mulhall's Dictionary of Statistics,* London, 1884.

C.F. Partington, *An Historical and Descriptive Account of the Steam Engine,* London, 1822.

C.F. Partington, *A Course of Lectures on the Steam Engine,* London, 1826.

W.J.M. Rankine, *A Manual of the Steam Engine and other Prime Movers,* London and Glasgow, 1859.

H. Reid, *The Steam Engine,* Edinburgh, 1838.

J. Ross, *A Treatise on Navigation by Steam, Comprising a History of the Steam Engine,* London, 1828.

J. Rumsey, *A Short Treatise on the Application of Steam . . .* Philadelphia, 1788.

J.S. Russell, *A Treatise on the Steam Engine,* Encyclopaedia Brit. 7th Ed. 1841.

Samuel Smiles, *Lives of Boulton and Watt, principally from the Original Soho Mss.,* 1865.

T. Tredgold, *The Steam Engine,* London, 1827.

F. Trevithick, *Life of Richard Trevithick, with an Account of his Inventions,* London, 1872.

T. Wickstead, *An Experimental Enquiry Concerning Cornish and Boulton and Watt Engines,* London, 1841.
T. Wickstead, *Further Elucidations of the Useful Effects of Cornish Engines,* London, 1859.
George Williamson, *Memorials of the Lineage, Early Life, Education and Development of the Genius of James Watt,* 1856.

(2) *Books 20th Century*

T.S. Ashton, *Iron and Steel in the Industrial Revolution,* 1924 and 1951.
T.S. Ashton and J. Sykes, *The Coal Industry of the Eighteenth Century,* Manchester, 1929 and 1964.
D. Bradford Barton, *The Cornish Beam Engine,* Truro, 1965.
Hedwig Behrens, *Mechanikus Franz Dinnendahl,* Köln, 1970.
R.H. Campbell, *Carron Company,* Edinburgh and London, 1961.
D.S.L. Cardwell, *Steam Power in the Eighteenth Century,* London and New York, 1963.
D.S.L. Cardwell, *From Watt to Clausins: The Rise of Thermodynamics in the Early Industrial Age,* London, 1971.
S.D. Chapman, *The Early Factory Masters,* Newton Abbot, 1967.
H.W. Dickinson, *Robert Fulton, Engineer and Artist: His Life and Works,* London, 1913.
H.W. Dickinson and R. Jenkins, *James Watt and the Steam Engine,* 1927 and Hartington, 1981.
H.W. Dickinson and A. Titley, *Richard Trevithick: The Engineer and the Man,* Cambridge, 1934.
H.W. Dickinson, *James Watt, Craftsman and Engineer,* Cambridge, 1936 and Newton Abbot, 1967.
H.W. Dickinson, *Matthew Boulton,* Cambridge, 1937.
H.W. Dickinson, *A Short History of the Steam Engine,* London, 1938 and 1963.
B.F. Duckham, *A History of the Scottish Coal Industry,* Vol. i, Newton Abbot, 1970.
T.R. Harris, *Arthur Woolf: The Cornish Engineer, 1766-1837,* Truro, 1966.
W.S. Harvey and G. Downs-Rose, *William Symington, Inventor and Engine Builder,* London, 1980.
Richard L. Hills, *Power in the Industrial Revolution,* Manchester, 1970.
A.H. John, *The Walker Family, Iron Founders and Lead Manufacturers,* Council for the Preservation of Business Archives, 1951.

R.J. Law, *The Steam Engine,* Science Museum, HMSO, 1965.
R.J. Law, *James Watt and the Separate Condenser,* Science Museum, HMSO, 1969.
J. Lord, *Capital and Steam Power 1750-1800,* London 1923 and 1966.
P. Mathias, *The Brewing Industry in England, 1700-1830,* Cambridge, 1959.
A.E. Musson and Eric Robinson, *Science and Technology in the Industrial Revolution,* Manchester, 1969.
A.E. Musson (ed) *Science Technology and Economic Growth in the Eighteenth Century,* London, 1972.
J. Payen, *Capital et Machine à Vapeur au XVIII*[e] Siècle, Paris, 1969.
S. Pollard, *The Genesis of Modern Management: A Study of the Industrial Revolution in Great Britain,* London, 1965.
C.W. Pursell, *Early Stationary Steam Engines in America: A Study in the Migration of a Technology,* Washington, 1969.
A. Rastrick, *Dynasty of Ironfounders : The Darbys and Coalbrookdale,* London, 1953 and Newton Abbot, 1970.
Eric Robinson and A.E. Musson, *James Watt and the Steam Revolution,* London, 1967.
Eric Robinson and Douglas McKie, *Partners in Science, Letters of James Watt and Joseph Black,* London, 1970.
Eric Roll, *An Early Experiment in Industrial Organisation: Being a History of the Firm of Boulton and Watt, 1775-1805,* London, 1930 and 1968.
L.T.C. Rolt and J.S. Allen, *The Steam Engine of Thomas Newcomen,* Hartington and New York, 1977.
W.J. Rowe, *Cornwall in the Age of the Industrial Revolution,* Liverpool, 1953.
E.K. Scott, *Matthew Murray: Pioneer Engineer,* Leeds, 1928.
R.E. Schofield, *The Lunar Society of Birmingham,* Oxford, 1963.
C.J. Singer, E.J. Holmyard, A.R. Hall and T.I. Williams (eds) *A History of Technology,* Oxford, 1954-8.
J. Tann, *The Development of the Factory,* London, 1970.
Barrie Trinder, *The Industrial Revolution in Shropshire,* London and Chichester, 1973.
G.N. von Tunzelmann, *Steam Power and British Industrialisation to 1860,* Oxford, 1978.

(3) Articles

J. Atack, F. Bateman and T. Weiss, the Regional Diffusion and Adoption of the Steam Engine in American Manufacturing, *Journ. Econ. Hist.* 40, 1980.

D.S.L. Cardwell, Early Development of the Concepts of Power, Work and Energy, *Brit. Journ. Hist. of Science,* 3, 1967.
W.H. Chaloner, The Cheshire Activities of Matthew Boulton and James Watt . . . 1776-1817, *Trans. Lancs. & Chesh. Antiq. Soc.* 61, 1949.
W.H. Chaloner, Robert Owen, Peter Drinkwater and the Early Factory System in Manchester, *Bull. John Rylands Library,* 37, 1954.
W.H. Chaloner, The Stockdale Family, the Wilkinson brothers and the Cotton Mills at Cark-in-Cartmel, c.1782-1800, *Trans. Cumb. and Westmorland Antiq. and Arch. Soc.*, 64, 1964.
S.D. Chapman, The Pioneers of Worsted Spinning by Power, *Business Hist.*, 7, 1965.
S.D. Chapman, The Cost of Power in the Industrial Revolution, *Midland Hist.*, 1, 1970.
J.C. Cule, Finance and Industry in the Eighteenth Century: The Firm of Boulton and Watt, *Economic History,* Feb. 1940.
N. Deerr, The Early Use of Steam Power in the Cane Sugar Industry, *Trans. Newcomen Soc.*, 4, 1923-4.
H.W. Dickinson and A. Lee, The Rastricks — Civil Engineers, *Trans. Newcomen Soc.*, 4, 1923-4.
J.S. Enys, Remarks on the Duty of the Steam Engines Employed in the Mines of Cornwall at Different Periods, *Trans. Inst. Civil Engineers,* 3, 1842.
E.A. Forward, Simon Goodrich and his Work as an Engineer, *Trans. Newcomen Soc.*, 3, 1922-3 and 8, 1937-8.
W.K.V. Gale, Soho Foundry, Some Facts and Fallacies, *Trans. Newcomen Soc.*, 34, 1961-2.
D. Gilbert, On the Progressive Improvements made in the Efficiency of Steam Engines in Cornwall, *Phil. Trans.*, 120, 1830.
J. Glynn, Draining Land by Steam Power, *Trans. Soc. Arts and Manufactures,* 2, Derby, 1836.
J. Goodchild, On the Introduction of Steam Power into the West Riding, *South Yorks Journ.*, 3, 1971.
J.R. Harris, The Employment of Steam Power in the Eighteenth Century, *History,* 52, 1967.
J.R. Harris, The Early Steam Engine on Merseyside, *Trans. Historic Soc. of Lancs & Chesh.*, 106, 1954.
W.J. Henwood, Account of Steam Engines in Cornwall, *Edinburgh Journ. of Science,* 10, 1829.
R. Hills and A.J. Pacey, The Measurement of Power in Early Steam Driven Textile Mills, *Technology and Culture,* 13, 1972.
G.J. Hollister-Short, The Introduction of the Newcomen Engine into Europe, *Trans. Newcomen Soc.*, 48, 1976-77.
D.T. Jenkins, Early Factory Development in the West Riding

of Yorkshire, 1770-1800, N.B. Harte and K.G. Ponting (Eds), *Textile History and Economic History, Essays in Honour of Miss Julia de Lacy Mann,* Manchester, 1973.

R. Jenkins, Jonathan Hornblower and the Compound Engine, *Trans. Newcomen Soc.,* 11, 1930-1.

R. Jenkins, A Cornish Engineer: Arthur Woolf 1766-1837, *Trans. Newcomen Soc.,* 13, 1932-3.

M. Kerker, Science and the Steam Engine, *Technology and Culture,* 2, 1960.

J.D. Marshall, Early Applications of Steam Power: The Cotton Mills of the Upper Leen, *Trans. Thoroton Soc.,* 60, 1956.

J. Mokyr, The Industrial Revolution in the Low Countries in the First Half of the Nineteenth Century, *Journ. Econ. Hist.,* 34, 1974.

W.A. Morshead, Junr., On the Relative Advantages of Steam, Water and Animal Power, *Journ. Bath and West of England Soc.,* 5, 1856.

A.E. Musson and E. Robinson, The Origins of Engineering in Lancashire, *Journ. Econ. Hist.,* 19, 1959.

A.E. Musson and E. Robinson, The Early Growth of Steam Power, *Econ. Hist. Rev.,* 11, 1958-9.

A.E. Musson and E. Robinson, Science and Industry in the late 18th Century, *Econ. Hist. Rev.,* 13, 1960.

A.E. Musson, Industrial Motive Power in the United Kingdom, *Econ. Hist. Rev.,* 29, 1976.

N. Mutton, Boulton and Watt and the Norfolk Marsh Land, *Norfolk Arch.,* 34, 1967.

F. Nixon, The Early Steam Engine in Derbyshire, *Trans. Newcomen Soc.,* 31, 1957-9.

R.A. Pelham, The Application of Steam Power to the Wiltshire Textile Industry in the Early 19th Century, *Wilts. Arch. & Nat. Hist. Mag.,* 54, 1952.

J-C. Perrier, Sur les Machines à Vapeur, *Bull. de la Société d'Encouragement pour L'Industrie Nationale,* 73, 1810.

K. van der Pols, Early Steam Pumping Engines in the Netherlands, *Trans. Newcomen. Soc.,* 46, 1973-4.

Fritz Redlich, The Leaders of the German Steam Engine Industry during the First Hundred Years, *Journ. Econ. Hist.,* 4, 1944.

W.G. Rimmer, Middleton Colliery near Leeds, 1779-1830, *Yorks. Bull. for Econ. and Soc. Research,* 7, 1955.

Eric Robinson, James Watt and the Tea Kettle. A Myth Justified, *History Today,* April, 1956.

Eric Robinson, The International Exchange of Men and Machines 1750-1800, *Business History,* I, 1958.

Eric Robinson, Training Captains of Industry, *Annals of Science,* 10, 1954.
Eric Robinson, James Watt and the Law of Patents, *Technology & Culture,* 13, 1972.
E. Robinson, The Diffusion of Steam Power, *Journ. Econ. Hist.,* 34, 1974.
N. Rosenberg, Selection and Adaptation in the Transfer of Technology: Steam and Iron in America, 1800-1870, *L'Aquisition des Techniques par les Pays Non-initiateurs,* Pont à Moulson, 1970.
F.M. Scherer, Invention and Innovation in the Watt-Boulton Steam Engine Venture, *Technology and Culture,* 6, 1965.
J. Tann, Richard Arkwright and Technology, *History,* 58, 1973.
J. Tann, Fuel Saving in the Process Industries during the Industrial Revolution — A Study in Technological Diffusion, *Business History,* 15, 1973.
J. Tann, The Employment of Power in the West of England Wool Textile Industry 1790-1840, N.B. Harte and K.G. Ponting (Eds), *Textile History and Economic History, Essays in Honour of Miss Julia de Lacy Mann,* Manchester, 1973.
J. Tann, Suppliers of Parts: The Relationship between Boulton and Watt and the Suppliers of Engine Components, *Centenary Trans. Birm. and Warwicks. Arch. Soc.,* 1974.
J. Tann, Research on the Boulton and Watt Papers: A Study in Methodology, *Business Archives,* 43, 1977.
J. Tann, Marketing Methods in the International Steam Engine Market: The Case of Boulton and Watt, *Journ. Econ. Hist.,* 38, 1978.
J. Tann and M.J. Breckin, The International Diffusion of the Watt Engine, *Econ. Hist. Rev.,* 31, 1978.
J. Tann, Boulton and Watt's Organisation of Steam Engine Production Before the Opening of Soho Foundry, *Trans. Newcomen Soc.,* 48, 1978.
J. Tann, Richard Arkwright and Steam Power, *Business History,* 21, 1979.
J. Tann, Makers of Improved Newcomen Engines in the late 18th Century, *Trans. Newcomen Soc.* 50, 1980.
J. Tann, Mr. Hornblower and His Crew: Watt Steam Engineer Pirates in the Late 18th Century, *Trans. Newcomen Soc.,* 51, 1981.
M. Teich, Diffusion of Steam, Water and Air Power to and from Slovakia during the 18th Century and the Problem of the Industrial Revolution, *L'Aquisition des Techniques par les Pays Non-Initiateurs,* Pont à Molson, 1970.
P. Temin, Steam and Water power in the Early Nineteenth

Century, *Journ. Econ. Hist.*, 26, 1966.
J. Thomas, Josiah Wedgwood as a Pioneer of Steam Power in the Pottery Industry, *Trans. Newcomen Soc.*, 17, 1936-7.
J. Thomas, The Pottery Industry and the Industrial Revolution, *Economic History,* Feb. 1937.
A. Titley, Trevithick and Rastrick and the Single-Acting Expansive Engine, *Trans. Newcomen Soc.*, 7, 1926-7.
A. Titley, Richard Trevithick and the Winding Engine, *Trans. Newcomen Soc.*, 10, 1929-30.
G.N. von Tunzelmann, Technological Diffusion during the Industrial Revolution: The Case of the Cornish Pumping Engine, in R.M. Hartwell (Ed), *The Industrial Revolution,* Oxford, 1970.
Rex Wailes, James Watt — Instrument Maker, *Chartered Mechanical Engineer,* March 1962.

Index – Persons

Index – Subject and Place